Contents

Dedication

To Ross Cunningham for his extraordinary wisdom, generosity, and encouragement.

To all our colleagues who are working hard to conserve the planet's biodiversity.

Acknowledgments

This book could only be written because of the support of our friends and colleagues at the Centre for Resource and Environmental Studies at the Australian National University in Canberra, Australia. In particular, Ross Cunningham, Adrian Manning, and Ioan Fazey have made major contributions to the development of our thinking on some of the issues we have tackled. We have also benefited greatly from our work with Jerry Franklin, Mark Burgman, Hugh Possingham, Denis Saunders, Mike McCarthy, Sue McIntyre, and Richard Hobbs. We thank Kimberlie Rawlings for discussions on vegetation restoration.

Comments on the manuscript by Rebecca Montague-Drake, Annika and Adam Felton, Nicki Munro, and Kara Youngentob were useful. Andrew Bennett, Yrjö Haila, Reed Noss, and Denis Saunders made highly professional critiques of the book and vastly improved its content—although it was not always possible to accommodate all of their excellent suggestions.

Financial support for this book was provided by the Kendall Foundation, Land and Water Australia, the Natural Heritage Trust, the Australian Research Council, and the Pratt Foundation.

Barbara Dean, Laura Carithers, and the rest of the team at Island Press assisted with steering this book through to completion. Nick Alexander, John Manger, and Briana Elwood from CSIRO Publishing also assisted with finalizing this book.

Nicki Munro and Rebecca Montague-Drake helped admirably with gathering the enormous amount of literature that characterizes research on landscape change and habitat fragmentation.

Clive Hilliker did an outstanding job of drawing figures and other images for the book. We are greatly indebted to those colleagues who provided photographs for this book. The Photographic Unit at The Australian National University helped with scanning images.

Preface

The world is experiencing a major extinction crisis. Losses of species over the coming decades are forecast to parallel those of the past five mass extinction events recorded in geological history. The modification of landscapes by humans and the associated loss of habitats are major drivers of this crisis. In an effort to slow rates of species loss, investigations of the effects of habitat loss and habitat fragmentation and ways to mitigate them are becoming significant research fields in ecology, conservation biology, landscape ecology, and natural resource management.

There are many books on related topics such as landscape mosaics, wildlife corridors, species–area relationships, and island biogeography. In addition, there are edited books with contributed chapters on many topics associated with habitat fragmentation. However, we are unaware of any books that have taken a similar approach to ours in deriving a synthesis of the various topics collected under the theme of "fragmentation." On this basis, this book was designed to be a broad overview of the array of topics that are directly and indirectly linked with the effects of habitat loss and habitat fragmentation on biota. We emphasize that this book is a synthesis of material rather than a detailed assessment of key topics associated with landscape change.

The task of writing this book was enormous—perhaps too ambitious—because the literature on habitat loss and habitat fragmentation is vast. It is so large (and expanding at a phenomenal rate) that it is impossible for any one, two, or even handful of people to stay on top of it. As a result, we have undoubtedly made mistakes, overlooked some key topics, and misinterpreted the findings of others' work. We apologize for these mistakes and also to our colleagues if they feel we have short-changed their work. This was by no means our intention and we look forward to the criticisms from them that will undoubtedly flow from having attempted to write a book like this.

We have, as much as possible, tried to include examples and insights from a range of ecosystems from around the world. However, we readily acknowledge that we have been heavily influenced by our own experiences and empirical studies in southeastern Australia and that, as a result, there are more examples from this region than from elsewhere. We also have focused primarily on terrestrial habitats, but we acknowledge that habitat loss and habitat subdivision can apply equally to aquatic and marine landscapes. Much of the book concerns the responses of animals to landscape alteration—which again reflects our own biases toward work on animals rather than plants. Finally, we are aware that there is a bias in our book toward places where major forms of human-derived landscape change such as land clearing have been relatively recent (e.g., the Americas and Australia), whereas places dominated by cultural landscapes (such as Europe) have received less attention.

The science of conservation biology and related fields of enquiry into landscape change, habitat loss, and habitat fragmentation are young but rapidly evolving. Perhaps our work will be outdated within the next decade or two or even more quickly. Despite this, we believe that a book on landscape change, habitat loss, and habitat fragmentation is timely to demonstrate the threats these processes pose—and to highlight the opportunities to conserve as much of the world's biota as possible within human-modified landscapes, both for its own sake and for future generations of our own species.

David Lindenmayer
Joern Fischer
December 2005

A modified farming and woodland agricultural landscape in southeastern Australia (photo by David Lindenmayer).

CHAPTER 1
Introduction

Why This Book Was Written

This book was written for two main reasons. First, themes associated with landscape change have become a major focus of conservation biology and landscape ecology (McGarigal and Cushman 2002; Fahrig 2003; Hobbs and Yates 2003). A review of all papers published in 2001 in the journals *Conservation Biology, Biological Conservation,* and *Biodiversity and Conservation* found that landscape change and habitat fragmentation were the two most frequently studied processes threatening species persistence (Fazey et al. 2005a). Similarly, a database search of journal articles prior to writing this book produced over 2000 published papers where the abstract or keywords contained the words "habitat loss" or "habitat fragmentation." Because the literature on landscape change and habitat fragmentation is so large and complex (and becoming increasingly so), we believe there is an important role for a book that provides an overview of the varied and interrelated topics encompassed by it.

The second reason we wrote this book was because the term "habitat fragmentation" has become vague and ambiguous due to its imprecise use, thereby limiting its practical value for conservation managers (Haila 2002; Regan et al. 2002). The term is losing its meaning because it is frequently used as an umbrella term for a wide range of interacting processes, including habitat loss, the subdivision of remaining habitat, an increase in edge effects, and altered species interactions or ecological processes. In essence, there are multiple processes grouped under the term "fragmentation," and there are multiple consequences of those processes; thus it is difficult to determine which consequences relate to which process. As a result, the "blizzard of ecological details" characteristic of the fragmentation literature, coupled with the way the term "habitat fragmentation" often means "all things to all people," has made the study of fragmentation a "panchreston" (Bunnell 1999a; see discussion in the next section). Given this, it is perhaps not surprising that even some major reviews of fragmentation (e.g., McGarigal and Cushman 2002) have failed to identify "clear insights into system dynamics" (Bissonette and Storch 2002). The "panchreston problem" is clearly hampering progress in the study of landscape change and ways to mitigate its negative impacts on species and ecosystems.

A possible way to gain a better understanding of the field is to disentangle the subcomponent parts and themes that have been collected under the umbrella of "fragmentation" research. We have tried to do this in our book by *summarizing* and identifying links between what we believe are key topics associated with landscape change, habitat loss, and habitat fragmentation. We emphasize the word "summarizing" here because we have attempted to provide a brief overview and synthesis of key topics, rather than an exhaustive review of each and every topic related to fragmentation.

"Disentangling" Habitat Fragmentation

Bunnell (1999a) argued that the generic term "habitat fragmentation" is often used loosely to encompass a myriad of processes and changes that accompany landscape alteration—making it a panchreston; which he defined as a

> proposed explanation intended to address a complex problem by trying to account for all possible contingencies but typically proving to be too broadly conceived and therefore oversimplified to be of any practical use.
>
> *The Random House Dictionary of the English Language* [undated] cited in Bunnell (1999a).

Bunnell (1999a) argued that a solution to the problems created by the fragmentation panchreston was to acknowledge that multiple processes take place and to work in ways to separate them so that consequences can be better assigned to particular processes (see also Figure 1.1). Unfortunately, the fragmentation literature rarely does this. Instead, typically what occurs is that one pattern (the spatial distribution of a species or suite of species) is correlated with another pattern (the spatial distribution of patches of native vegetation), with little emphasis on the ecological processes that link the two. In this book, we argue that tackling the panchreston will help: (1) key questions to be better defined—a major issue because problem-definition is something that ecologists often do poorly (Peters 1991); (2) ensure better and more focused studies that contribute an improved understanding of the effects of landscape change on biota; and (3) more rapidly evolve improved conservation and resource management strategies (Box 1.1). For example, although habitat loss and habitat subdivision often go together (see Chapter 4), the distinction between them is important. This is because dealing with habitat loss will require formulating quite different kinds of hypotheses, quite different types of studies, and, ultimately,

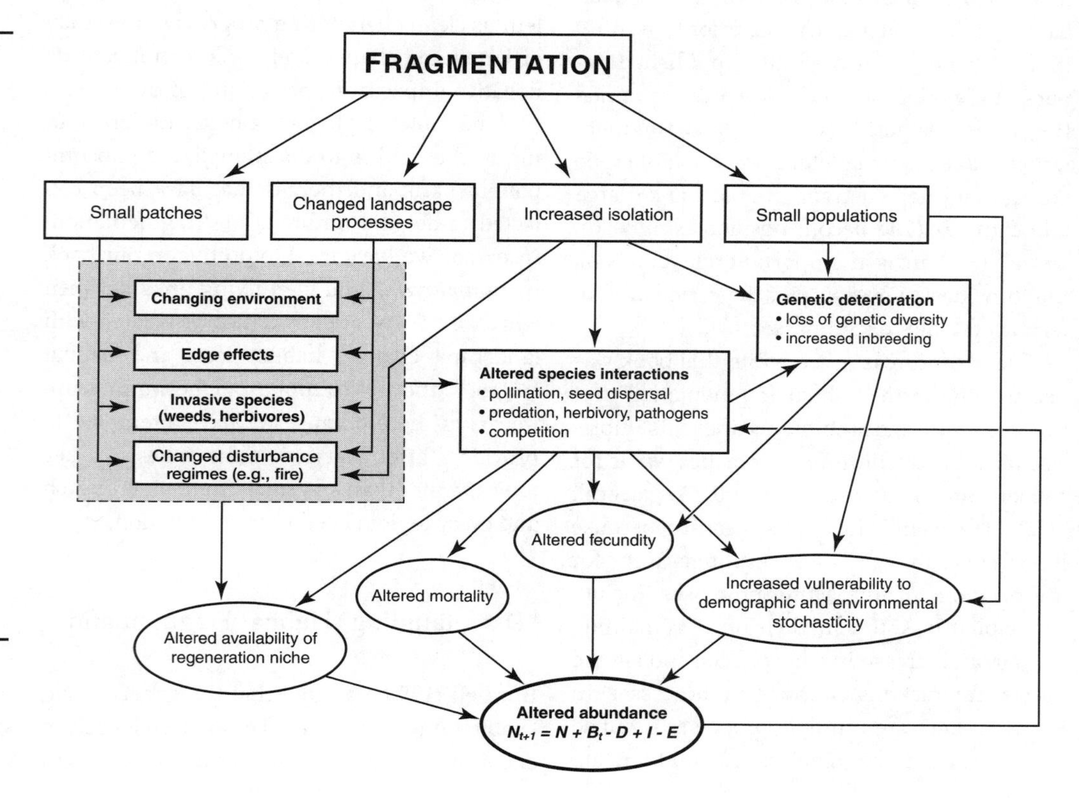

Figure 1.1. A conceptual model of landscape change effects on plants (redrawn from Hobbs and Yates 2003: reproduced with permission from the *Australian Journal of Botany*, **51**, 471–488 [Hobbs, R.J. & Yates, C.J.]. Copyright CSIRO 2003. Published by CSIRO PUBLISHING, Melbourne VIC, Australia [http://www.publish.csiro.au/journals/ajb]).

different kinds of mitigation strategies than if the focus is on habitat subdivision.

While we argue that an important way forward is to focus on particular subcomponents of landscape modification, we also acknowledge the importance of recognizing the complexity that is typically associated with multiple biotic responses to multiple interacting processes of landscape change (Crome 1994). This is why we dedicate an entire section of this book (Part V) to approaches and general principles for mitigating the multiple threatening processes and negative effects on species and assemblages that often accompany landscape change.

Box 1.1. Why Disentangling Threatening Processes Is Important for Conservation: The Case of the Brown Treecreeper

Identifying which of the processes associated with landscape change are most threatening to a given species of conservation concern is critical for the development of efficient and effective conservation strategies. This is illustrated by the case of the southern subspecies of the brown treecreeper *(Climacteris picumnus picumnus)*, a declining bird species in many woodland areas in southeastern Australia (Cooper and Walters 2002). The species is absent from, or has declined severely, in many remnants of apparently suitable woodland. For example, in Mulligan's Flat Nature Reserve near Canberra, a previously stable population has declined over the last decade to a single individual in mid-2006 (J. Bounds, pers. comm.). Developing strategies to conserve the brown treecreeper depends on identifying the underlying reasons for its decline and low levels of patch occupancy.

Work on the brown treecreeper indicates that a lack of habitat connectivity and hence disrupted dispersal is the most likely reason for the decline of the species in landscapes subject to major human modification (Walters et al. 1999). Where woodland patches are isolated and the landscape between them is devoid of suitable habitat, brown treecreepers are often absent—even though some woodland patches are large and contain all the essential habitat components needed by the species (e.g., fallen logs and tree hollows; Cooper et al. 2002). Therefore, if a small population in a patch is lost, sources of dispersing animals may be too remote to reverse the local extinction. Conservation of the species will depend on ways to promote habitat connectivity. These may include revegetation programs to link remnants and promote natural dispersal, or "artificial dispersal" through translocation programs of birds to empty patches or patches with only a few remaining birds.

The Scope of This Book: Definitions and Key Themes

Our overarching goal is to tackle the habitat fragmentation panchreston. To achieve this goal, we provide a conceptual framework for the study and management of modified landscapes. We also provide many empirical examples from around the world to give tangible support to this conceptual framework. An important part of tackling the fragmentation panchreston is to carefully define appropriate terms such as "habitat" and "fragmentation." This is because the imprecise use of terminology can lead to the inappropriate use of concepts and theories in management (Murphy 1989) or inappropriate testing of theory (Fazey 2005). Given this, we have attempted to be as precise as possible in our use of terms throughout this book—although errors have undoubtedly crept in.

Definitions

Key terms that we use frequently throughout the text and that have a specific meaning in this book (sometimes different from standard usage in the literature) are outlined in Table 1.1. From an ecological perspective, what constitutes a landscape will usually be a function of the scales over which a given species moves and how it perceives its surrounds (Wiens 1997; Manning et al. 2004a). However, for the practical purposes of this book, we define the term "landscape" from a human perspective and consider it as an area that covers hundreds to thousands of hectares. When landscapes are changed by vegetation clearing or other kinds of anthropogenic modification, we use the

Table 1.1. The definition of key terms used widely throughout this book

Term	*Definition*	*Chapter where discussed in detail*
Species perspective of a modified landscape	The perception of a landscape as it is assumed to be experienced by a given (nonhuman) species; important features are assumed to include sources of food and shelter, sufficient space, and appropriate climatic conditions	Chapter 3
Human perspective of a modified landscape	The perception of a landscape as experienced by humans; features typically noted include patches of different types of land cover and their spatial arrangement (including native vegetation)	Chapter 3
Habitat	The environments suitable for a particular species—a species-specific entity	Chapter 4
Habitat loss	Loss of suitable habitat for a given species, making the area unsuitable for that species to occur there	Chapter 4
Habitat degradation	The reduction in quality of an area of habitat for a given species. The species may still occur in the area but, for example, may not be able to successfully breed there	Chapter 5
Habitat subdivision	Subdivision of a single large area of habitat into several smaller areas	Chapter 6
Habitat isolation	The isolation of habitat areas for a given species—the opposite of habitat connectivity	Chapter 6
Habitat connectivity	The connectedness of habitat patches for a given species—a species-specific entity	Chapter 6
Vegetation loss	The loss of native vegetation cover	Chapter 9
Vegetation deterioration	The deterioration in condition of the cover of native vegetation	Chapter 10
Edge effect	The change in biological and physical conditions that occur at an ecosystem boundary and within adjacent ecosystems	Chapter 11
Landscape connectivity	The connectedness of areas of native vegetation cover within a landscape from a human perspective	Chapter 12
Ecological connectivity	The connectedness of ecological processes at multiple spatial scales	Chapter 12

interchangeable terms "landscape change," "landscape alteration," or "landscape modification." We use the term "habitat" to mean the environments suitable for a particular species. Following this definition, habitat therefore is a species-specific entity (see Table 1.1). "Habitat loss" refers to the loss of suitable habitat for a given species such that the particular species no longer occurs in that area. We do not consider the term "habitat loss" to be synonymous with the loss of native vegetation. This is because a landscape extensively altered by humans where vegetation loss has been substantial may effectively experience no loss of suitable habitat for some species. Conversely, a landscape supporting a complete cover of native vegetation may contain no suitable habitat for some species (e.g., because of a lack of naturally suitable environmental conditions) (Table 1.2). Given this, much of this book labors the point about the importance of understanding what constitutes suitable habitat as a prelude to mitigating habitat loss. Habitat subdivision is the process of subdividing a single large area of habitat into several smaller areas, a practice that is also referred to as habitat fragmentation. We follow the logic of Fahrig (2003) in recognizing that the spatial process of habitat subdivision (habitat fragmentation) is distinctly different from the process of habitat loss. Moreover, we rarely use the term "habitat fragmentation" in the remainder of this book because of the confusion that its loose application can create.

Finally, the term "biodiversity," which is a

Table 1.2. Examples of nonanthropogenic factors influencing the distribution of individual species and patterns of species richness

Factor	*Explanation*	*Reference*
Climate	Climate shapes the broadscale distribution patterns of species	MacArthur 1972; Woodward 1987
Elevation	Some species are restricted to high elevations; species richness tends to be lower at higher elevations	Huggett and Cheesman 2002; Gaston and Spicer 2004
Latitude	Most species are restricted to certain latitudes; species richness tends to be lower at higher latitudes	Armesto et al. 1998
Productivity	At a global scale, species richness increases with levels of energy and nutrient availability; at a regional scale, for many animal groups there is a hump-shaped relationship between species richness and productivity—species richness first increases and then declines as productivity increases	Rosenzweig 1995; Gaston and Spicer 2004
Biogeographic history	Changing sea levels, isolation of land masses, and orogeny isolate populations and/or lead to the speciation of new taxa	Mayr 1942; Croizat 1960
Natural disturbance	Different species are adapted to different disturbance regimes (e.g., hurricanes, fires, windstorms); species richness is sometimes highest at intermediate levels of disturbance	Connell 1978

contraction of the words "biological diversity," is used occasionally throughout this book. There are many definitions of "biodiversity" (Bunnell 1998). For the purposes of this book, the term encompasses genes, individuals, demes, populations, metapopulations, species, communities, ecosystems, and the interactions between these entities (Box 1.2). This definition stresses both the numbers of entities (genes, species, etc.) and the differences within and between those entities (see Gaston and Spicer 1998, 2004). Bunnell (1998) reviewed approximately 90 interpretations of the biodiversity concept. Many of these definitions were very abstract, making it difficult to use them in management applications. For instance, definitions that include the maintenance of genetic diversity and the maintenance of ecosystem processes or functions are difficult to apply. On this basis, Bunnell et al. (2003) argued that the best surrogate for sustaining the diversity of biological entities is the maintenance of species richness (sensu Whittaker et al. 2001; Gaston and Spicer 2004), although there are some important caveats with this approach and it cannot be applied uncritically to all assemblages in all landscapes and under all circumstances (see Box 1.2).

The Importance of Spatial Scale

Humans can modify environments at several spatial scales (Angelstam 1996). For example, at a regional scale, massive changes can be caused by deforestation or urban expansion. Second, within landscapes completely covered by native vegetation, formerly continuous areas of distinct vegetation types or successional stages (e.g., old-growth forest stands) can be lost or become subdivided. Finally, within given areas of particular kinds of vegetation, structural and floristic elements can be lost (e.g., large fallen logs; Angelstam 1996). The appropriate scale of an investigation or explanation is also related to the species of interest (Wiens 1989; Box 1.3). Importantly, all species are affected by ecological phenomena at multiple spatial scales (Forman 1964; Diamond 1973; Mackey and Lindenmayer 2001). This means there is no single correct scale at which to study landscape change, or at which to mitigate its effects on ecosystems. Throughout the book, we have attempted to explicitly highlight the importance of considering multiple scales.

Box 1.2. Diversity Concepts

"Biodiversity" encompasses genes, species, ecosystems, and their interactions (Noss 1990). Several related concepts are frequently used in conservation biology. The most widely used diversity concept is species richness, which is simply the number of species in a given area. A likely reason why species richness is widely used is that it is often relatively straightforward to measure and directly interpretable. Some authors use "species diversity" as an interchangeable term for "species richness," but others refer to species diversity as a weighted index of the number of species and their relative abundances. Given this ambiguity, we do not use the term "species diversity" in this book (see also Whittaker et al. 2001).

Diversity concepts are scale dependent. For example, two forest patches in a given modified landscape each may have a species richness of X (e.g., 12) forest-dependent bird species. However, depending on the degree to which these patches support the same species, species richness at the landscape scale could be between X and $2X$ (e.g., between 12 and 24).

Unfortunately, the scale dependence of species richness is not always acknowledged in conservation management. However, the scale at which species richness should be maximized needs to be carefully considered (Murphy 1989; Gilmore 1990). For example, in some forest environments, maximum bird species richness at a local scale may occur following timber harvesting as a result of invasions by bird species more typically associated with open vegetation types (Shields and Kavanagh 1985). Although a higher number of species might occur under such a management regime, species that depend on intact forest ecosystems may be eliminated from such modified ecosystems. Hence, species richness at a broader scale (e.g., across entire forest landscapes) may be reduced because taxa sensitive to logging operations are lost (Noss and Cooperrider 1994).

An additional problem with the adoption of species richness as a primary conservation goal is that all taxa are inadvertently assigned equal status. In the worst case, introduced pest species may be assigned the same value as threatened species. For example, in southeastern Australia, the introduced common starling (*Sturnus vulgaris*) may inadvertently be given equal weighting as the threatened and forest-dependent sooty owl (*Tyto tenebricosa*; Milledge et al. 1991). To avoid such oversimplification, it is important to consider the identity of the taxa that comprise an assemblage of species. As Gilmore (1990, p. 384) noted: "we do not conserve diversity indices," which, if used simplistically, can "obscure the taxonomic information content of plant and animal assemblages."

Putting Landscape Change into Context

A key problem in studies of modified landscapes is that landscape change has been regarded by some workers to be the single dominant reason for why species and populations come to be where they are in a landscape, and the single dominant reason for the loss or decline of species. However, we believe that landscape change needs to be put into context as only one of several factors which influence ecosystem processes, species richness, and the distribution of particular species in a landscape (see Table 1.2). In some cases, factors not directly related to habitat loss, habitat degradation, or habitat subdivision will be the most important ones. For example, the ecological literature is replete with examples of relationships between high latitude or high elevation and low levels of species richness in largely natural landscapes (reviewed by Gaston and Spicer 2004; Figure 1.2). In another example, in parts of Central America, many areas of forest remain relatively intact, yet many species of vertebrates have been lost because of intensive hunting pressure by humans (Redford 1992). Hunting effects have also had major negative impacts on a wide range of North American species, including North American elk *(Cervus elaphus)*, pronghorn *(Antilocapra americana)*, mountain goat *(Oreamnos americanus)*, big-horned sheep *(Ovis canadensis)*, gray wolf *(Canis lupus)*, and grizzly bear *(Ursus horribilis)* (Mech 1970; Noss and Cooperrider 1994; Cole et al. 1997). Similarly, in some Australian landscapes where patterns of vegetation cover have not undergone major alteration, species loss and population decline have been substantial as a result of the impacts of introduced species such as the cane toad (*Bufo marinus;*

Bennett 1997) and red fox (*Vulpes vulpes;* Kinnear et al. 2002). The decline of forest birds on the island of Guam is another classic example where landscape change has not been the primary cause of species loss (see Box 1.4).

Single Species or Multiple Species?

This book provides an overview of the wide range of topics associated with habitat loss, changes in landscape patterns, and other changes that typically accompany human landscape modification. A recurring theme throughout this text is that there is an inherent tension between approaches that explore the effects of landscape change on aggregate measures of (multi-) species occurrence, such as species richness or species composition (e.g., island biogeography theory [MacArthur and Wilson 1963, 1967] or nested subset theory [Patterson and Atmar 1986]), and other approaches that focus on the responses of individual species to landscape change (e.g., Lamberson et al. 1994; Ferreras 2001). While the former approach of focusing on aggregate measures of species occurrence groups several species together, single-species investigations often highlight that each individual species responds uniquely to landscape change, habitat loss, and habitat subdivision (e.g., Robinson et al. 1992; Lindenmayer et al. 2002a). Such findings emerge naturally from niche and habitat theory (Hutchinson 1958), which recognize that what constitutes suitable habitat (and hence what constitutes habitat loss) and what constitutes habitat subdivision will be species-specific. However, a detailed focus on the response of an individual species may tell us little about the overall pattern of change in larger assemblages of species, and the management reality is that it is rarely possible to consider more than a handful of individual species in any given area. Conversely, the identification of general patterns involving many species is often particularly useful from a management perspective. A potential problem is that management tools that attempt to predict general species distribution patterns from a few individual species rely on a range of assumptions about the nature of species co-occurrence patterns (e.g., indicator species [Landres et al. 1988; Lindenmayer et al. 2000; Rolstad et al. 2002], focal species [Lambeck 1997; Lindenmayer and Fischer 2003], or umbrella species [Caro 2001, 2004; Sergio et al. 2005]). In all cases there is an inherent tension

Box 1.3. Why an Appreciation of Scale Is Important

An appreciation of scale is important for many types of ecological investigations (Wiens 1989). The general notion of scale refers to the area and time frame over which phenomena take place. For natural phenomena, large scales tend to be associated with relatively slow changes over large areas (e.g., global climate change; Thomas et al. 2004), whereas small scales are often associated with more rapid changes over small areas (e.g., temperature changes in partially shaded areas of a small lizard's habitat; Avery 1979). Two concepts related to scale are extent and grain (Kotliar and Wiens 1990; Whittaker et al. 2001). The extent of a study relates to the area over which comparisons are made (e.g., the size of the study area), and the grain of a study is the finest unit of observation or analysis. Appreciating these concepts can be important for conducting and interpreting ecological work. Many ecological phenomena are best understood if multiple scales are considered simultaneously. Multiscaled investigations have led to useful insights on the habitat requirements of single species, including plants (Forman 1964), mammals (Lindenmayer 2000; Johnson et al. 2004), lizards (Rubio and Carrascal 1994), and birds (Luck 2002a). For example, Forman (1964) demonstrated how factors at a hierarchy of spatial scales influenced the distribution of the moss *Tetraphis pellucida* from global climate to the microhabitat of an individual log, with different factors being the most critical ones at a particular spatial scale. Entire communities can also be studied at multiple spatial scales (Kinnunen et al. 2001; Weibull et al. 2003). Finally, the consideration of multiple spatial scales is important for successful biological conservation. Forest management, for example, should consider the likely effects of management actions at multiple scales, from individual trees to landscapes and regions (Tang and Gustafson 1997; Lindenmayer and Franklin 2002).

Figure 1.2. Torres del Paine, Chile. A rugged and mountainous area where elevation and high latitude play important roles in influencing the distribution and abundance of plant and animal taxa (photo by David Lindenmayer).

between finding generalities that apply broadly across many landscapes and many species on the one hand, and accurately predicting the effects of landscape change for single species on the other hand (Harper 1982; Figure 1.3).

A key issue is to determine when it is appropriate to use aggregate measures of multispecies occurrence and when a more detailed focus on single species is more instructive. There are no simple answers to this problem or generic recipes that can be applied uncritically in all circumstances. Rather, the study of processes driving single species' responses to landscape change and the study of aggregate patterns of species distribution provide complementary insights, which can both be valuable from a management perspective. Throughout the book, we have attempted to distinguish clearly where we focus on single species and where we focus on aggregate measures of species occurrence.

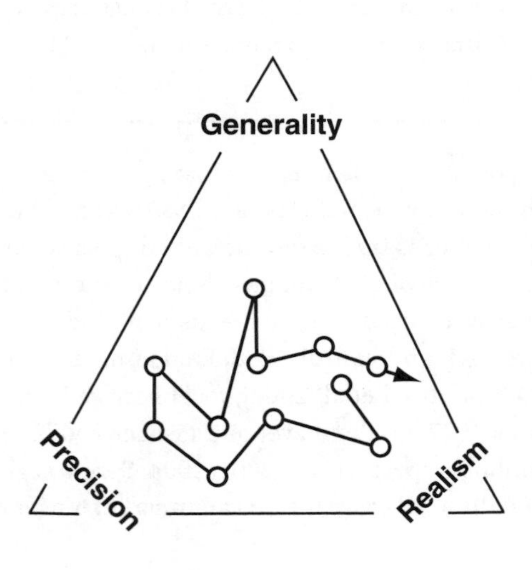

Figure 1.3. Harper's triangle (reworked from Whelan, R.J. 1995: *The Ecology of Fire.* Cambridge Studies in Ecology, Cambridge University Press, Cambridge).

Pattern versus Process

A second major tension that arises in the study of altered landscapes is that landscape modification involves both altered ecological processes and altered patterns of vegetation cover

(Cale 1999). Both processes and patterns can be significantly correlated with various measures of biota. As part of tackling the fragmentation panchreston, we highlight explicitly where we are examining altered landscape patterns as opposed to ecological processes. We are acutely aware that examining both patterns and processes is important. However, there is a clear asymmetry in documenting patterns versus processes. Patterns are directly observable because they consist of the configuration of one or more entities. By contrast, processes can only be inferred indirectly through the patterns they produce (Y. Haila pers. comm.). Nevertheless, a better understanding of the processes that give rise to emergent patterns may clarify how to extrapolate findings from one landscape to another previously unstudied one.

Box 1.4. Forest Birds and the Introduction of the Brown Tree Snake on Guam

Island faunas are particularly susceptible to introduced species (Simberloff 1995). The bird assemblage on the island of Guam in the North Pacific Ocean experienced a massive collapse during the second half of the 20th century (Wiles et al. 2003). Of the 18 native species of birds on Guam, populations of all but one have been either extirpated or severely reduced (Wiles et al. 2003). Many other species of mammals and reptiles have also declined. The decline of Guam's bird fauna is a clear case where landscape changes such as vegetation loss, vegetation subdivision, and edge effects are largely irrelevant factors influencing species loss. Rather, the primary cause of bird declines is predation by the introduced brown tree snake (*Boiga irregularis*). The species is believed to have been accidentally introduced to Guam in the 1940s or 1950s with cargo transported to the military base on the island (Savidge 1987).

A series of studies by Savidge (1987) provided good tests of the hypothesis that predation by the brown tree snake was responsible for the collapse in native bird populations. Bird populations on similar nearby islands without snakes were reasonably stable. The timing of expansions of the brown tree snake's range (in the early 1960s) correlated closely with dramatic declines in the ranges of 10 species of forest-dependent birds on Guam. The northern area of the island was the last known habitat supporting all 10 species, and it was the last place to be invaded by tree snakes (Savidge 1987). Snake control measures are under way, such as aerial baiting with mice injected with the human painkiller Tylenol, which is toxic to snakes. However, the complete loss of so many bird species means that the prospects for avifaunal restoration on Guam are limited (Wiles et al. 2003). Hence key conservation strategies on other islands are to develop predator-proof fences and use sniffer dogs and snake traps (Vice et al. 2005) at air terminals and shipping ports to limit accidental introductions of snakes to other islands via shipments of cargo (Engeman and Linnell 1998; Engeman et al. 1999).

The Structure of This Book

Given that one of our overarching aims was to break the fragmentation panchreston into its subcomponents, this book is organized as a series of parts (Fig. 1.4)—some significantly longer and more complex than others. Variation in the lengths of chapters and parts reflects variation in the richness of material we considered appropriate for discussion. In addition, we have organized chapters and parts so that they build on one another (Figure 1.4). The earlier chapters of the book in particular are basic building blocks for our discussion of more complex material in later chapters. We encourage readers not to skip the early chapters entirely just because they seem somewhat basic—it is precisely because terms like "habitat" or "habitat patch" are often applied carelessly that "fragmentation" has become a panchreston. Because each chapter is only a short foray into what, in most cases, are massive topics (that deserve books in their own right), we provide a Further Reading summary at the end of every chapter.

Chapter 2 in Part I reviews some of the typical ways in which landscapes are altered by human activities. In Chapter 3, we develop a conceptual framework for how individual species and aggregate measures of species occurrence are related to landscape change. This framework forms the basis of Parts II and III of the book.

Part II (Chapters 4–8) examines the responses of individual species to landscape

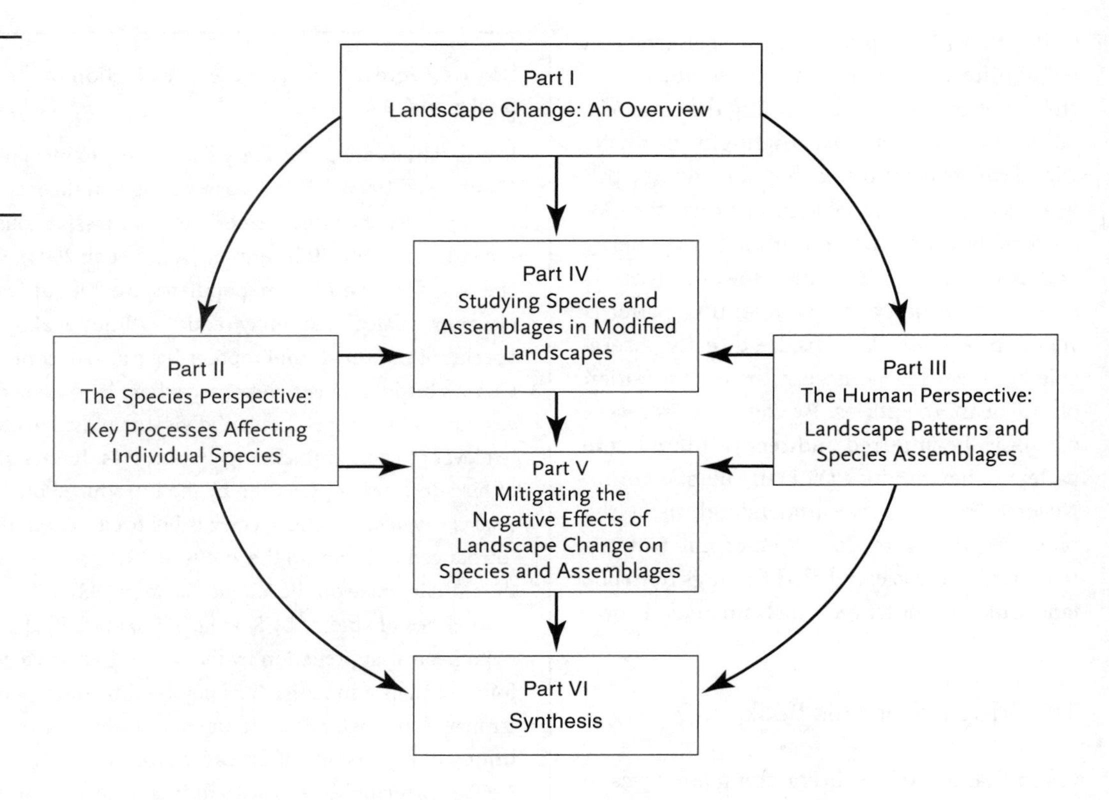

Figure 1.4. Topic interaction diagram showing links between the different parts of the book.

change; particularly the impacts of key threatening processes associated with landscape change—habitat loss, habitat degradation, habitat subdivision, and altered biological processes. Chapter 8 synthesizes information presented in Part II and discusses how various interacting effects of landscape change may lead to population declines for some species.

While Part II corresponds to the ecological processes associated with landscape modification and their effects on single species, Part III focuses on patterns of landscape modification and their relationships with aggregate measures of species occurrence. These themes of processes and patterns are partitioned between Parts II and III in explicit recognition of the fact that landscape modification changes both ecological processes and landscape pattern (Cale 1999). Chapters 9 through 15 cover various themes frequently discussed in the context of pattern-oriented research, including the size of remnant vegetation patches, landscape connectivity, edge effects and vegetation structure, the role of the matrix, and landscape heterogeneity.

Following the overview of how individual species and aggregate measures of species occurrence are affected by landscape change (Parts II and III), Part IV explores methods to study, explain, and predict species distribution patterns in modified landscapes—including empirical investigations, mathematical models, and quantitative as well as qualitative methods of reviewing existing information.

Part V examines approaches and general principles for mitigating the negative impacts of landscape modification on individual species and assemblages of species. The various chapters address the key problems that affect individual species and assemblages of taxa outlined in Parts II and III, respectively; particularly strategies to limit habitat loss and losses in native vegetation cover, ways to buffer edge effects, and ways to maintain or increase landscape connectivity. The final chapter in Part V is an overarching framework for mitigating the impacts of landscape change on biota.

Part VI closes the book with a short synthesis and discusses some future directions and key knowledge gaps.

How to Read This Book

We have tried to make the text accessible to as many readers from different backgrounds as possible. Our intended audience is wide—undergraduate and postgraduate university students, academics and teachers, natural resource managers, conservation biologists, ecologists, and decision makers in natural resource management. We have assumed readers will have a reasonable understanding of basic ecology and conservation biology concepts.

We anticipate that different readers will use this book in different ways. Some readers will dip in and out according to their interests and requirements. Others will want to read the book from beginning to end, whereas some will only want to skim each chapter. Therefore, we provide a short Summary at the end of each chapter. Finally, a complicating problem in synthesizing information on landscape change and its impact on biota is that many topics are intimately associated with other topics. Although we have attempted to minimize repetition in this book, the intimate connectedness of many aspects of landscape change has meant that there is inevitably some overlap in the themes and ideas between the different chapters and parts of this book. Drawing boundaries between them can be difficult, and assigning material to one chapter and leaving it out of another can be somewhat arbitrary at times. Given this, most chapters in this book have a section, Links to Other Chapters, to help guide the reader to other closely related topics examined elsewhere in the book.

Further Reading

Numerous books and articles examine the many subtopics associated with landscape change, habitat loss, and habitat fragmentation. An outstanding set of studies in the highly modified woodlands of the wheatbelt of Western Australia has been published in the series of books by Saunders et al. (1987, 1993a) and Saunders and Hobbs (1991). Although more than a decade old, they are still highly relevant to many issues and topics examined in highly modified landscapes. In many respects, the highly modified prairies of midwestern North America provide a parallel set of circumstances to that of the Western Australian wheatbelt, and the edited book by Schwartz (1997) is a valuable one in this regard. A useful set of papers on modified tropical systems appear in the edited volumes by Laurance et al. (1997) and Bierregaard et al. (2001).

There are several useful reviews of habitat loss and habitat fragmentation that provide an informative entry into this massive and complex topic. Seven that give quite different but interesting perspectives are Davies et al. (2001a), Haila (2002), McGarigal and Cushman (2002), Bennett (2003), Fahrig (2003); Hobbs and Yates (2003), and Noss et al. (2005). Debinski and Holt (2000) provide a useful review of the limited number of fragmentation experiments that have been conducted to date.

PART I
Landscape Change: An Overview

This section contains two chapters (Figure I.1). The first briefly explores ways in which humans have typically changed landscapes in contemporary times. A prominent part of Chapter 2 is that landscape change is rarely a random process; the most productive parts of landscapes are typically those first targeted for modification. If areas of original vegetation cover remain in modified landscapes, they are typically in the least productive places. This phenomenon can have major implications for patterns of species occurrence, as well as attempts to study and mitigate the effects of landscape change, habitat loss, and habitat fragmentation.

In Chapter 3, we explore how the typical kinds of landscape changes brought about by human activities (see Chapter 2) affect single species and aggregate measures of multispecies occurrence. We present several conceptual landscape models—the landscape contour model, the island model, the patch–matrix–corridor model, and the variegation model. We show how these landscape models can assist our analysis of species distribution patterns in human-modified landscapes. Chapter 3 provides the conceptual background for Parts II and III of the book, which contain a detailed discussion of the responses of single species and aggregate measures of species occurrence to landscape change.

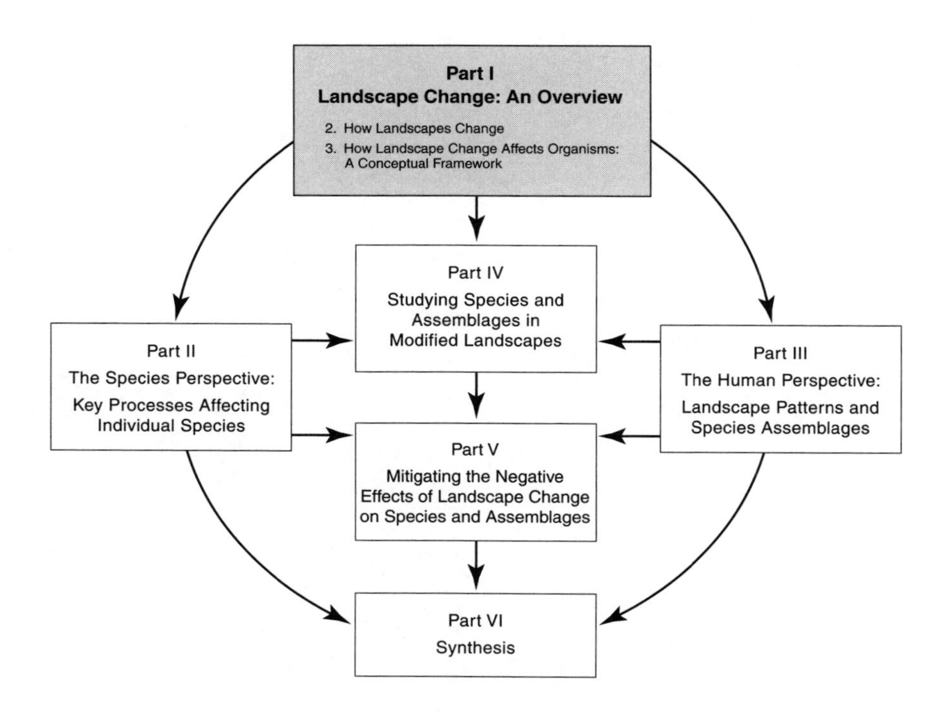

Figure I.1. Topic interaction diagram showing links between the different parts of the book, and the topics covered in the chapters of Part I.

Aerial photo of a modified landscape in the Brazilian Pantanal showing various spatial patterns of vegetation resulting from human land use change as well as natural patterns of vegetation cover resulting from seasonal flooding regimes (photo by David Lindenmayer).

CHAPTER 2
How Landscapes Change

Natural processes can alter landscapes and destroy habitat for some species. Examples are volcanic eruptions (Croizat 1960; Franklin et al. 1985), long-term climatic change (Cunningham and Moritz 1998; Thomas et al. 2004), and wildfires (Williams and Gill 1995; Agee 1999; Bradstock et al. 2002). However, landscape modification by humans is by far the most important modern cause of habitat loss and habitat fragmentation, reducing levels of biodiversity worldwide (Saunders et al. 1987; Primack 2001; Kerr and Deguise 2004). For example, a large part of the Amazon basin has been altered by human activities, largely by land clearing. By 2001, 14% of the Amazon basin or 5 million km^2 had been extensively altered (Skole and Tucker 1993; Ferraz et al. 2005). In the year ending August 2004, over 26,000 km^2 of Amazon rainforest were cleared, primarily for cattle ranches and soybean farms. This was the second highest rate of clearing on record (Government of Brazil 2005). Numerous other examples of extensive vegetation clearing come from Europe, North America, Australia, and Asia, and they parallel those of the Amazon (Wilcove et al. 1986; Saunders et al. 1993a, b; Webb 1997; Box 2.1). However, landscapes are rarely altered and changed in a continuous and unidirectional manner. This chapter explores some of the ways in which landscapes are typically modified by humans.

Typical Patterns of Landscape Change

Landscape modification takes place for numerous and often quite different reasons. Among the most common ones are agricultural expansion (Landsberg 1999; Daily 2001) and urbanization (Luck et al. 2004). In general, landscape change has followed similar patterns in many different parts of the world. Here, we discuss two classifications of landscape change as a useful basis from which to explore how assemblages and individual species respond to landscape change (see Chapter 3).

Forman's Patterns of Landscape Change

Forman (1995) identified five main ways in which humans can alter landscapes spatially: perforation, dissection, fragmentation, shrinkage, and attrition (Figure 2.2). These changes result in different spatial patterning of landscapes and can alter ecological processes and the distributions of plants and animals. Table 2.1 presents some examples of the ways in which these forms of landscape change might manifest.

> **Box 2.1. Vegetation Loss in the Brazilian Cerrado**
>
> A large amount of the native vegetation that is cleared around the world each year is in Brazil, and much of it is in Brazilian Amazonia (Ferraz et al. 2005). However, the extent of land clearing is substantial in other parts of Brazil. The second-largest biome in Brazil is the Cerrado, which is a mosaic of grassland, palm stands, and forests (Figure 2.1). The Cerrado supports more than 7000 species of endemic plants and an array of threatened vertebrates. The amount of land clearing is as high and perhaps higher than in the Amazon and is taking place to establish crops, including soybeans, rice, wheat, and maize. Only ~2% of the Cerrado is protected (compared with more than three times that level for the Amazon) and the area, although rich in endemic species, has received limited attention from conservation scientists (Klink and Machado 2005).

Figure 2.1. Brazilian Cerrado (photo by David Lindenmayer).

Collinge and Forman (1998) empirically tested the effects of different patterns of landscape change on organisms by quantifying the response of grassland insects in a mowing experiment conducted in an artificial "micro-ecosystem." They found significant differences in the insect communities subject to different spatial patterns of landscape change. Species richness declined where shrinkage had occurred, but increased where the landscape pattern was fragmented. In fragmented systems, species richness increased in smaller, separated patches, possibly as a result of crowding via the influx of animals from the surrounding modified areas. Many other effects were identified, but overall the investigation demonstrated that insects responded strongly to the different spatial patterns created by different kinds of landscape change (Collinge and Forman 1998).

The McIntyre and Hobbs Model of Landscape Change

An additional description of landscape change to Forman's (1995) was presented by McIntyre and Hobbs (1999). These authors suggested that landscape change tended to have a temporal component, with landscape modification often increasing through time. According to their classification, four broad classes of

Table 2.1. Examples of Forman's (1995) patterns of landscape change (also see Figure 2.2)

Pattern of landscape change	*Example*
Perforation	Mine sites in remote areas (e.g., West Papua)
Dissection	Roading in remote areas (e.g., Pan-American Highway in Nicaguara)
Fragmentation	Remnant vegetation in grazing lands
Shrinkage	Patch size reduction due to realignment of fences around pastures
Attrition	Lowest productivity patches cleared last in heavily cleared landscapes

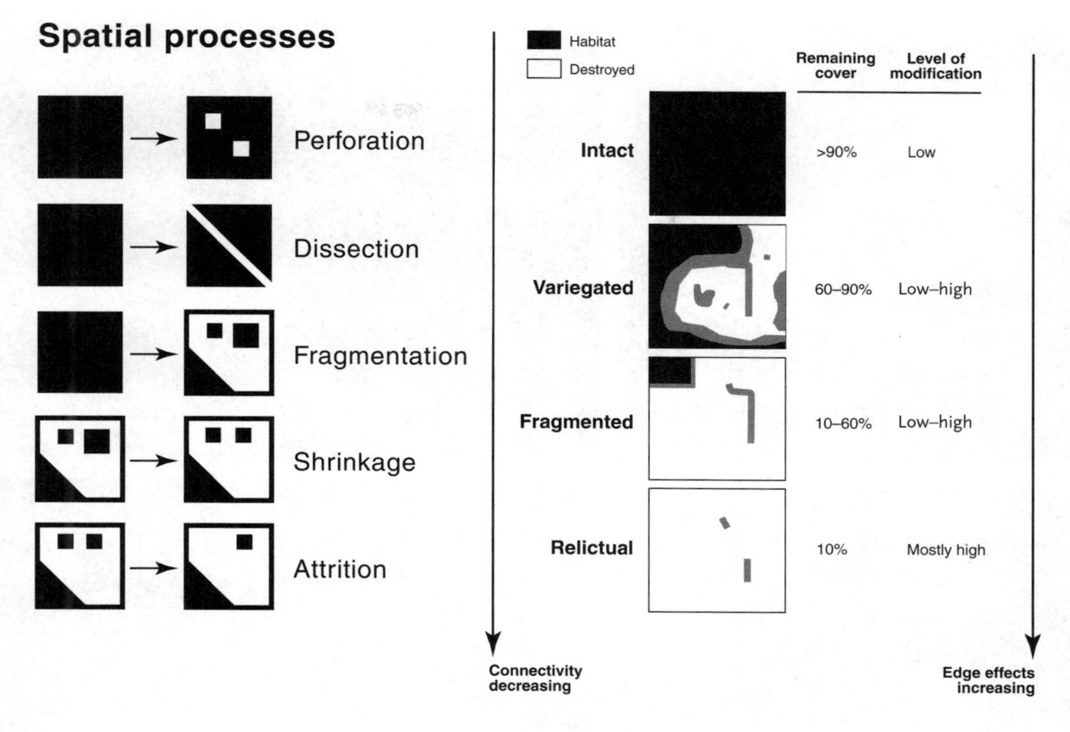

Figure 2.2. (Left) Five ways in which landscapes can be modified by humans (redrawn from Forman, R.T. 1995: *Land Mosaics. The Ecology of Landscapes and Regions.* Cambridge University Press, New York).

Figure 2.3. (Right) Landscape alteration states (following McIntyre, S. and Hobbs, R. 1999: *Conservation Biology*, **13**, 1282–1292, A framework for conceptualizing human effects on landscapes and its relevance to management and research models, with permssion from Blackwell Publishing).

landscape condition could be identified along a continuum of human landscape modification. Landscapes could be intact, variegated, fragmented, or relictual (Figures 2.3, 2.4). As in Forman's (1995) model, these different classes corresponded to different spatial patterns in the landscape. McIntyre and Hobbs (1999) suggested that as human landscape modification increased, the amount of intact habitat would decrease, and habitat degradation would increase. In addition, relictual and fragmented landscapes were characterized by more sharply defined patch boundaries than variegated or intact landscapes.

Aspects of the model by McIntyre and Hobbs (1999) had previously been explored by McIntyre and Barrett (1992), and, similarly to Forman's (1995) model, parts of their framework have been examined empirically. For example, Ingham and Samways (1996) demonstrated that the distribution patterns of macroinvertebrates in a modified landscape in South Africa changed gradually through space. Hence they argued that, in the specific context of macroinvertebrates, the alteration state of the landscape they investigated could be best described as variegated, rather than fragmented or relictual (Ingham and Samways 1996).

The models of Forman (1995) and McIntyre and Hobbs (1999) could be seen as complementary because they both recognize that landscape change often results in discontinuities in remaining land cover and a decrease in the amount of native vegetation. Landscape modification typically results in a decrease in the average size of remaining vegetation patches, an increase in the average distance between these patches, a decrease in landscape connectivity between patches, and an increase in the ratio of patch edges to patch sizes (see Part III). The impacts of such changes on biota may be exacerbated by changes in fluxes of wind, water, radiation, and nutrients (Laurance et al. 1997; Harper et al. 2005).

Figure 2.4. States of landscape condition: Intact (top left), Variegated (top right), Fragmented (bottom left), and Relictual (bottom right). (photos by Richard Hobbs).

Examples of Landscape Change

Figure 2.5 gives examples of typical patterns of anthropogenic landscape change. The figure includes trends in vegetation cover studied at Naringal, Victoria, southeastern Australia (Bennett 1990a, b). These patterns provide a useful example of how landscapes can change and are indicative of those seen in many other places elsewhere around the world.

The Naringal region was originally covered by "thick forest" and was first settled by Europeans in the late 1830s. The forests were used to graze domestic livestock, and extensive clearing began some 30 to 40 years after European settlement. By 1947, approximately half the native vegetation in the Naringal region had been cleared, and the rate of forest removal accelerated substantially during the following decades (see Figure 2.5a). By 1966, a total of 19% of the original vegetation cover remained, and in 1980 only 8.5% of the original vegetation cover remained. Not surprisingly, there was a marked change in the size of remnant blocks of vegetation. In 1947, about 90% of remnant vegetation occurred as patches exceeding 100 ha in size. In 1980, 92% of remnants measured less than 20 ha and none was larger than 100 ha (Bennett 1990b). The Naringal region now supports numerous small dairy farms. The small patches of native vegetation are either surrounded by pasture or occur as remnant riparian strips or

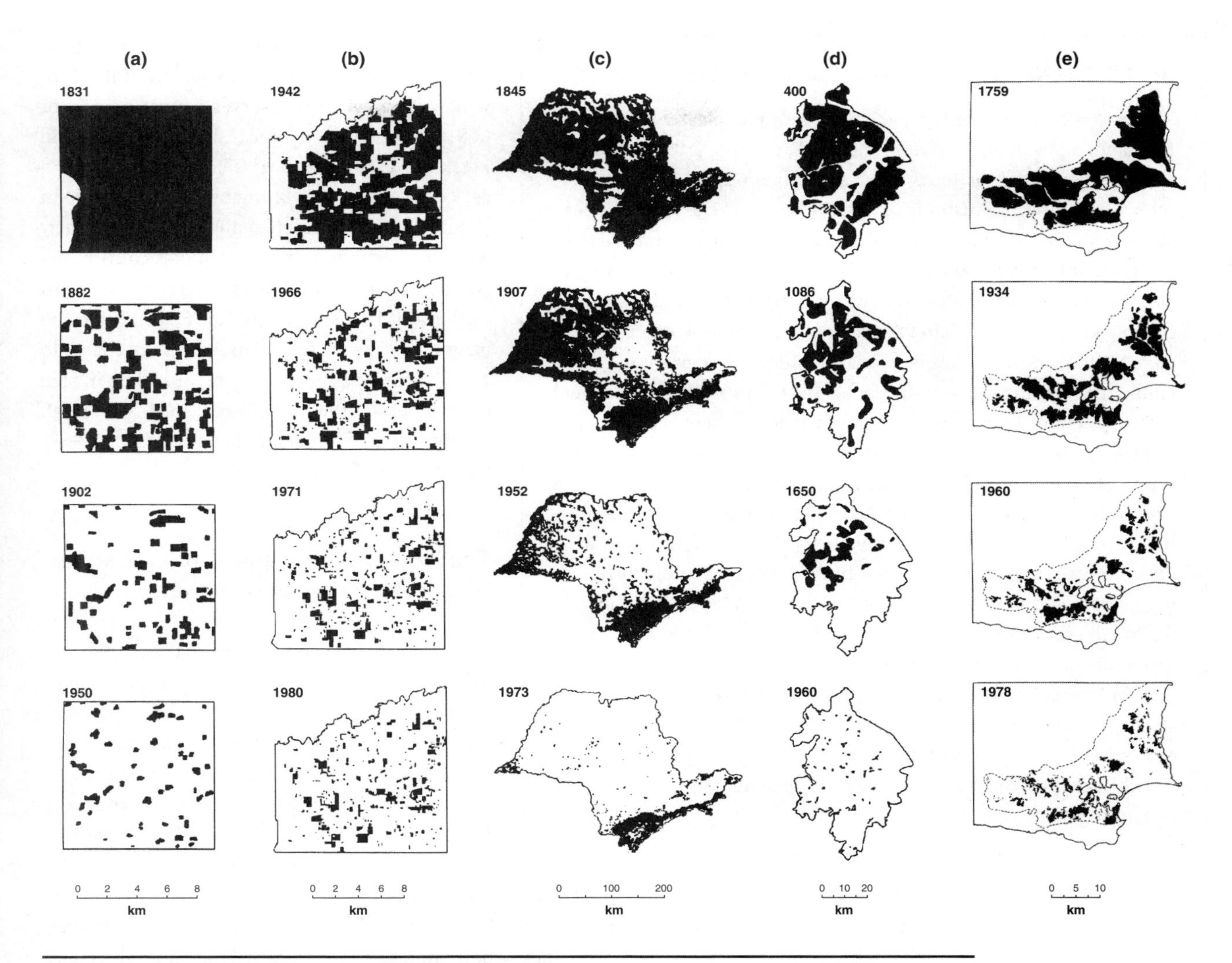

Figure 2.5. Land clearance patterns and habitat fragmentation in: (a) Naringal, Victoria; (b) Cadiz township, Green County, Wisconsin, USA; (c) Poole Basin, Dorset, England; (d) São Paulo, Brazil; and (e) Warwickshire, England (after Webb, N.R. and Haskins, L.E. 1980. Reprinted from *Biological Conservation,* **17**, 281–296, An ecological survey of heathlands in the Poole Basin, Dorset, England in 1978, with permission from Elsevier; after Bennett, A.F. 1990b. Reproduced with permission from *Australian Wildlife Research,* **17**, 325–347 [Bennett, A.F.]. Copyright CSIRO 1990. Published by CSIRO PUBLISHING, Melbourne VIC, Australia [http://www.publish.csiro.au/journals/ajb]; after Oedekoven, K. 1980. Reprinted from *Environmental Policy and Law,* **6**: 184–185, The vanishing forest, with permission from IOS Press; after Smith, D.S. and Hellmund, P.C. 1993: *Ecology of Greenways: Design and Function of Linear Conservation Areas,* University of Minnesota Press, Minneapolis).

Box 2.2. The Nonrandom Pattern of Landscape Change

The ecological literature is replete with examples clearly illustrating that landscape change is rarely random. Landsberg (1999) studied the vegetation cover patterns of the Australian Capital Territory, the region surrounding Australia's capital, Canberra. The region spans approximately 235000 ha, and more than 50% of it is in formal reserves. Subalpine woodlands are largely intact, with just 1 to 12% cleared, whereas lower elevation, warmer, and more productive foothill woodlands are 85 to 98% cleared. Formal protection of vegetation communities is in reverse order to the extent of landscape modification—alpine woodlands are strongly represented in the system of protected areas, but lowland woodlands are extremely poorly reserved (Landsberg 1999). This phenomenon of nonrandom patterns of landscape change and protection is repeated countless times around the world. In northwestern North America, highly productive sites were occupied by private landowners more than a century ago, and so they are heavily modified and poorly represented in forest reserves. Consequently, protected areas in this region are dominated by steep, high-elevation lands with low vertebrate diversity (Harris 1984; Scott 1999). Similar circumstances exist in Scandinavia (Virkkala et al. 1994), North America (Scott et al. 2001), and South America. For example, in Chile, areas below 43 degrees latitude support the highest numbers of species but are heavily modified and have virtually no representation in the reserve system (Armesto et al. 1998).

roadside reserves. The changes in vegetation cover at Naringal have had profound impacts on the vertebrates of the region—which we describe in more detail in the section on observational studies in Part IV (Chapter 16).

The patterns of landscape alteration documented by Bennett (1990a, b) are paralleled by others documented elsewhere around the world. For example, Webb (1997) described the processes of landscape alteration in the heathlands of Dorset in England. Between 1759 and the mid-1980s, the area of heathland was reduced to approximately 6000 ha from an original 40000 ha. The initial extent of heathland in the mid-1700s was in 10 large blocks separated by rivers. In the following 2 centuries, heathland cover was reduced by ~85% and the number of patches increased from 10 to more than 600. In 1978, 14 patches exceeded 100 ha, of which three were in reserves. In contrast, over 608 patches were 4 ha or smaller (Webb and Haskins 1980; Webb 1997). The loss of forest cover in the Cadiz township in Wisconsin (USA) shows a similar pattern. Green County was characterized by an almost continuous cover of native forest at the time of European settlement in 1831 (Curtis 1956). By 1978, the original cover of forest had been reduced to 111 forest remnants averaging only 9 ha in size (Sharpe et al. 1987). Between 1831 and 1950, the total area of forest had been reduced by 96% (Curtis 1956; Burgess and Sharpe 1981).

The Nonrandom Patterns of Landscape Change and Vegetation Cover

Patterns of landscape change, habitat loss, and habitat fragmentation are rarely random processes. Some vegetation and habitat types survive relatively unscathed, whereas others may be eliminated. In many cases, the most productive land is extensively modified (e.g., Landsberg 1999; Box 2.2). Remnant vegetation is often located in areas that are unsuitable for other land uses; for example, because they are too steep, too rocky, or located at high elevations (a phenomenon sometimes called the "worthless land hypothesis" [Hall 1988]; see Boxes 2.2, 2.3).

Some Implications of Nonrandom Patterns of Landscape Change

The nonrandom nature of landscape change has some important implications. First, the spatial distribution of vegetation types reflects spatial heterogeneity in landscape productivity. Remnant vegetation will not be a representative sample of the unmodified landscape (Norton et al. 1995; Huggett and Cheesman 2002). Usually, different faunal assemblages associate with different vegetation types (Bennett et al. 1991; Morrison et al. 1992). For example, some theoretical and empirical evidence suggests that species diversity of plants and animals may be positively associated with productivity (Srivastava and Lawton 1998; reviewed by Waide et al. 1999; Mittelbach et al. 2001), although for some groups and at some spatial scales (e.g., the regional scale), hump-shaped relationships are apparent, with species richness increasing initially and then declining at higher levels of productivity (Rosenzweig 1995; Gaston and Spicer 2004). More generally, upland and lowland areas in a given landscape are likely to support different sets of species (Sabo et al. 2005). Thus populations of particular species may be smaller in low productivity areas, such as those on poor soils (Braithwaite 1984), on steep terrain (Lindenmayer et al. 1991a; Braithwaite 2004), or at high elevations (Harris 1984)—although this does not mean that low productivity areas (such as heathlands or places with limited soil fertility) are without conservation value (Whittaker 1954a, b; Keith et al. 2002a). Nevertheless, previously abundant species characteristic of productive vegetation types may be those most susceptible to decline because of habitat loss and may be the first to be lost. This also means that current patterns of species distribution and abundance may not be closely correlated with those that existed prior to landscape change (Seabloom et al. 2002). Godwin (1975) provides an interesting example of the short-leaved lime *(Tilia cordata)* from southern England. The species was most probably an important tree in forest environments on lighter soils before they were cleared to establish agricultural fields. The species is now virtually confined to hedgerows in such agricultural areas. Notably, the species was probably largely absent from native forests and woodlands that remained unconverted because such areas occur primarily on heavy clay-dominated soils, which are unsuitable for the short-leaved lime (Whittaker 1998).

Box 2.3. The Nonrandom Nature of Patterns of Human Population Density and the Distribution of Biota

Problems created by nonrandom patterns of vegetation clearing and nonuniform patterns of species distribution are not ones that are simply legacies of past eras. Unfortunately, there is considerable potential for future spatial "conflict" between the expansion of the world's human population and ongoing losses of species. For example, Luck et al. (2004) found that human population densities in North America and Australia tended to be high where the inherent species richness of birds, mammals, amphibians, and butterflies also was high. Similar problems have been found by other workers, such as Cincotta et al. (2000), who showed that the world's 25 hotspots where high concentrations of species (and threatened species) occur have human population densities (~73 people per km^2) almost twice as high as the global average (~ 42 people per km^2). Human population growth rates and increases in per capita consumption in hotspots are also faster than elsewhere and are forecast to continue that way for several decades to come (e.g., Liu et al. 2003).

A second reason that nonrandom patterns of landscape change are important relates to attempts to protect formally representative samples of original vegetation cover that existed prior to contemporary human disturbance. "Representativeness" is a term often applied in reserve design and selection (Margules and Pressey 2000), and it relates to attempts to conserve examples of the full range of environments, vegetation types, or species (Lindenmayer and Burgman 2005). In areas that are highly modified by humans, remaining productive parts of landscapes may be small or highly degraded (and poorly represented in reserves). Nevertheless, such areas may still be very important for conservation even though they are small and/or in relatively poor condition. There are many examples from around the world where this is the case, including the tallgrass prairie and savannah communities of the midwestern United States (Robertson et al. 1997; Schwartz and Mantgem 1997; Figure 2.6), Florida scrub in the southeastern United States (McCoy and Mushinsky 1999), and the heavily

Figure 2.6. Tallgrass prairie in Illinois (photo by K. R. Robertson, Illinois Natural History Museum).

grazed temperate woodlands of southeastern and southwestern Australia (Abensperg-Traun and Smith 2000; Gibbons and Boak 2002).

Finally, nonrandom patterns in landscape change can be important for studies that attempt to quantify the response of biota to landscape change. Many studies of habitat fragmentation compare the fauna (and less often the flora—see Hobbs and Yates 2003) of patches of remnant vegetation of varying sizes. This is undoubtedly a useful endeavor, but nonrandomness in landscape change means that the environmental conditions and habitat types that characterize large vegetation remnants are likely to be quite different from those in small ones. Therefore, there can be confounding effects of remnant size and remnant characteristics—larger remnants are often located in less productive areas. This can mean that some species may be absent from particular remnants not because of the effects of habitat loss or habitat subdivision but because conditions within unproductive areas may never have been suitable for them. This underscores one of the key themes of this book—the importance of understanding the habitat requirements of organisms as part of determining the impacts of landscape change on them (see Part II).

Dynamism in the Patterns of Vegetation Cover and Landscape Change

Many perspectives on landscape alteration are unidirectional, with continued alteration assumed to reduce both the size of individual patches ("shrinkage" in Figure 2.2) and the overall number of patches ("attrition" in Figure 2.2). However, there are many cases where trends in landscape change have been reversed (Vellend 2003). For example, in some areas of wood production forest, cover patterns change rapidly because of regeneration of logged areas (Fahrig 1992). In the northeastern United States, extensive forest clearing for agricultural development accompanied European settlement. In the state of Massachusetts, less than 30% of the original forest cover existed by the mid-1800s, and the total amount of remaining old-growth forest was less than 500

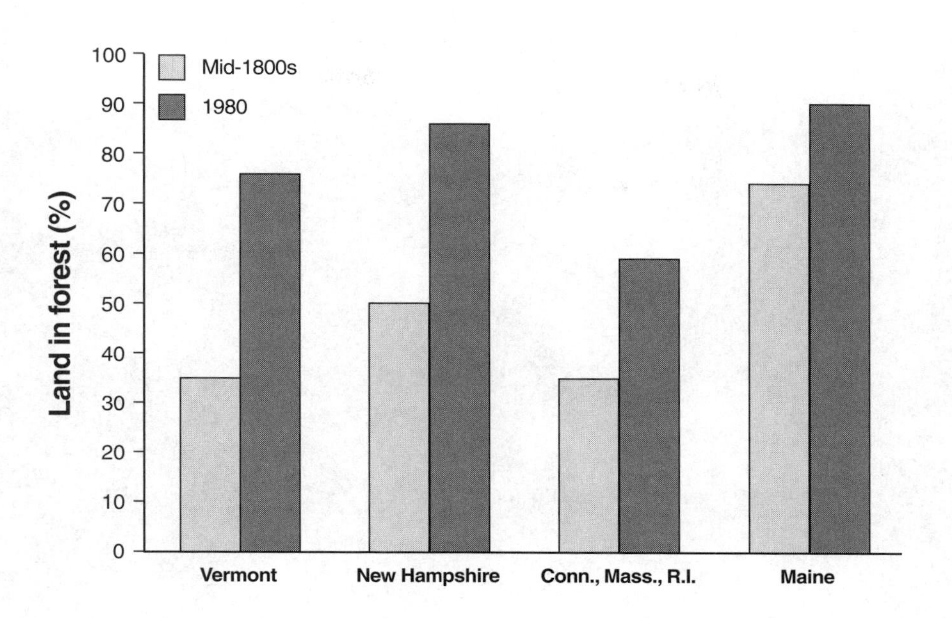

Figure 2.7. Trends in forest cover, mid-1800s and 1980 across several states in northeastern United States (redrawn from MacCleery, D.W. 1996: *American Forests: A History of Resiliency and Recovery,* Forest History Society Issues Series, Durham, North Carolina, reproduced with permission from The Forest History Society).

ha, primarily in patches 1.5 ha or smaller (Hall et al. 2002). Following the industrial revolution, many workers left agricultural employment and began working in factories and service industries. Abandoned agricultural fields regenerated and forest cover increased rapidly. Extensive areas of Massachusetts are now covered in forest, although pressure on the forest has returned, this time from urban expansion (Figures 2.7, 2.8). However, the structure of these recovering forests remains greatly altered in comparison with pre-European conditions and, for example, old-growth features such as very large cavity trees remain rare and may take another several centuries to develop (D. Orwig, pers. comm.).

There are many other examples of native vegetation recovery following land clearing. For example, some temperate woodlands in southeastern Australia have been cleared at least three times in the last 150 years and have recovered to some extent each time (Lindenmayer et al. 2001). New forests are also emerging on abandoned agricultural land that was previously extensively cleared in the Canadian province of Quebec (Burton et al. 2003) and on the island of Puerto Rico (Lugo and Helmer 2004). In each of these cases, although vegetation cover has increased, the composition and structure of that vegetation cover may take considerable additional time to revert to preclearing conditions.

The foregoing examples indicate that trends in landscape cover and associated patterns of species distribution and abundance are not necessarily unidirectional. Mather (1990) summarized broadscale patterns of forest change and associated human perceptions based on historical forest cover in Great Britain in a simple conceptual model. The model encompasses cases where forest cover was depleted and then recovered and how such changes were perceived by human communities (Table 2.2).

There are many other important examples where the dynamics of cultural landscapes can have a significant impact on biota such as those in parts of Europe where there has been a long history of human modification (Peterken and Francis 1999). For example, the loss of cultural landscape elements (such as hedgerows,

Figure 2.8. Forest cover in western Massachusetts that was formerly entirely cleared for agricultural production (photo by David Lindenmayer).

ditches, and footpaths) from Danish agricultural landscapes has had negative impacts on many native species typically associated with cultural landscapes containing these elements (Agger and Brandt 1988). Similarly, forest encroachment on open vegetation types such as grasslands and heather can have negative impacts on a range of plant and animal biota in the United Kingdom as well as other parts of Europe such as France, Sweden, and Norway (Peterken and Ratcliffe 1995; Aanderaa et al. 1996). In the Wyre Forest National Nature Reserve in the midlands of England, areas of heath are deliberately kept open because the invasion of trees and shrubs would degrade habitat quality for reptiles such as the adder *(Vipera berus)*. In the forests of the Swiss Jura, a reduction in the cover of trees and shrubs is also considered critical for the survival of populations of the asp viper *(Vipera aspis)* (Jäggi and Baur 1999). Finally, in Spain and Portugal, the maintenance of open woodland cork plantations is important for some key elements of the biota, including the highly endangered Iberian lynx *(Lynx pardinus)* (Ferreras 2001).

In their model of landscape change, McIntyre and Hobbs (1999; see also Wiens 1994; Pearson et al. 1996) recognized that landscape change could be dynamic and was not necessarily a unidirectional process of vegetation loss or deterioration (see Figure 2.3). They believed that by recognizing different landscape alteration states, conservation strategies could be better focused. Thus, for numerous elements of the biota, it may be possible to return degraded landscapes to a variegated or even intact condition (Hobbs et al. 1993; McIntyre et al. 1996), although it may take a long time for this to occur (Franklin et al. 1981).

Summary

In modern times, human activities have had far greater impacts on landscapes, assemblages, and individual species than have nat-

Table 2.2. Sequential model of forest change (modified from Mather 1990)

Phase	*Description*	*Trend in forest area*	*Public perception of change*
1	"Unlimited resource"	Contraction	Positive or neutral
2	Depleting resource	Contraction	Negative
3	Expanding resource	Restoration/expansion	Neutral or positive
4	Equilibrium (?)	Stability (?)	Positive/neutral/negative

ural disturbances. Human-derived landscape changes are not random. High-productivity areas are typically those modified first and most extensively. Because high-productivity areas tend to support different sets of species (and potentially a larger number of species), the bias in human landscape modification has major implications for: (1) which species are lost and which species remain; (2) interpreting studies of the effects of landscape change on biota, particularly in the context of contrasts between different vegetation remnants (e.g., large versus small areas), and between modified and unmodified areas; and (3) strategies that attempt to conserve representative samples of vegetation in modified landscapes. Patterns of landscape change are not always unidirectional, and there are several examples of ecosystem recovery in areas subject to extensive human modification. Importantly, there is no single "correct" way in which to think about ecological patterns in modified landscapes. In different situations, different models of landscape change can be useful to communicate patterns of landscape change—including patterns of ecological degradation as well as recovery.

Links to Other Chapters

Chapter 3 provides a conceptual overview of how single species and species assemblages respond to the patterns of landscape change discussed in this chapter. The effects of various changes in landscape pattern on assemblages are examined in more detail in Part III. Part V of the book introduces approaches for the mitigation of the negative effects of landscape change on biota. Several of the strategies discussed are related to reversing changes in landscape pattern.

Further Reading

Forman (1995) gives an excellent outline of the ways landscapes can change in his seminal book *Land Mosaics*. An empirical examination for insect populations of the Forman (1995) model can be found in Collinge and Forman (1998). Hunter (1997) provides a valuable discussion of the Forman (1995) model in a forest management context. McIntyre and Hobbs (1999) discuss their conceptual framework of human landscape modification, and Lindenmayer et al. (2003a) apply it in a native forest/exotic plantation context. Saunders et al. (1987) and Norton et al. (1995) discuss many of the problems associated with the fact that landscape change is a nonrandom process. Useful examples of the problem are provided for vegetation types by Landsberg (1999) and Scott et al. (2001) and for ecological processes by Norton et al. (1995) and Laurance et al. (2004). The book edited by Schwartz (1997) provides a range of valuable examples of conservation issues associated with small vegetation remnants in highly modified landscapes. Peterken and Francis (1999) and Jäggi and Baur (1999) give some interesting examples of where dynamic landscape change in Europe can have negative impacts on biota.

CHAPTER 3

How Landscape Change Affects Organisms: A Conceptual Framework

In Chapter 2, we described some typical ways in which human activities can change landscapes. But how do the patterns described in the previous chapter affect the distribution of organisms? This chapter provides the conceptual basis from which to address this question in more detail in later parts of the book. In this chapter, we introduce four different landscape models. We loosely define a landscape model as a conceptual tool that provides terminology and a visual representation that can be used to communicate and study how organisms are distributed through space. In theory, landscape models could be applied at many organizational levels—from genes to ecosystems. In practice, the species is the most widely accepted organizational unit in both scientific and land management contexts (Box 3.1). For this reason, we begin this chapter by focusing on the processes that underlie the distribution of individual species, and we present a landscape model for individual species. Because it is impossible to know everything about the spatial distribution of all individual species, we also discuss landscape models that focus on the relationship between landscape pattern and aggregate measures of species occurrence (species richness and species composition). We argue that a focus on the relationship of single species and ecological processes is complementary to a focus on the relationship between landscape pattern and species richness or species composition.

Processes Affecting the Distribution of Individual Species

Much research on how landscape change affects organisms launches directly into an assessment of how species are distributed in relation to human-defined vegetation patches. However, different species occur in many different places, and for many different reasons—and the distribution pattern of some species may not be closely related to human-defined patches (see Table 1.2). An alternative is to consider the various requirements of a species that need to be met for it to complete its life cycle. For example, species can be limited in their spatial distribution by climatic conditions, food or shelter availability, insufficient space, or the presence of competitors, predators, and mutualists (Fischer et al. 2005a; Table 3.1).

Because different species have different requirements, they can coexist in a given location (Schoener 1974). For example, Pianka (1973) reviewed how different lizard species could coexist in North American deserts by having slightly different food requirements. In arid Australia, lizard species richness is particularly high compared to deserts in other parts of the world. Here, different species use different microhabitats with different thermal properties, have different food requirements, and forage at different times of the day (Pianka 1973; see also Avery 1979).

In addition to some species co-occurring at a single location, different species' unique requirements often mean that they occur at very different locations. In his seminal paper on plant distribution patterns, Gleason (1939) discussed his "individualistic concept of plant association." According to Gleason (1939), different plant species have unique biophysical requirements and their spatial co-occurrence is largely a function of these requirements being met. The idea of species responding uniquely to their biophysical environment was further elaborated by advocates of the continuum concept (Austin and Smith 1989), which recognizes that species composition changes gradually "along environmental gradients, with each species having an individualistic and independent distribution" (Austin 1999, p. 171). The "Gleasonian" school of thought highlights that a human-defined vegetation patch may not necessarily equate to suitable habitat for all species—for example, for some species it may lack suitable locations to find shelter, or it may be climatically unsuitable (see Table 3.1). On this basis, we believe it is important to remember that it is likely that all species in a given landscape respond differently to landscape change, and that each species may have its own unique spatial distribution.

The Landscape Contour Model

The way animals perceive a landscape can be very different from how humans perceive it. Manning et al. (2004a) attempted to capture the concept that each species perceives a landscape in its own particular way by reapplying the notion of Umwelt (von Uexküll 1926) to animal ecology. "Umwelt" is the German word for "environment" and literally means the "surrounding world"; the concept emphasizes species' unique perceptions of their environment. A similar realization led Fischer et al. (2004a) to develop the landscape contour model. This model incorporates multiple species and their unique habitat requirements and can characterize gradual habitat change across multiple spatial scales. The conceptual foundation comes from Wiens (1995), who suggested

Box 3.1. The Species Concept and Its Usefulness in Conservation

Traditionally, most biologists think of species as "interbreeding natural populations that are reproductively isolated from other such populations" (Noor 2002, p. 153). This definition is generally considered the "biological species concept" (Mayr 1942; reviewed by Mallet 1995). Over the last decade, there has been much debate about the appropriateness of the biological species concept (Noor 2002). For example, Mallet (1995) pointed out that the notion of species being genetically isolated from one another contradicted the necessity of gradually changing gene flow, which is widely regarded as a major driver of speciation. For this and other reasons, several additional definitions of what constitutes a species have recently been suggested, and the debate on how best to define a species is far from resolved (see Mallet 1995; Noor 2002; Isaac et al. 2004). Indeed, as indicated in the quote below, this problem was identified by Charles Darwin (1859, p. 101) and remains today:

> No one definition has as yet satisfied all naturalists; yet every naturalist knows vaguely what he means when he speaks of a species.

Despite the limitations of the species concept, focusing on individual species can be a useful starting point to understand the effects of landscape change on biota. First, many studies have been conducted on individual species, and these studies provide a valuable basis from which to understand and mitigate species decline. Second, the biological species concept is useful in practice because it is widely understood outside the scientific community (Caughley and Gunn 1996). Third, the species concept is deeply embedded in a wide range of legislation and conservation strategies (e.g., threatened species recovery plans; Department of Environment and Heritage 2005).

It is important to note that the individual species is a convenient taxonomic unit, rather than an inherently more worthwhile unit to conserve than, for example, genetic diversity (but see Moritz 1994, 1999 for a discussion of evolutionarily significant units) or the diversity of genera (Gaston and Spicer 2004). This is because biodiversity conservation, by definition, is concerned with all levels of ecological organization (see Box 1.2). Protecting individual species, or maintaining species richness, should be seen as one particularly tangible means to achieve the overriding objective of biodiversity conservation.

Table 3.1. Examples where food, shelter, space, climate, and interspecific processes are significantly related to the spatial distribution of a species

Example	*Reference*
Food	
Ontario, Canada: The ovenbird *(Seiurus aurocapillus)* nests in territories with significantly higher prey biomass than at randomly selected sites	Burke and Nol 1998
Southern Australia: Honeyeaters are limited by food availability in summer and autumn	Paton 2000
Shelter	
Iberian peninsula, Spain: The Iberian rock lizard *(Lacerta monticola)* is restricted to areas with many large rocks	Martín and Salvador 1997
Saskatchewan, Canada: Forest structure was shown to be related to forest age, and bird species sheltering in different habitat features (shrubs, canopy, cavities) occurred in different age classes	Hobson and Bayne 2000
Space	
Central United States: Populations of the bald eagle *(Haliaeetus leucocephalus)* in Yellowstone National Park include individuals that have moved in from the surrounding source areas outside the national park	Swenson et al. 1986
Northern Territory, Australia: To gather sufficient food, frugivorous birds are reliant on a large mosaic of rainforest patches and regularly move between patches	Price et al. 1999
Meso- and microclimate	
Switzerland: Climate variables are reliable predictors of the occurrence of several reptile and plant species at the mesoscale	Woodward 1987; Guisan and Hofer 2003
Southeastern Madagascar: Reptiles and amphibians inhabiting fragments of tropical forests respond strongly to microclimatic gradients at the edges of forest patches	Lehtinen et al. 2003
Interspecific processes	
California Channel Islands, United States: By providing abundant food, the introduction of the feral pig *(Sus scrofa)* facilitated the colonization of previously uninhabited islands by the golden eagle *(Aquila chrysaetos)*	Roemer et al. 2002
New Zealand: Competition by southern beech (*Nothofagus* spp.) limits the spatial distribution of other widespread conifer and broad-leaved tree species	Leathwick and Austin 2001
Southern Spain: The distribution of the butterfly *Plebejus argus* is closely related to the distribution of its mutualist ant *(Lasius niger)*	Gutierrez et al. 2005

viewing landscapes as "cost–benefit contours," and Lindenmayer et al. (1995a), who highlighted how spatially explicit habitat models are similar to contour maps.

Contour maps provide a familiar graphical representation of complex spatial information. The landscape contour model represents a landscape as a map of habitat suitability contours overlaid for different species (Figure 3.1). The landscape contour model is based on the following premises (after Fischer et al. 2004a):

- Habitat is a species-specific concept (Hutchinson 1958; Block and Brennan 1993).
- The spatial grain at which species respond to their environment varies between species (see Kotliar and Wiens 1990; Manning et al. 2004a). This is translated onto a contour map through different contour spacing. Species may have densely spaced contours with many peaks and troughs (fine spatial grain) or widely

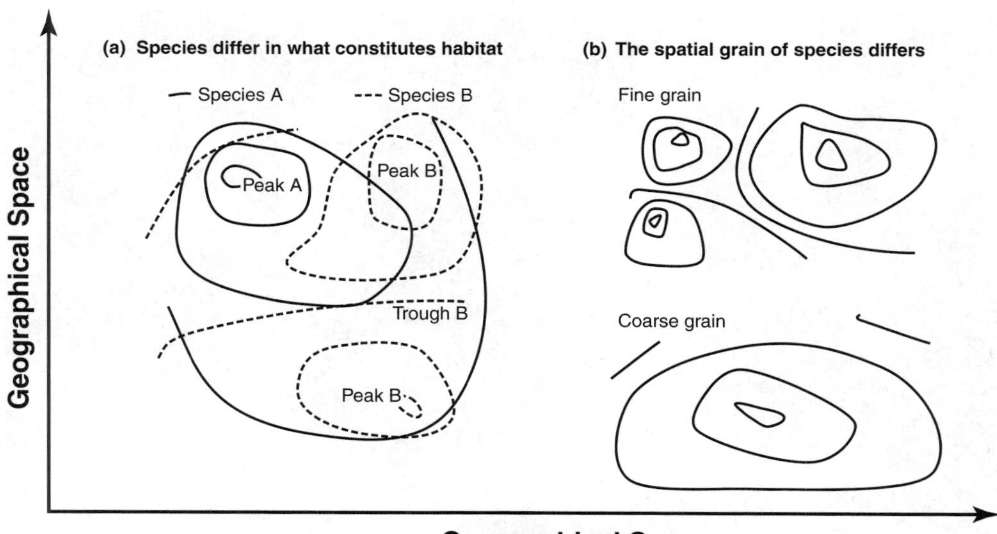

Figure 3.1. Graphical presentation of a contour-based conceptual landscape model. The model recognizes gradual changes in habitat suitability through space. Habitat contours are the emergent spatial pattern resulting from a myriad of ecological processes, including availability of suitable food, shelter, climate, and sufficient space; as well as competition, predation, and other interspecific processes (see Table 3.1).

spaced contours with few peaks and troughs (coarse spatial grain).

- Different spatial resolutions can be used to represent responses at different spatial scales. For example, at the continental scale, the interval between contours can be coarse.
- A habitat contour map for a single species emerges from many ecological processes operating at many spatial scales. The map may not directly reflect the processes causing the pattern of distribution or abundance.
- A contour map does not have a temporal component. Multiple "snapshots" of contour maps at different times could reflect temporal dynamics such as changing habitat requirements at different stages of a life cycle or changes in habitat suitability with ecological succession (e.g., Palomares et al. 2000).

Fischer et al. (2004a) combined data (tree age and hollow tree abundance, with mapped information on these variables) on the habitat requirements of the greater glider *(Petauroides volans)* (Figure 3.2), an Australian arboreal marsupial in the montane ash forests of the Central Highlands of Victoria. The resulting "contour map" of the species' predicted distribution is shown in Figure 3.3. The empirical application of the contour approach depends on quantification of habitat requirements of the species and maps of the habitat features.

Limitations of Single-Species Approaches

Fundamental to the landscape contour model is the recognition that different species occupy different habitats, and that no two species will respond to landscape change in precisely the same way. Although, strictly speaking, this may be true, it poses a difficult challenge to land management (Simberloff 1998). First, the identity of all species in a given landscape is often not known. Globally, Gaston and Spicer (2004) estimated there are about 13.5 million species, of which only 1.75 million are currently described. Second, even for those species that are known, it is often not known what constitutes habitat for most of them (Morrison et al. 1992). Hence, in the foreseeable future, detailed studies of the processes affecting individual species' distribution patterns will probably be impossible for the vast majority of spe-

Figure 3.2. The greater glider (photo by David Lindenmayer).

cies and will most likely occur only for common or charismatic species or species of particular conservation concern. This recognition means that some level of generalization across multiple species is often necessary. One way to do this is to emphasize the value of natural landscape heterogeneity. Natural landscape heterogeneity can be broadly defined as the diversity, size, and spatial arrangement of native vegetation patches (Lindenmayer and Franklin 2002) and it is known to have an important influence on the diversity of many assemblages (e.g., Halley et al. 1996; Turner et al. 1997; Saab 1999; Debinski et al. 2001). For example, Benton et al. (2003) reviewed species distribution patterns in European agricultural systems. They argued that the loss of landscape heterogeneity was the most likely cause of species loss in these landscapes. Weibull et al. (2003) compared organic and conventional farms in central east Sweden and reached similar conclusions to those of Benton et al. (2003). They also found a significant relationship between various measures of biodiversity and landscape heterogeneity (Weibull et al. 2003). These findings result because species differ in their habitat requirements, and many different habitat types are available in naturally heterogeneous landscapes (see also Chapter 14).

An alternative way to generalize across species is to consider situations where different species have broadly similar habitat requirements, and then focus management efforts on these species. Such groups of species may include forest birds (Recher et al. 1987; McGarigal and McComb 1995), woodland birds (Watson et al. 2001, 2005), arboreal marsupials (Lindenmayer 1997), or forest fungi (Berglund and Jonsson 2003). The distribution patterns of such groups can be analyzed as aggregate measures of species occurrence, such as species richness (e.g., in relation to the size and spatial juxtaposition of forest remnants in a cleared landscape).

Throughout this book, we argue that a focus on single species and their unique habitat requirements (Part II) is complementary to a fo-

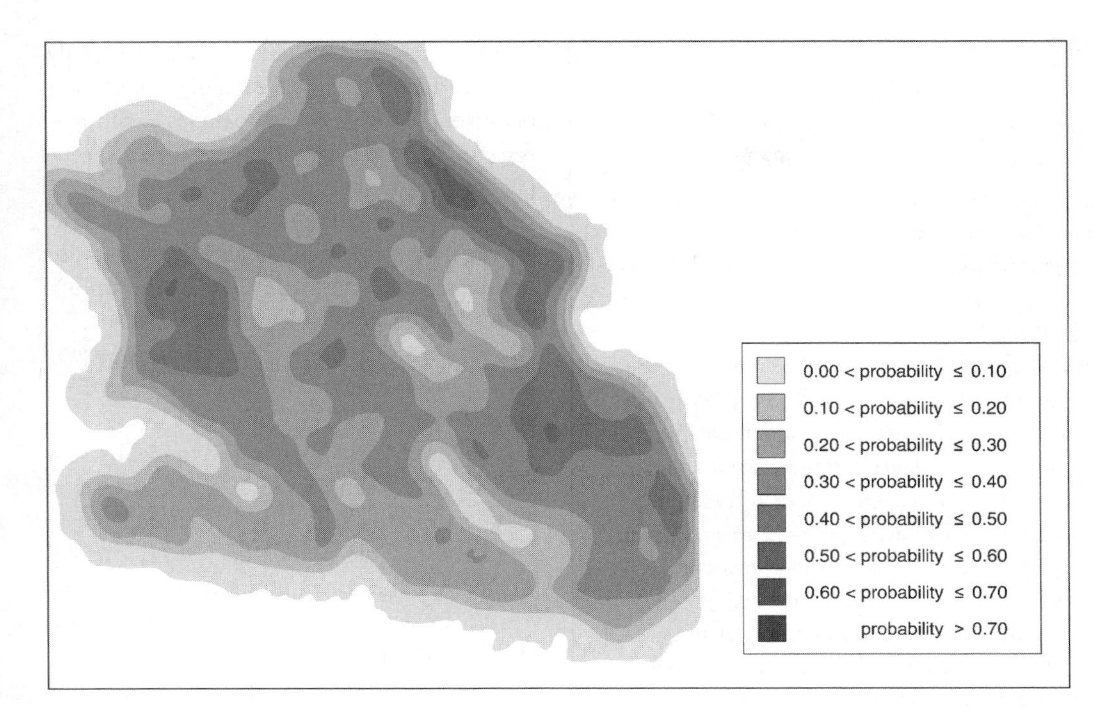

Figure 3.3. Contour map showing the predicted spatial distribution of the greater glider in a block of Victorian montane ash forest, southeastern Australia.

cus on the relationship of landscape pattern with species richness or species composition (Part III). Which of the two approaches is most useful in a land management context will depend on the particular situation and will need to be carefully considered on a case by case basis (Part V).

Pattern-Based Landscape Models

Landscapes can be described in relation to the habitat requirements of selected species (see earlier discussion), or from the perspective of humans (Chapter 2). Investigating the relationship between human-defined landscape patterns and species occurrence patterns has been a popular, albeit controversial, research area in applied ecology and conservation biology (Haila 2002). Three landscape models are used particularly frequently, both implicitly and explicitly: the island model, the patch–matrix–corridor model, and, to a much lesser extent, the variegation model.

The Island Model

The underlying premise of the island model is that fragments of original vegetation surrounded by cleared or highly modified land are analogous to oceanic islands in an "inhospitable sea" of unsuitable habitat (reviewed by Haila 2002). There are three broad assumptions made under the island model. They are: (1) islands (or vegetation patches) can be defined in a meaningful way for all species of concern; (2) clear patch boundaries can be defined that distinguish patches from the surrounding landscape; and (3) environmental, habitat, and other conditions are relatively homogeneous within an island or patch.

Much empirical work has shown that large areas support more species than smaller ones, and the number of species can often be predicted (albeit crudely) from species–area functions (e.g., Arrhenius 1921; Rosenzweig 1995; Scheiner 2003; see Chapter 9). The theory of island biogeography (MacArthur and Wilson

Box 3.2. Limitations of the Island Model

Several empirical studies illustrate the limitations of the island model. One of these is the Biological Dynamics of Forest Fragments Project in Brazil, which has been examining species and ecosystem responses to landscape change and rainforest clearing since 1979. In that study, edge effects and matrix conditions had substantially greater impacts on some groups of animals than did remnant size (Brown and Hutchings 1997; Gascon et al. 1999). For example, there was an increase in frog, small mammal, and butterfly species richness in rainforest patches following their isolation (Gascon and Lovejoy 1998). An increase in species richness in fragments may be due to the influx of organisms displaced from adjacent areas (Bierregaard and Stouffer 1997; see also Darveau et al. 1995) or the addition of generalist taxa from surrounding cleared areas (Gascon and Lovejoy 1998; Ås 1999). Such species are sometimes called invaders (see also Halme and Niemelä 1993; Saurez et al. 1998).

Other studies, such as those of invertebrates and plants (Ås 1999; Cook et al. 2002), have also produced results that contradict predictions from the island model. Estades and Temple (1999) recorded increased bird species richness in small rather than large *Nothofagus* spp. remnants embedded within an exotic pine (*Pinus* spp.) plantation in Chile. As in the case of the work in Brazil, species richness was elevated by an influx of taxa capable of using the changed matrix.

1963, 1967) was developed to explain species–area phenomena for island biotas. Part of this theory considers aggregate species richness on islands of varying size and isolation from a mainland source of colonists (Shafer 1990). The balance between the rates of extinction and colonization as influenced by island size and isolation is considered to produce an equilibrium number of species (Macarthur and Wilson 1967; see Chapter 9).

The literature on the theory of island biogeography is immense and it is well beyond the scope of this book to review it (e.g., see Simberloff 1988; Shafer 1990; Whittaker 1998). Much of it deals with the adaptation of island biogeography theory to reserve design (see Doak and Mills 1994), another topic that is beyond the scope of this book. As a model of landscape dynamics, the island model can fail (Gilbert 1980) because: (1) areas between patches of remnant vegetation are rarely nonhabitat for all species (Daily 2001; Lindenmayer and Franklin 2002) and, (2) the island model does not account for important interactions between vegetation remnants and the landscapes surrounding them (Janzen 1983; Manning et al. 2004a; Box 3.2; Figure 3.4) and the species that can live in areas surrounding patches (e.g., Ås 1999; Cook et al. 2002).

The limitations of the island model become obvious given consideration of the drivers of the dynamics on real oceanic islands. That is, even true oceanic islands do not exist in isolation from the surrounding marine environment. For example, the surrounding oceans provide food (e.g., for seabirds that nest on islands) and generate edge effects (e.g., wind-blown salt driven on-shore). These limitations will be magnified for modified terrestrial landscapes where areas adjacent to patches of native vegetation may be habitat for some species (Daily 2001; Chapter 14) and a source of negative or positive edge effects for others (Janzen 1983; Laurance et al. 1997; Chapter 11).

Despite the inherent problems of the island model, the broad notion of islands in an inhospitable sea has spawned the development of many related theories and concepts (see Rosenzweig 1995; Whittaker 1998) that are used widely in work on landscape modification (Doak and Mills 1994; Pullin 2002). These include wildlife corridors (Chapter 12), nested subset theory (Chapter 13), and the notion of vegetation coverage thresholds (Chapter 15). Notably, the island model could be seen as a good example of the progress of science and the extent to which it has evolved over the past 30 years. Indeed, in their seminal book, MacArthur and Wilson (1967) noted that much of the theory they presented was likely to be falsified when tested with empirical data.

Figure 3.4.
A rainforest remnant in the Biological Dynamics of Forest Fragments Project in Brazil (photo by Rob Bierregaard).

The Patch–Matrix–Corridor Model

Rather than conceptualizing landscapes as "islands in a sea of nonhabitat," Forman (1995) developed a model in which landscapes are conceived as mosaics of three components: patches, corridors, and a matrix (Figure 3.5). The focus is not so much on aggregate species richness, but rather the geographical composition of landscapes, with the different components having different characteristics, shapes, and functions. Forman (1995) defined the three components of his conceptual model as follows:

- Patches are relatively homogeneous nonlinear areas that differ from their surroundings.
- Corridors are strips of a particular patch type that differ from the adjacent land on both sides and connect two or more patches.
- The matrix is the dominant and most extensive (and often most modified) patch type in a landscape. It is characterized by extensive cover and a major control over dynamics.

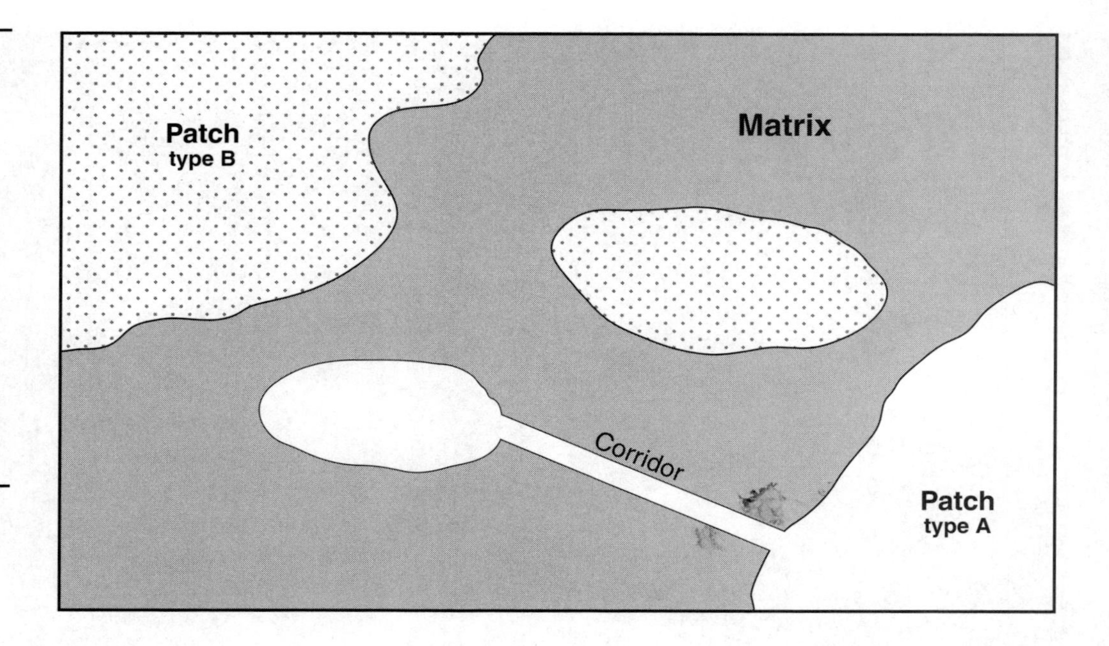

Figure 3.5. A landscape perspective based on the patch–matrix–corridor model (sensu Forman 1995). The landscape shown here has two patch types (A and B) and an extensive background matrix, and the patches of type A are connected by a corridor.

The name "patch–matrix–corridor model" is often used for Forman's (1995) conceptualization of landscape cover patterns. In the patch–matrix–corridor model, the matrix is intersected by corridors or perforated by smaller patches. Patches and corridors are readily distinguished from the background matrix (Forman and Godron 1986; Kotliar and Wiens 1990; Figure 3.5). Forman (1995) noted that every point in a landscape was either within a patch, corridor, or the background matrix, and that the matrix could be extensive to limited, continuous to perforated, and variegated to nearly homogeneous.

The patch–matrix–corridor model has been widely adopted in conservation biology. It helps land managers and researchers to translate their ideas into a spatial context. It is an extension of the island model, which often oversimplifies landscapes into areas of habitat and nonhabitat. However, the patch–matrix–corridor model also makes some simplifying assumptions. First, it does not generally deal with spatial continua (apart from edge effects; see Laurance et al. 1997; Fischer et al. 2004a). Second, it implicitly assumes that a single classification of landscape pattern can work for all species. This can be an important limitation because many organisms do not perceive landscapes in the same way as humans (Bunnell 1999a, b; Lindenmayer et al. 2003a). Therefore, patterns of landscape cover seen from a human perspective may not always provide a useful framework for interpreting biotic response to landscape conditions (Manning et al. 2004a).

The Variegation Model

Some authors have objected to the sharp boundaries and discrete classes prevalent in the island model and patch–matrix–corridor model (Harrison 1991). In response, McIntyre and Barrett (1992) developed the variegation model (see also McIntyre and Hobbs 1999). In many landscapes, the boundaries between patch types are diffuse, and differentiating them from the background matrix may not be straightforward. The term "variegated landscape" was coined to incorporate gradual spatial changes or gradients in vegetation cover (McIntyre and Barrett 1992; McIntyre 1994).

The variegation model was originally proposed for semicleared grazing and cropping landscapes in rural eastern Australia. These landscapes are characterized by small patches of woodland, relatively isolated native trees scattered throughout grazing lands, and areas of native ground and understory cover with no overstory trees (McIntyre and Barrett 1992; McIntyre et al. 1996). Here, from a human perspective, patches and corridors are difficult to identify among the loosely organized and spatially dispersed trees and other ecological communities such as native grasslands. For instance, numerous trees scattered across a landscape collectively provide habitat for some species (e.g., for some woodland birds; Barrett et al. 1994; Fischer and Lindenmayer 2002a, b). The variegation model takes account of small habitat elements that might otherwise be classified as unsuitable habitat in the background matrix (Tickle et al. 1998).

Limitations of Pattern-Based Landscape Models

One of the common goals of pattern-based landscape models is to reduce the complexity created by having to analyze every single species in its own right, which would pose insurmountable challenges to landscape management. Some level of simplification of reality in landscape models is necessary, and in fact, desirable (Burgman et al. 2005). However, irrespective of the original intentions of the architects of pattern-based landscape models, these models are sometimes used uncritically and can oversimplify ecological patterns. Sometimes, biologists represent landscapes as universally suitable habitat patches contrasting markedly with remaining areas of nonhabitat, without carefully assessing if it is appropriate to aggregate multiple species in this way. Mapping tools (like geographical information systems) are sometimes used to define habitat patches, assuming that species perceive patches in the same way and at the same scale as humans (Bunnell 1999a). Often, approaches based on landscape pattern do not consider the habitat requirements, movement patterns, and other important ecological attributes of the organisms of interest.

In some cases, a possible solution to ensure that ecological complexity is not oversimplified is to avoid using pattern-based landscape models and instead to apply concepts such as natural landscape heterogeneity or the landscape contour model (see earlier discussion). In other cases, pattern-based landscape models can be improved by carefully assessing how species should be aggregated. For example, some workers have attempted to overcome the problem of species-specific responses to landscape conditions by classifying species according to their use of a given modified landscape. Terms frequently used include "forest-interior species" (Tang and Gustafson 1997; Villard 1998), "edge species" (Howe 1984; Bender et al. 1998; Euskirchen et al. 2001) and "generalist species" (Andrén 1994; Williams and Hero 2001). Others have classified patches by the extent to which species use them or by their propensity to provide dispersers (the source–sink concept; Pulliam et al. 1992; Breininger and Carter 2003; Kreuzer and Huntly 2003). Some concepts discriminate original vegetation cover from relictual vegetation cover (McIntyre and Hobbs 1999; see Chapter 2) or describe the matrix in more detail (Gascon et al. 1999). All these approaches are potentially useful refinements to what might otherwise be an overly simplistic or inappropriate anthropocentric classification of landscape pattern.

The Link between Single Species and Multiple Species

The effects of landscape change can be assessed for a single species or for multiple species simultaneously. Single-species investigations tend to be more detailed and tend to have a reasonable grasp of the ecological processes that limit the distribution and abundance of a given species (e.g., see Table 3.1). In contrast, investigations on multiple species often need to aggregate species into groups and may need to make several assumptions about how landscape patterns are related to a given group of species. A common, but problematic, assumption is that human-defined patches correspond to habitat for a group of species. For example, Watson et al. (2001) implicitly assumed that human-defined patches of native vegetation provided essential habitat for woodland birds in the Australian Capital Territory. Later, Watson et al. (2005) demonstrated that the nature of the matrix (i.e., the area between woodland vegetation patches) actually had a major influence on woodland birds, most likely because some species could use the matrix for foraging or breeding or could readily move through it.

Given this, when is it necessary to study every single species, and when is it reasonable to aggregate species and focus on landscape patterns? This dilemma is (often implicitly) a fundamental source of many debates in conservation biology; for example, about indicator species (Landres et al. 1988; Lindenmayer et al. 2000), umbrella species (Lambeck 1997; Roberge and Angelstam 2004), small reserves (e.g., Gilpin and Diamond 1980; Simberloff and Abele 1982), and wildlife corridors (Noss 1987; Simberloff et al. 1992).

The short answer to this dilemma is that there is no simple solution. Throughout this book, we recognize that detailed studies on single species (which often elucidate ecological processes) are complementary to studies that relate human-defined landscape patterns to aggregate measures like species richness or species composition. Neither approach is inherently superior because both approaches have their own unique limitations. Single-species approaches are limited mainly in their practical utility because it will be impossible to study every single species. Multispecies approaches are limited because aggregation usually relies on simplifying assumptions about species co-occurrence patterns.

To emphasize the complementary nature of single-species approaches (which often focus on ecological processes) and multispecies approaches (which often focus on landscape pattern), we have structured this book accordingly (Table 3.2). In Part II, we focus on the processes accompanying landscape change and how they affect single species. In Part III, we focus on various pattern-based notions applicable to multispecies investigations—many of which are based on the patch–matrix–corridor model (e.g., edge effects, patch sizes, corridors, matrix management).

Some of the concepts that we discuss at the single-species level in Part II are mirrored by related concepts at the multispecies level in Part III (see Table 3.2). For example, habitat loss for single species (Chapter 4) can translate into relationships between species richness and patch sizes of remnant vegetation if such vegetation is the primary habitat for most species under investigation (Chapter 9). Similarly, the processes of habitat subdivision and habitat isolation discussed in the context of single species in Part II (Chapter 6) have strong parallels to our exploration of landscape connectivity in vegetation cover for multispecies assemblages in Part III (Chapter 12). Further parallels between single-species concepts and multispecies concepts are outlined in Table 3.2. Given these parallels, some readers may find it counterintuitive that we have separated topics in this way. Notably, the distinction between single-species approaches that emphasize ecological processes and multispecies approaches that emphasize landscape patterns is not absolute, but rather represents a continuum of research approaches. However, we believe that to overcome Bunnell's (1999a) "fragmentation panchreston" (Chapter 1), it is important to make clear distinctions at the conceptual level between processes and patterns, and landscape

Table 3.2. Parallel concepts for processes influencing single species, and landscape patterns frequently related to multiple species, and where they are discussed in the book

Species perspective (Part II: Chapters 4–8)	*Human perspective (Part III: Chapters 9–15)*
Areas of emphasis:	*Areas of emphasis:*
• Ecological processes	• Land cover patterns
• Mechanistic view of species occurrence patterns (biology, ecology, interspecific processes)	• Phenomenological view of species occurrence patterns (species richness, community composition)
• Biotic considerations (e.g., population viability, habitat suitability)	• Abiotic considerations (e.g., patch sizes, corridors)
• Variable spatial grain (depending on organism, multiple scales per organism)	• Relatively constant spatial grain (human definition of patches and landscape patterns)
Threatening processes:	*Key patterns related to threatening processes:*
• Habitat loss (Chapter 4)	• Land clearing (Chapter 9)
• Habitat patch shrinkage (Chapter 4)	• Vegetation patch shrinkage (Chapter 9)
• Habitat degradation (Chapter 5)	• Vegetation deterioration, loss of structural complexity (Chapter 10); edge effects (Chapter 11)
• Habitat isolation, loss of habitat connectivity, and subdivision (Chapter 6)	• Loss of landscape connectivity (Chapter 12)
• Biological changes and immediate species interactions (Chapter 7)	• [no directly parallel concept]
• [no directly parallel concept]	• Community composition (Chapter 13)
• [no directly parallel concept]	• Roles of the matrix and landscape heterogeneity (Chapter 14)
• Extinction proneness and interactions of threatening processes (Chapter 8)	• Cascading ecosystem changes (Chapter 15)

models relying on few assumptions about species co-occurrence patterns, versus approaches that rely on many assumptions.

Summary

For effective biological conservation, it is critical to determine and understand how landscape change affects organisms. Landscape models can be used to conceptualize the effects of landscape change on biota. Landscape models can (1) attempt to consider the perspective of a given species, or (2) take a human perspective of landscapes. Considering the effects of landscape change from a single species' perspective is a useful starting point to understand key ecological processes. The landscape contour model is a potentially useful tool to conceptualize the distribution of individual species in landscapes. However, it is impossible to study in detail every single species and every associated landscape change process that impacts upon it. In contrast, the island model, the patch–matrix–corridor model, and the landscape variegation model usually focus on responses to predetermined (and human-defined) patterns of vegetation cover. Studies using these pattern-based landscape models often rely on aggregating groups of species and make simplifying assumptions about commonality of responses to landscape change among species in such groups. Because both types of landscape models have advantages and disadvantages, a focus on single species and the ecological processes affecting them is complementary to a focus on landscape pattern in relation to assemblages of species. Such issues

> **Box 3.3. Why Landscape Models Are Important**
>
> Conservation biologists are often called upon to make practical decisions about landscape and conservation planning. The landscape model they use to characterize landscapes can make a significant difference to the practical conservation recommendations they make. If the patch–matrix–corridor model is the favored model, then for example, one of the conservation strategies applied might be the maintenance or establishment of wildlife corridors to link remaining vegetation patches surrounded by what is deemed to be an unsuitable landscape matrix. Corridor establishment would aim to reduce patch isolation and increase species richness (Saunders and Hobbs 1991) and would be appropriate for species that disperse along connecting vegetation links and make only limited movements across open areas (Brooker and Brooker 2002). However, such corridors may also attract nest predators or parasites (Temple and Cary 1988) or edge-using species that are aggressive to other taxa (Piper et al. 2002). Conversely, if the landscape variegation model is used, a different set of management prescriptions might arise. For example, in some cases, it may be more appropriate to develop stepping stones that may better facilitate the movement of species across a landscape in which the surrounding matrix retains habitat value for some species and is "permeable" to the movement for others (Franklin et al. 1997; Fischer and Lindenmayer 2002b) as was recommended in the case of the northern spotted owl (*Strix occidentalis caurina*) in the Pacific Northwest of the United States (Murphy and Noon 1992).
>
> The contrast in conservation outcomes that arise from different landscape models shows that it can be important to be explicit about both which conceptual model is applied and the objectives of its application.

associated with landscape models are important because they can influence the kinds of conservation strategies that are implemented (Box 3.3).

Links to Other Chapters

This chapter is the conceptual basis for Part II and Part III of the book. The various chapters in Part II discuss the processes associated with landscape change and their effects on single species. Part III discusses the relationships between landscape pattern and aggregate measures of biodiversity such as species richness.

Further Reading

Gaston and Spicer (2004) provide an excellent discussion on why species are the typical currency of discussions on biodiversity conservation. Manning et al. (2004a) discuss how species may differ from humans in their perceptions of the environment. The landscape contour model is presented in Fischer et al. (2004a). Our outline of the island model is built in part on the vast literature on island biogeography theory. The original text on this theory by MacArthur and Wilson (1967) is a good entry to that topic. The books by Rosenzweig (1995) and Whittaker (1998) are valuable and balanced overviews of the many aspects of island biogeography theory. Shafer (1990) provides a review of many aspects of island biogeography theory, but Gilbert (1980), Burgman et al. (1988), Simberloff (1988), and Doak and Mills (1994) give useful critiques of it; as do Haila (2002) and Manning et al. (2004a) in a landscape ecology context. Zimmerman and Bierregaard (1986) and Cook et al. (2002) are two of many useful examples illustrating problems with the island model using empirical data. A practical application of island theory to landscape planning and management is the book by Harris (1984), which has a particular focus on old-growth forest conservation. Forman and Godron (1986) and Forman (1995) contain valuable discussions of a raft of topics associated with landscapes and landscape ecology, including the patch–matrix–corridor model described in this chapter. McIntyre and Barrett (1992) and McIntyre and Hobbs (1999) outline the landscape variegation model. Ingham and Samways (1996) provide an interesting comparison of different landscape models in an empirical study of South African invertebrates.

PART II

The Species Perspective: Key Processes Affecting Individual Species

Landscape modification can affect the distribution patterns of species. Some species can benefit from landscape modification. For example, in the wheatbelt of western Australia, clearing of eucalypt *(Eucalyptus)* woodlands for agriculture has led to an increase in several bird species, including the crested pigeon *(Ocyphaps lophotes)*, galah *(Cacatua roseicapilla)*, Australian wood duck *(Chenonetta jubata)*, and pied butcherbird *(Cracticus nigrogularis)* (Saunders 1989; Saunders and Ingram 1995). However, for most species, human landscape modification typically leads to local or regional declines and can even result in the extinction of species (e.g., Saunders 1989). In Part II of this book, we are largely concerned with the negative effects of landscape change on individual species, although we acknowledge that landscape change can have positive effects for some species.

In this part of the book, we discuss key threatening processes that are often associated with landscape change (Figure II.1). Habitat loss is discussed in Chapter 4. It is the primary driver of species' declines and extinctions throughout the world. However, habitat loss rarely oc-

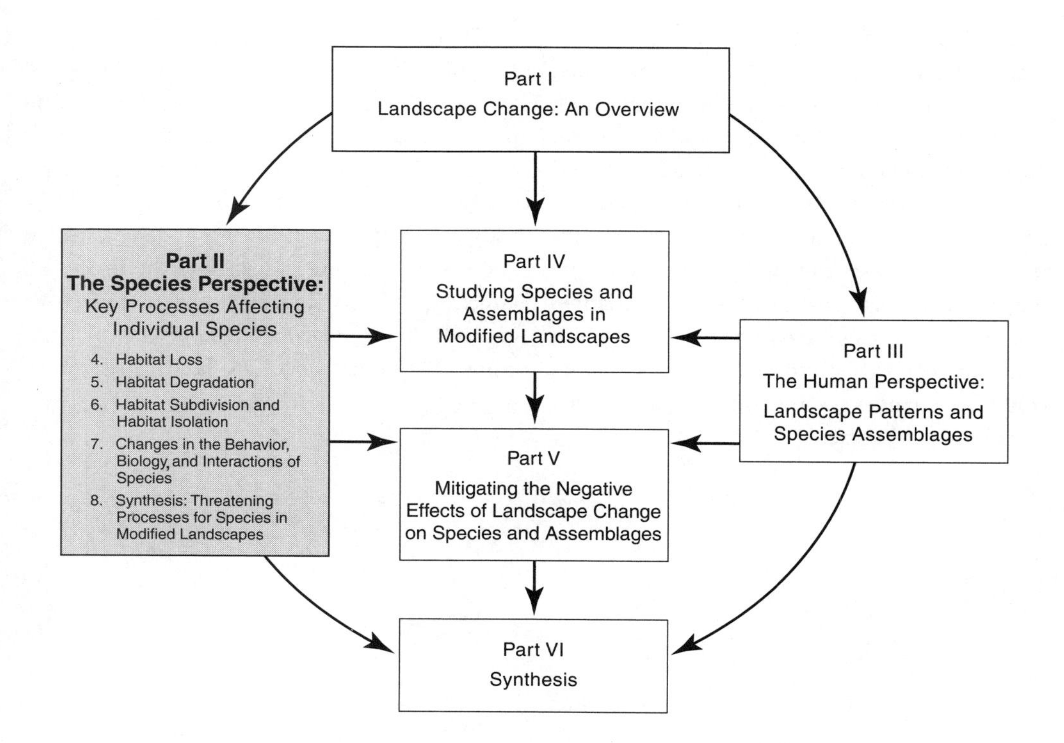

Figure II.1. Topic interaction diagram showing links between the different parts of the book, and the topics covered in the chapters of Part II.

Aerial photo of a human-modified landscape showing scattered paddock trees that are used as stepping stones by a wide range of native birds and insects. This landscape can be described as variegated under the model articulated by McIntyre and Hobbs (1999; see Chapter 3) (photo by David Lindenmayer).

curs in isolation from other major processes of landscape change. It is typically accompanied by habitat degradation and habitat subdivision, which we discuss in Chapter 5 and Chapter 6. Together, habitat loss, habitat degradation, and habitat subdivision often lead to changes in a species' biology and behavior. Also, species interact with other species, and these interactions may be altered as a result of landscape change. Chapter 7 discusses changes to the biology and behavior of species and also describes some common changes in the immediate interactions between species.

Chapter 8 is the final chapter in Part II. It synthesizes information from Chapters 4 through 7 and explains how the threatening processes discussed in the previous chapters can interact and lead to population declines, reduced genetic variability, and extinction. This is a logical end point to Part II of the book, which is devoted to single species and the processes affecting their distribution and abundance. We expand our discussion of the effects of landscape change to a focus on landscape pattern in relation to multiple species in Part III.

CHAPTER 4
Habitat Loss

Access to suitable habitat is essential for the survival and successful reproduction of all species. It is not surprising then that habitat loss is a major factor influencing species loss all over the world (Fahrig 1999; Craig et al. 2000; Thomas et al. 2004). The importance of habitat loss as a process threatening the persistence of species in the United States is illustrated in Figure 4.1. Similar trends are also apparent in many other parts of the world (Fahrig 2003).

Clearly Defining Habitat as a Prelude to Understanding Habitat Loss

A potential problem with data like those shown in Figure 4.1 is that the term "habitat loss" is often used in a generic sense as being synonymous with the loss of native vegetation (e.g., Kerr and Deguise 2004). In many cases the two are not the same, and the differences between native vegetation cover and habitat can be important (Box 4.1). This is because what constitutes suitable habitat is species-specific (Morrison et al. 1992). Hence habitat loss is also species-specific.

Simply equating habitat loss with loss of vegetation cover as a general rule across all terrestrial taxa can be misleading for at least three reasons. First, as outlined in Chapter 2, landscape modification is not a random process. Species differ in their habitat requirements and occupy different locations along environmental gradients. Notably, most native vegetation,

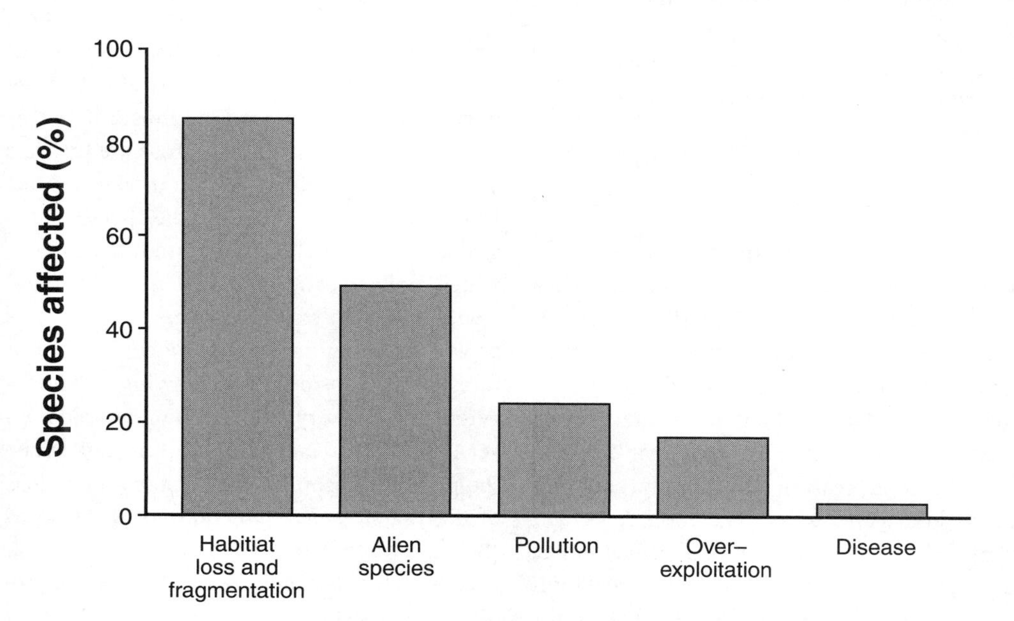

Figure 4.1. The relative impacts of different threatening processes on species in the USA (redrawn from Population Action International 2000 and based on data in Wilcove et al. 1997). Note that percentages do not add to 100% because many species are influenced by several threatening processes.

Box 4.1. Defining Habitat

Block and Brennan (1993, p. 36) defined "habitat" as the "subset of physical environmental factors that permit an animal (or plant) to survive and reproduce." We assume that habitat is associated with a geographic location (Krebs 1985). Differences in the habitat requirements of different species allow them to coexist in the same area (Lindenmayer 1997). In contrast, a niche is not tied to a geographic location (Hutchinson 1958). The fundamental niche of a species is the set of physical limits within which a species can live and reproduce. The realized niche is the environment to which a species is limited, both by tolerance of physical variables and by historical events and social and biotic interactions, including predation and competition. Concepts related to habitat include microhabitat, critical habitat, habitat use, habitat selection, habitat suitability, habitat requirements, habitat preference, and habitat association (Block and Brennan 1993). All of these terms indicate some type of relationship between an organism and its environment (Morrison et al. 1992).

Access to habitat influences the distribution and abundance of all organisms (Elton 1927; Andrewartha and Birch 1954; Morrison et al. 1992), and it significantly influences survival, reproduction and long-term persistence (Krebs 1978; Block and Brennan 1993). A species of plant or animal cannot survive without access to suitable habitat. Suitable habitat may vary from extensive relatively pristine forest stands for area-sensitive organisms like some wide-ranging carnivores (Milledge et al. 1991; Bart and Forsman 1992) to the moisture and decay conditions provided by individual logs for invertebrates (Økland 1996; Meggs 1997).

particularly in protected areas, remains in low productivity areas that have little value for other uses like agriculture or forestry (Braithwaite et al. 1993; Scott et al. 2001). Despite intact vegetation cover, unmodified low productivity sites may be unsuitable habitat for some species (Braithwaite et al. 1993; Armesto et al. 1998).

The second reason why simply equating habitat loss with loss of vegetation cover can be misleading is that a landscape may either be (1) completely covered in native vegetation (e.g., forest) but support no suitable habitat for a given species (e.g., for old-growth–dependent species; Marcot 1997; Lindenmayer et al. 1999a; Figure 4.2), or (2) entirely cleared of native vegetation, yet still completely suitable for a given species (although highly unsuitable for many others). Indeed, for some species such as those that use disturbed environments, the amount of suitable habitat may actually increase as the area between remnants of original vegetation cover expands (Box 4.2; Ås 1999).

The third reason why simply equating habitat loss with loss of vegetation cover can be misleading is that simple maps of native vegetation versus cleared land ignore gradients and the variation in distribution and abundance patterns of species (see Chapter 3). Such variation in habitat suitability matters because managing habitat loss depends on understanding what constitutes habitat for a given species. This knowledge is, in turn, essential for understanding why species respond to landscape change in the way they do (Brooker and Brooker 2001, 2002).

The scientific literature contains many examples of species that have declined when the amount of habitat suitable for them has been reduced (e.g., Caughley and Gunn 1996). The case of the hipsid hare or Assam rabbit *(Caprolagus hispidus)* is just one of many examples that illustrate the negative effects of habitat reduction and habitat loss. The hipsid hare was formerly widely distributed across the foothills of the Himalayas from Assam to Nepal (MacDonald 2001). The habitat of the species is tall grassland that would have naturally been maintained by floods and fires (Bell et al. 1990) and would have occurred near rivers and wetlands as well as being an understory component of forests. The hipsid hare is now highly endangered and only very rarely seen in the wild (Oliver 1984, 1987). Habitat loss is considered to be the primary reason for the decline and/or loss of the majority of populations of the species (Bell and Oliver 1992). Habitat loss is occurring because tall grasslands are (1) cut down to produce roof thatching, (2) burned to create green pick for domestic livestock, and (3) destroyed as

Figure 4.2. (Above) Aerial view of wet eucalypt montane ash forest in the Victorian Central Highlands, southeastern Australia. The image shows continuous native forest cover. However, because there is no suitable old-growth forest habitat for the yellow-bellied glider *(Petaurus australis)* (Below, left) and the sooty owl *(Tyto tenebricosa)* (Below, right), both species are absent (photos by David Lindenmayer (landscape and glider) and Esther Beaton (owl) ©.

part of agricultural development or the creation of dams and flood mitigation works (Bell and Oliver 1992).

Habitat Loss Is a Deterministic Process

In Box 4.1, we broadly defined "habitat" as the environments suitable for a particular species. In other words, species need habitat to survive. Thus the loss of habitat is a highly deterministic process leading to population declines (although some of the underlying causes of habitat loss, such as wildfires, may be stochastic). Many studies have shown that the distribution and abundance of individual species decline with habitat loss (e.g., Kattan et al. 1994; Hinsley et al. 1995; Best et al. 2001). At a landscape scale, populations of a given species often decline proportionally with habitat loss (Andrén 1994). However, at low levels of remaining suitable habitat, other effects may exacerbate

Box 4.2. Vegetation Clearing and Range Expansion in the Galah

The Australian continent supports about one-sixth of the world's parrot species. One of these is the galah (*Cacatua roseicapilla*). Unlike many other native animals, this species has benefited from changes in the environment brought about by human activities (Rowley 1990; Saunders and Ingram 1995; Forshaw 2002). The galah's diet is predominantly seeds, especially those from cereal crops and agricultural weeds. It is probably the most abundant species of Australian parrot, and flocks exceeding 1000 birds are not uncommon, particularly in places where food is abundant (Forshaw 2002).

Prior to 1750, the galah was associated with watercourses in arid and semiarid areas (Saunders et al. 1985). Now, its distribution spans most of the continent (Blakers et al. 1984; Forshaw 2002; Barrett et al. 2003). This range expansion followed clearing of native vegetation for agriculture, provision of dams as watering places for stock (and inadvertently for the galah), and the widespread establishment of pasture and cereal crops, which provide major food sources for the species. The example of the galah illustrates why it is important not to simply equate the loss of native vegetation cover with habitat loss.

the negative effects of habitat loss (e.g., habitat subdivision and isolation), and this may further accelerate the decline of populations (Andrén 1994; Chapter 8).

At a local scale, many different factors can lead to small patches of habitat being less suitable than large habitat patches (Bender et al. 1998). For example, availability of food, shelter, and space may be more limited in smaller patches, leading to smaller populations of individual species (Burke and Nol 1998; Box 4.3). Such smaller populations may, in turn, be at greater risk of local extinction (Berger 1990). Smaller populations are at greater risk of extinction through stochastic events, including demographic stochasticity (Pimm et al. 1988; McCarthy et al. 1994), environmental stochasticity (Thomas 1990; Tscharntke 1992), and genetic stochasticity (Lacy 1987, 1993; Young et al. 1996; see Chapter 8 for more details).

Habitat Loss as a Temporal Phenomenon

The loss of habitat in a landscape may or may not be a permanent and unidirectional problem. It is likely to be so for species dependent on old-growth forest if such a forest is converted to a high-density urban settlement. Such spatial conflicts between human settlements and species occurrence are common. For example, Luck et al. (2004) demonstrated that areas which have inherently provided habitat for many species of birds, mammals, amphibians, and butterflies have often become major centers of urbanization in both North America and Australia.

In other situations, degraded or lost habitat may eventually recover. For example, a logged and regenerated forest may eventually regain part or all of its habitat suitability status for some forest-dependent species (see Fahrig 1992), depending on factors such as the intensity of the harvesting operation, the success of attempts to retain key structures on a cutover unit (such as cavity trees, large pieces of woody debris, etc.), and the interval elapsed and the pace of successional processes before the area is logged again (Franklin et al. 1997; Hunter 1999). Other examples such as postfire vegetation succession (Agee 1993; Bradstock et al. 2002) and ecosystem recovery following flooding (Gregory 1997) also demonstrate that habitat loss can be a temporary phenomenon. Similarly, one key motivation of restoration activities throughout the world is to increase the amount of habitat for native species (Hobbs 2003).

Habitat Loss and Other Threatening Processes

Despite the importance of habitat loss as a key threatening process, few studies in conservation biology directly and specifically focus on habitat loss (Fazey et al. 2005a). Forman's (1995) landscape modification model, which considers various patterns of landscape change (e.g., perforation and shrinkage; see Chapter 2) shows that, theoretically, habitat loss can occur without habitat subdivision or habitat degradation. However, in the majority of cases, it will be extremely difficult to separate the effects of habitat loss from the effects of habitat subdivision (Andrén 1997; Harrison and Bruna 1999) and habitat degradation (Thiollay 1998; Bunnell 1999a). As a result, in many studies, several threatening processes are confounded (Simberloff 1992). Although theoretical models assessing populations in relation to habitat cover attempt to separate habitat loss and habitat subdivision (Fahrig 1997, 1999), from a practical perspective it is essential to be aware that landscape modification is nearly always closely accompanied by several threatening processes (Bunnell 1999a; Fahrig 2003). This confounding may partly explain why so few studies have focused specifically on habitat loss (Fazey et al. 2005a). The case study of the giant panda *(Ailuropoda melanoleuca)* in Box 4.4 illustrates how habitat loss is typically linked with habitat degradation and habitat subdivision.

Caveats

Although habitat loss holds primacy as the key driver of the loss of individual species, interrelationships between habitat loss and other threatening processes highlight a need for caution in studying habitat associations of a declining species. That is, the distribution of a species and hence correlates of attributes of the places where it occurs (e.g., for habitat relationships; see Part IV) may not in fact be a good reflection of the true habitat requirements of a given species (van Horne 1983). Rather, such relationships may reflect where other threatening processes associated with landscape change, like predation, are less intense (Caughley and Gunn 1996).

> Box 4.3. Problems with Small Habitat Patches
>
> Many studies have found that populations of individual species decline with habitat loss. Ecological theory predicts this relationship from simple generalizations of per capita consumption of resources and the finite nature of those resources (Hutchinson 1958; MacArthur and Wilson 1967). Zanette et al. (2000) examined this phenomenon in a study of the eastern yellow robin (*Eposaltria australis*) in forest remnants in northern New South Wales in eastern Australia. They found that, because small remnants supported a lower biomass of insects than large remnants, nesting female eastern yellow robins in small habitat remnants (1) left their nests more often to find food, (2) had a shorter breeding season, (3) laid eggs that were lighter in weight, and (4) reared smaller nestlings. Zanette et al. (2000) concluded that per capita food limitations explained why the eastern yellow robin was less abundant in smaller remnants than in larger ones.

Summary

The primary threat to species associated with landscape change is the loss of habitat. Access to suitable habitat is fundamental to the survival of all species. The habitat requirements of any given taxon are species-specific. Hence habitat loss is also a species-specific entity. Notably, native vegetation loss and habitat loss are not synonymous because what constitutes habitat loss is different for each species; for some generalist species, the amount of suitable habitat may even increase with the reduction in native vegetation cover. Habitat loss can occur at multiple spatial scales. Habitat loss rarely occurs in isolation from other threatening processes associated with landscape change such as habitat subdivision, habitat isolation, habitat degradation, and changed species interactions, although these are largely regarded as having secondary (and hence less severe) impacts on the survival of individual species than habitat loss. Understanding the intimate links between multiple threatening processes is important for mitigating the negative effects of landscape change on species.

Box 4.4. The Intimate Links between Habitat Loss, Habitat Degradation, and Habitat Isolation: The Case of the Giant Panda

The giant panda *(Ailuropoda melanoleuca)* inhabits bamboo (subfamily Bambusoideae) forests in southern China (Carter et al. 1999; Wei et al. 2000), and was deliberately constructed as a "flagship" for conservation by the World Wide Fund for Nature. In 1975, the Wolong Nature Reserve was established in Sichuan Province to protect the species. Despite its status as a protected area, the giant panda's habitat in Wolong Nature Reserve is now more degraded than when the reserve was established. Highly suitable habitat for the species has become increasingly degraded, and remaining habitat patches are smaller and more isolated than prior to the establishment of the reserve (Liu et al. 2001).

The reasons for the loss, degradation, and subdivision of the giant panda's habitat are directly related to human activities in the reserve. The human population inhabiting the reserve nearly doubled between 1975 and 1995, and the number of households more than doubled over this period. In addition, the Wolong Nature Reserve is now a popular tourist destination. Demand for agricultural produce, fuelwood, and timber for construction work has increased, resulting in the logging, degradation, and subdivision of bamboo forest (Liu et al. 1999, 2001).

This example illustrates that the processes of habitat loss, habitat degradation, and habitat isolation often occur simultaneously in the real world. Thus, while it can be useful to differentiate between these processes from a theoretical perspective (Fahrig 2003), in reality, they are often closely linked. The example also highlights that nature reserves can sometimes be limited in their effectiveness to protect biota. This underlines one of the key themes in this book; that is, the importance of biological conservation in human-modified landscapes, outside protected areas.

Links to Other Chapters

Habitat loss rarely occurs in isolation from habitat degradation or habitat subdivision, which are discussed in Chapters 5 and 6. Chapter 9 discusses the link between the loss of native vegetation and assemblages of species. Chapter 16 discusses some of the most widely used methods to quantify the habitat of a given species. Mitigation approaches to habitat loss are discussed in Part V of this book.

Further Reading

Given the central role of habitat in understanding species decline, useful books and articles on habitat include those by Saunders et al. (1987), Saunders and Hobbs (1991), Morrison et al. (1992), Block and Brennan (1993), and Lindenmayer and Burgman (2005). Guisan and Zimmerman (2000), Elith (2000), and Burgman et al. (2005) contain useful reviews of the raft of statistical and other analytical methods that can be employed to quantify the habitat of a species. The reviews by Fahrig (1999, 2003) explain why it is important to recognize habitat loss as a different process from habitat subdivision.

CHAPTER 5
Habitat Degradation

The previous chapter highlighted the fundamental importance of access to suitable habitat for the persistence of an individual species. For a given species, the loss of suitable habitat will threaten its survival. Habitat can be lost rapidly or it can degrade in quality over time—the process of habitat degradation. Habitat degradation is common in landscapes subject to human modification (Ambuel and Temple 1983; Saunders et al. 2003), and it can be broadly defined as the slow decline or attrition of habitat suitability. It is a process that can eventually lead to habitat loss if it is not reversed (New 2000). This chapter describes two main types of habitat degradation—the decline in food resources and the decline in shelter availability. Together, these processes may lead to the reduced abundance of a species in a given area, or they may prevent it from successfully reproducing. In keeping with the theme of Part II of this book, our primary focus is on individual species and the processes that influence their distribution and abundance in response to landscape modification. However, we recognize that some types of landscape modification can lead to habitat degradation not only for individual species but also for assemblages of taxa (see Chapter 10).

Case Studies of the Effects of Habitat Degradation

A wide range of factors influence the distribution of plants and animals, and many of these operate at different spatial and temporal scales (Gaston and Spicer 2004). They include abiotic processes, biologically mediated processes, and processes dominated by biotic interactions (Mackey and Lindenmayer 2001). The availability of food and access to shelter sites are two of the key multiscaled factors that shape the distribution and abundance of animals (Elton 1927). Together, they are critical components of the habitat requirements of animal taxa and limit the spatial distribution of many species (Morrison et al. 1992; Chapter 3).

Decline in Food Resources

Felton et al. (2003) studied the effects of low-intensity hand-logging on the orangutan *(Pongo pygmaeus pygmaeus)* in the rainforests of West Kalimantan, Indonesia (Figure 5.1, Figure 5.2). Logging led to a significant reduction in large food trees and a substantial increase in canopy gaps, resulting in habitat degradation. Increased numbers of large canopy gaps led to increased travel times between large fruit-producing food trees. Although the species persisted in logged areas, indicating that habitat had not been totally lost, the density of orangutan nests was more than 20% lower in logged than in unlogged areas. Adult females were particularly affected, which raised concerns about the reproductive viability of the remaining population.

Decline in Shelter and Nest Availability

A prime example of the effects of habitat degradation on the availability of shelter sites comes from the temperate woodland remnants of the Western Australian wheatbelt. Although these

Figure 5.1. Orang-utan (photo by Adam Felton).

areas of temperate woodland currently provide habitat for a range of native birds (Saunders and Ingram 1995), the ongoing degradation in their condition (Hobbs 2001; Saunders et al. 2003) will make them increasingly unsuitable habitat for some taxa. A monitoring study by Saunders et al. (2003) has shown there has been no regeneration of young trees or saplings in the woodland remnants they studied since major grazing and cropping practices began in the region in the 1920s. The lack of recruitment of younger cohorts of trees means there is little or no understory and midstory vegetation. Perhaps more importantly, it also means there are no young trees that can eventually replace the old-growth trees, which currently dominate woodland stands. Old trees in this region are often characterized by having many cavities. A lack of recruitment of large trees with cavities will eventually make these woodland stands unsuitable habitat for a range of cavity-dependent woodland bird species. For example, the decline of suitable nesting sites is one of the factors that negatively influence the occurrence of Carnaby's cockatoo (*Calyptorhyncus funerus latirostris;* Figure 5.3) in remnants of native vegetation in the Western Australian wheatbelt (Saunders and Ingram 1987). Limited tree regeneration is a common problem in woodland ecosystems in many Australian agricultural landscapes (Spooner et al. 2002; Dorrough and Moxham 2005; Fischer et al. 2005b).

Some kinds of environments can be highly susceptible to habitat degradation. Caves are an important example because they provide roosting, hibernating, and hiding places for many individual species, including some highly specialized organisms found nowhere else. Culver et al. (2000) calculated there were 927 species in the U.S.A.'s 48 contiguous states that were limited entirely to caves and associated subterranean habitats. Human activities at the surface can lead to habitat degradation in caves in several ways (Brown 1985). For example, small changes in the amount of organic materials

Figure 5.2. Rainforest logging in Kalimantan (photo by Annika Felton).

entering a cave through logging operations or human recreation can substantially degrade habitat suitability for many individual resident species (Wilson 1978; Brown 1985).

Other Examples of Habitat Degradation

The ecological literature contains numerous other examples of particular species being negatively affected by habitat degradation. Hazell et al. (2004) discussed the impacts of the degradation of aquatic ecosystems on amphibians in southeastern Australia since European settlement. As a result of land clearing, overgrazing by domestic livestock, and mining operations, deeply incised, fast-flowing streams now characterize areas that were formerly a series of loosely connected chains of ponds (Hazell et al. 2003; Figure 5.4). Habitat is still available for species such as Bibron's toadlet *(Pseudophryne bibronii)*, but populations are now declining. Bibron's toadlet lays large eggs with a substantial amount of yolk so that tadpoles develop to an advanced stage within the egg capsule. Young animals need to remain in shallow ponds to complete their development, but habitat degradation of aquatic zones means that the creation of fast-flowing, deeper water has greatly impaired reproduction in this species. Other potential effects of habitat degradation include egg and tadpole desiccation and the isolation of populations (Hazell et al. 2003, 2004). Notably, several other species of frogs that have different reproductive strategies have not been as severely affected by stream incision as Bibron's toadlet (Hazell et al. 2003). The example

Figure 5.3. Carnaby's cockatoo (photo by Graeme Chapman).

of Bibron's toadlet highlights two key points: that habitat degradation can be a species-specific process and, as a result, that habitat degradation can occur somewhat independently of vegetation deterioration. There are many similar cases of habitat degradation and its effects in aquatic ecosystems from around the world. For example, in a Canadian study of riparian buffer strips, a lack of tree cover was found to result in changed microclimatic conditions (like temperature regimes) within streams, with corresponding negative impacts on the suitability of sheltering and breeding areas for native rainbow trout (*Oncorhynchus mykiss;* Barton et al. 1985).

Chronic Degradation and Extinction Debts

A particular problem with habitat degradation is that it is not always easy to detect. Because habitat has not been lost completely, some individuals of a given species of concern usually remain, albeit at a reduced population density (Box 5.1). This problem is exacerbated by two factors. First, some types of habitat degradation take a long time to have a noticeable effect on a given species. For example, the loss of large trees with cavities is a major problem in many forest and woodland ecosystems worldwide, and many individual species are threatened by it (Fischer and McClelland 1983; Gibbons and Lindenmayer 2002). Habitat degradation through the loss of mature trees can be gradual, such as in the cases of repeated low intensity selective logging (Lindenmayer and Franklin 2002), long-term firewood collection in woodlands and forests (Wall 1999), or high-intensity grazing that precludes the recruitment of new trees and alters tree species composition (Vera 2000; Saunders et al. 2003; see earlier discussion). Because the recruitment of trees with cavities is slow, reversing the loss of such a key element of the habitat of many species can also be slow and requires prolonged and highly targeted management efforts.

A second problem of detecting the impacts of habitat degradation is that some species have very slow life cycles—several species of large cockatoos and other parrots are examples (Forshaw 2002). Therefore, individuals may persist for a prolonged period in an area of degraded habitat, although they fail to breed, or they produce only very few offspring. Rapid "snapshot" studies (sensu Diamond 1986) may confirm the presence of such species but fail to detect problems with limited reproductive success. Long-term population declines as a consequence of habitat degradation may therefore go undetected or be extremely difficult to reverse once they are identified (Caughley and Gunn 1996). The delay in a species' extinction following landscape change is sometimes termed an extinction debt (Tilman et al. 1994; McCarthy et al. 1997; see Chapter 9).

Figure 5.4. Chain-of-ponds (photo by Donna Hazell).

Box 5.1. The Decline of the Ethiopian Bush Crow

Habitat degradation can cause the gradual decline of a species. Repeated long-term monitoring is often needed to detect gradual changes in population sizes as a result of habitat degradation. Borghesio and Giannetti (2005) documented how habitat degradation has led to the gradual decline of the Ethiopian bush crow *(Zavattariornis streseinanni)*. The species is endemic to southern Ethiopia, and its range covers approximately 5000 km^2. A large part of the species' range is within the Yabello Wildlife Sanctuary, which receives little official protection. Borghesio and Giannetti (2005) compared abundance data on the Ethiopian bush crow gathered in 1989, 1995, and 2003 and found that the abundance of the crow had decreased by 80% over this time. Although they acknowledge the potential for natural population fluctuations, they highlighted that changes in crow abundance occurred at the same time as major structural changes to the species' habitat. Thus the authors reasoned it was likely that habitat degradation had led to the observed decline of the species (Borghesio and Giannetti 2005). Land cover changes outside the Yabello Wildlife Sanctuary were related to agricultural intensification and a thinning of native vegetation. In contrast, within the reserve, vegetation thickness had increased due to the encroachment of unpalatable thorny shrubs. Shrubby thickets were less suitable habitat for the Ethiopian bush crow (i.e., structural changes corresponded to habitat degradation for this species). Bush encroachment is common in African savannahs and also threatens many species other than the Ethiopian bush crow (e.g., Meik et al. 2002). It typically results from overgrazing by domestic livestock, the suppression of wildfires, and the local extinction of native herbivores such as the African elephant *(Loxodonta africana)*, giraffe *(Giraffa camelopardalis)*, and African buffalo *(Syncerus caffer)*. Given the likely causal link between changes in land cover and the marked decline of the Ethiopian bush crow, Borghesio and Giannetti (2005) recommended the species be officially listed as endangered rather than vulnerable. Without long-term monitoring, it is likely that the decline of the Ethiopian bush crow would not have been detected in time to take conservation action.

The difficulties in detecting habitat degradation mean that mitigation strategies need considerable foresight. Depending on the species of interest, they may include adding nest boxes for cavity-dependent species (Munro and Rounds 1985; Newton 1994; Smith and Agnew 2002), weed removal (Brothers and Spingarn 1992), the control of grazing (Spooner et al. 2002), or the regulation of firewood harvesting (Driscoll et al. 2000).

Summary

"Habitat loss" and "habitat degradation" are terms that are often used interchangeably in the scientific literature. However, they can be quite different, with each having its own significant implications for species conservation. Habitat loss is final (for the time being at least); it means that a site is no longer suitable for a given species and it cannot survive there. Habitat degradation means that some of the attributes of the original habitat remain, but the quality of the habitat is reduced for the given species of interest. For example, the quality of the habitat may be diminished in ways that do not preclude individuals of a particular species from persisting, but prevent them from reproducing. Differentiating between habitat loss and habitat degradation can be critical for identifying appropriate strategies for tackling habitat degradation. For example, restoration strategies may be quite different if all of a particular species' habitat has been lost as opposed to cases where some habitat remains but is degraded and in poor condition. The case studies presented in this chapter highlight that mitigating habitat degradation requires a good understanding of the habitat requirements of particular species, as well as its response to changes in habitat condition. The distinction between habitat loss and habitat degradation can become blurred if habitat degradation continues over time, eventually leading to a site being totally unsuitable for a species (viz. habitat loss).

Links to Other Chapters

Habitat degradation is a process closely related to habitat loss (Chapter 4). Often the factors or human land use practices leading to the degradation of habitat for a given species will also degrade habitat for many other members of a species assemblage. Examples include grazing by domestic livestock, logging, and firewood collection. The effects of these practices on assemblages are discussed in more detail in Chapter 10.

Further Reading

The study of the orangutan in West Kalimantan described briefly in this chapter is one of many that illustrate where habitat remains but its quality as a suitable foraging resource is significantly degraded (Felton et al. 2003). Saunders et al. (2003) outline the impacts of long-term habitat degradation on the quality of nesting resources for Carnaby's cockatoo and other cavity-nesting birds. Chains-of-ponds ecosystems and their value for amphibians are discussed by Hazell et al. (2003). The book by Vera (2000) is a fascinating account of the relationships between grazing and vegetation change, including habitat degradation in the forests of Europe.

CHAPTER 6

Habitat Subdivision and Habitat Isolation

Landscape change often subdivides remaining areas of a given species' habitat. Habitat subdivision involves breaking a habitat patch into two or more smaller patches. This is congruent with various dictionary definitions of "fragmentation" (i.e., "to break apart"). Habitat subdivision is also the "fragmentation process" in Forman's (1995) model of landscape change (see Chapter 2). As discussed in Chapter 4, the smaller habitat patches resulting from habitat subdivision may be unable to support viable populations (Shaffer 1981; Shaffer and Samson 1985; Armbruster and Lande 1993; also see Chapter 8). In addition to breaking larger single patches into smaller ones, habitat subdivision can isolate remaining habitat patches from one another.

We make an explicit distinction between the different landscape perspectives of individual species and humans. Habitat isolation is a process that affects an individual taxon. It is the opposite of habitat connectivity, which refers to the connectedness of habitat patches for a given individual species. This chapter focuses on these species-specific concepts. In contrast, landscape connectivity is a human perspective of the connectedness of vegetation patterns that cover a landscape and is discussed in Chapter 12. This chapter outlines multiple scales at which habitat isolation can occur, discusses the concept of metapopulations, and highlights that what constitutes habitat isolation can differ strongly between different species.

Scales of Habitat Isolation

Some species have a naturally patchy distribution. Particularly good examples are plants and animals that are largely confined to rocky outcrops (Hopper 2000; Schilthuizen et al. 2005) or those that live on mountain tops (Mansergh and Scotts 1989) and in freshwater ponds (Sjorgen-Gulve 1994). However, many other species whose natural distribution is more continuous are negatively affected by habitat subdivision. Habitat subdivision can lead to larger distances between habitat patches and hence greater habitat isolation (Fritz 1979; Smith 1980; Hanski 1994a). For plants, habitat isolation may limit the movements of propagules such as spores, pollen, and seeds (Duncan and Chapman 1999; Cascante et al. 2002). For animals, habitat isolation may impair movements at several spatial and temporal scales: day to day movements, dispersal movements, movements of individuals in metapopulations, and larger-scale movements like nomadic movements or seasonal migration, or range shifts in response to climate change (Hunter 1994).

Day to Day Movements

Habitat isolation can impair day to day movements within the home ranges or territories of individual animals. For example, habitat isolation may affect foraging behavior (Box 6.1) or may prevent a species from moving between its nesting habitat and foraging areas (Saunders 1977, 1980). The spatial scale of day to day movements can vary greatly between species. For example, small species such as skinks may only move several dozen meters (Turner et al. 1969; James 1991). Other species' day to day movements may span much larger distances, in the order of hundreds to thousands of meters for many birds (Wegner and Merriam 1979; Ford and Barrett 1995; Graham 2001), bats (Lumsden et al. 1994, 2002; Galindo-González and Sosa 2003), or mammalian carnivores like the North American black bear (*Ursus americanus;* Klenner and Kroeker 1990).

Dispersal between Habitat Patches

Habitat isolation can impair dispersal movements between the natal territory and suitable habitat patches, which are typically made by juvenile or subadult animals attempting to establish new territories (Wolfenbarger 1946; Stenseth and Lidicker 1992a). This interruption to dispersal can reduce the genetic size of populations through impaired patterns of gene flow (Leung et al. 1993; Mills and Allendorf 1996; Saccheri et al. 1998; Epps et al. 2005). Importantly, effective dispersal involves not only the movement of an individual but also its successful reproduction in the receiving population (Mech and Hallett 2001). In some cases, males and females of a given species do not respond in the same way to habitat isolation (Davis-Born and Wolff 2000; Berry et al. 2005). For species whose dispersal is sex biased (Cockburn et al. 1985; Stenseth and Lidicker 1992a), isolation may result in single-sex patches (Temple and Cary 1988; Cooper and Walters 2002), thereby ultimately leading to nonviable populations. In addition, the recolonization of vacant territories in some habitat patches by individuals originating from other habitat patches is critical for maintaining the overall demographic size of a given species' population (Brown and Kodric-Brown 1977; Cooper and Walters 2002).

By affecting patterns of dispersal between patches, habitat isolation can have significant effects on the occupancy of otherwise suitable habitat patches (Villard and Taylor 1994; Pither and Taylor 1998), including protected areas like nature reserves (Burkey 1989). For example, population recovery after disturbance may be impaired by habitat isolation (e.g., Lamberson et al. 1994). Habitat isolation can therefore have profound impacts on population persistence (Mills and Allendorf 1996) and contribute significantly to the proneness of an individual species to extinction (Angermeier 1995; Chapter 8).

Given the key roles of dispersal and recolonization, understanding the effects of habitat isolation on these processes is important for sustainable landscape management (Lamberson et al. 1994), designing reserve networks (Burkey 1989), establishing wildlife corridors (Andreassen et al. 1996; Rosenberg et al. 1997; Haddad et al. 2003), and determining the ability of populations to recover from disturbance (e.g., Whelan 1995). Despite these needs, knowledge about dispersal and recolonization in modified landscapes is limited (Chepko-Sade and Halpin 1987; Dieckmann et al. 1999). Few empirical studies have examined how patches are

Box 6.1. Habitat Isolation and Impaired Foraging

The spatial isolation of food resources can affect whether individuals of a given species are able to utilize those resources. For example, scattered fruit trees occur in many agricultural areas in Central America (Guevara et al. 1998; Harvey and Haber 1999). These trees provide food for many native species of frugivorous birds. However, not all trees are equally accessible. In Costa Rica, Luck and Daily (2003) showed that the visitation rate of fruit trees by small birds declined with increasing distance from a large area of relatively unmodified native rainforest. In Mexico, a similar set of observations was made by Guevara and Laborde (1993). These authors noted that individual birds visiting scattered fruit trees were more likely to arrive at these trees from directions where trees were relatively densely spaced. Thus the actual isolation of fruit trees for frugivorous birds in Central American agricultural landscapes appears to be related to the physical distance of trees from native rainforest as well as the density of scattered trees in a given landscape.

recolonized (but see Middleton and Merriam 1981; Johnson and Gaines 1985), and there are few rigorous empirical tests of the prediction that recolonization probability depends on the distance from a source population (Fritz 1979; Johnson and Gaines 1990; Hanski 1994a; Lindenmayer et al. 2005a).

Irrespective of the difficulties of studying dispersal in modified landscapes, a better understanding of it is essential for determining how to best mitigate the negative effects of landscape modification (Bullock et al. 2002). An excellent example is the case of the blue-breasted fairy-wren *(Malurus pulcherrimus)* in the wheatbelt of Western Australia. Brooker and Brooker (2001, 2002) studied banded fairy-wrens in woodland and heath remnants of different sizes over 6 years (1993–1998). They demonstrated that low levels of habitat connectivity reduced dispersal recruitment to unoccupied patches below the levels required to replace mortality. Subsequent calculations by Smith and Hellman (2002) using data from Brooker and Brooker (2002) showed that annual population growth rates of +5% characterized well-connected networks of patches. These data highlighted the importance of minimizing isolation between remaining habitat patches as a key conservation strategy for the species.

Metapopulations

Habitat isolation may shift a formerly contiguous and interacting population into a series of loosely connected subpopulations (i.e., a metapopulation). A metapopulation is defined as: "a set of local populations which interact via individuals moving between local populations" (Hanski and Gilpin 1991, p. 7).

The term "metapopulation" is sometimes applied to all species distributed across a number of vegetation remnants (Arnold et al. 1993). However, true metapopulations are defined by dynamic properties, including (after Hanski 1997, 1999a):

- Suitable habitat must be restricted to particular patches that can be differentiated from the surrounding unsuitable matrix.
- The populations in most patches must be at risk of extinction at some stage.
- There must be interpatch dispersal and colonization, but not too much so that the local dynamics of populations within a patch are not synchronous with the dynamics of populations in other patches.

Metapopulations can be important in modified landscapes because an ensemble of patchy populations supports a larger number of individuals than a single isolated population and, as a result, is likely to be less extinction-prone (Lindenmayer and Possingham 1995; Hanski 1999a). In addition, under true metapopulation conditions, localized extinction in a given patch (e.g., due to a wildfire) may be reversed by recolonization from dispersing individuals born in another patch (Stacey and Taper 1992). Such spatial discontinuities in distribution can therefore be important in promoting population persistence. Thus a metapopulation as a whole may persist much longer than any individual population (Harrison 1991; Hanski 1999a).

There are two extreme forms of metapopulation dynamics (Hanski and Gyllenberg 1993; Chesson 2001; Figure 6.1). The mainland–island structure includes a large "mainland" area in which populations are secure and very rarely (if ever) become extinct, and an array of small patches in which extinctions are more likely. The mainland provides a source to recolonize the patches and reverse localized extinctions (Hanski 1999a). Several species of butterflies are thought to have a mainland–island metapopulation structure (Harrison et al. 1988; Hanski and Thomas 1994). In many respects because populations in mainland areas rarely if ever suffer localized extinction, the mainland–island structure is more akin to a source–sink population (sensu Pulliam et al. 1992; Kreuzer and Huntly 2003) than a true metapopulation. The Levins structure (also called the classical model) takes its name from Levins (1970) and assumes patches are (approximately) the same size. Local extinctions are frequent, and empty patches are recolonized by migration (Hanski and Gyllenberg 1993). Most metapopulations have patch size distributions intermediate between the mainland–island and Levins models.

Arnold et al. (1993) examined populations of the euro *(Macropus robustus)*, a large macropod (kangaroo) species, in highly modified fragmented landscapes in the wheatbelt of Western Australia. They identified a metapopulation including several subpopulations comprising just a few individuals (Figure 6.2). Animals moved between several habitat patches scattered throughout the landscape. Adults in the largest patch were sedentary but their offspring dispersed into other patches. Several other authors have recorded similar patterns for other groups of organisms (Table 6.1; see also Hill et al. 1996; Carlson and Edenhamn 2000).

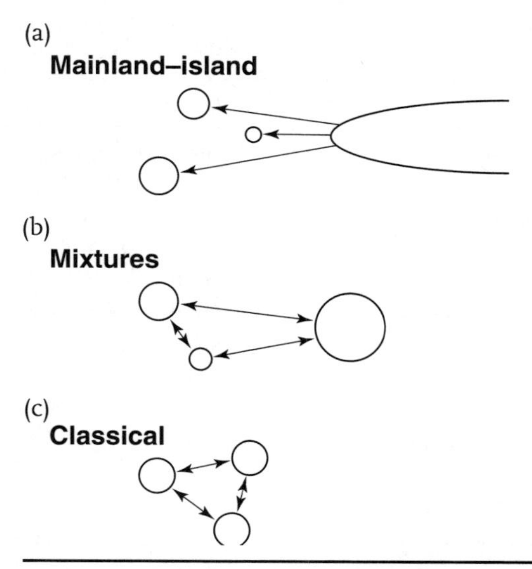

Figure 6.1. Different kinds of metapopulation structure: (a) mainland–island populations, (b) mixtures, metapopulations composed of habitat patches of different sizes, and (c) classical metapopulations with equal patch sizes (Levins structures) (after Hanski, I. and Gyllenberg, M. 1993: Fig. 1. Distribution of patch sizes in the simple model, p. 18, *The American Naturalist*, **142**, 17–41, Two general metapopulation models and the core-satellite hypothesis, published by, and reproduced with permission from, the University of Chicago).

Not All Patchy Populations Are Metapopulations

Patchily distributed populations of a species do not always conform to a true metapopulation structure (Bradford et al. 2003). For example, the areas between vegetation patches may be so hostile that they preclude movement so that populations are confined to isolated habitat patches. Sarre et al. (1995) compared two species of gecko that occupied woodland patches in Western Australia. One species, the tree dtella *(Gehyra variegata)*, probably exhibited a metapopulation structure in which individuals were able to move between woodland patches. In contrast, movements between patches by the other species, the reticulated velvet gecko *(Oedura reticulata)*, were extremely uncommon or absent. Thus, although both species had patchy populations restricted to woodland remnants, one species was distributed as a metapopulation while the other was not (Sarre et al. 1995).

At the other extreme, species in landscapes that appear patchy from a human perspective also may not be distributed as metapopulations if areas between patches are sufficiently suitable to facilitate frequent movements between them (Price et al. 1999), or conditions allow organisms to forage or live in areas between patches (Tubelis et al. 2004). The latter can be the case in some agricultural landscapes. In a study of patchy populations of the eastern chipmunk *(Tamias minimus)* in fencerows in Canada, Bennett et al. (1994) found that movements of animals were so frequent that the species in their study area functioned as a single dynamic population rather than as semi-isolated sub-

Table 6.1. Examples of species or species groups where field data indicated that they were likely to be distributed as metapopulations

Taxon	*Location*	*Citation*
Cougar *(Puma concolor)*	New Mexico, USA	Sweanor et al. 2000
Iberian lynx *(Lynx pardinus)*	Southwestern Spain	Palomares et al. 2000
Black-tailed prarie dog *(Cynomys ludovicianus)*	Colorado, USA	Roach 2001
Pika *(Ochotona princeps)*	California, USA	Peacock and Smith 1997
Euro *(Macropus robustus)*	Southwestern Western Australia	Arnold et al. 1993
Tree dtella *(Gehyra variegata)*	Southwestern Western Australia	Sarre et al. 1995
Several salamander species	Southwestern France	Joly et al. 2001
Several frog species	Sweden	Sjorgen-Gulve 1994
Pond-dwelling plant *(Ranunculus nodiflorus)*	Northern France	Kirchner et al. 2003
Carabid beetle *(Abax parallelepipedus)*	Northern France	Petit and Burel 1998
Checkerspot butterflies	Finnish Islands	Wahlberg et al. 2002
Glanville fritillary butterfly *(Melitaea cinixia)*	Finnish Islands	Hanski 1999a, b
Bog fritillary butterfly *(Proclossiana eunomonia)*	Southern Belguim	Schtickzelle and Baguette 2003
Several butterfly species	North America	Harrison et al. 1988
Two species of midges	Germany	Berendonk and Bonsall 2002
Bush cricket *(Metrioptera bicolor)*	Sweden	Kindvall 1999

populations, each susceptible to extinction (see also Bennett 1998). In another example, the superb parrot *(Polytelis swainsonii)* in southeastern Australia is not confined to woodland patches but frequently nests in scattered "paddock trees" in agricultural land (Manning et al. 2004b). In these cases human-defined vegetation patches do not coincide with habitat patches for the species of interest. Hence the metapopulation concept—which describes a species' population dynamics in relation to the spatial distribution of its habitat—does not apply.

If a species has a metapopulation structure, there should be a characteristic spatial distribution pattern in which locations close to occupied patches are more likely to be occupied than more distant ones (Hanski 1994a; Smith 1994; Koenig 1998). The metapopulation concept is most useful when successful interpatch dispersal is infrequent and migration distances are limited (Hanski and Simberloff 1997). Hastings (1993) suggested that populations should be considered 'independent' rather than part of a metapopulation if less than 10% of resident populations disperse.

Notably, although Table 6.1 gives some examples of metapopulations, demonstrating that patchy populations are truly metapopulations is often difficult because it requires the identification of effective dispersal. It remains to be seen whether some so-called metapopulations are in fact isolated or partially isolated subpopulations undergoing gradual decline.

Large-Scale Movements

In addition to day to day movements within home ranges and movements that link individuals within a genetically connected population, many species need to make less frequent movements over much larger distances. Examples include annual patterns of long-distance migration, which can span continents or hemispheres (Keast 1968; Flather and Sauer 1996). For example, roughly 35% of Britain's nonpasserine and passerine birds migrate. Other

Figure 6.2. Spatial structure in a metapopulation of the euro (after Arnold, G.W., Steven, D.E. and Weeldenburg, J.R. 1993: Reprinted from *Biological Conservation*, **64**, 219–230, Influences of remnant size, spacing pattern and connectivity on population boundaries and demography in Euros *Macropus robustus* living in a fragmented landscape, with permission from Elsevier). The map shows remnants of native vegetation in an agricultural landscape. The dashed line encloses patches of habitat in which euros were captured and patches postulated to be part of the metapopulation.

examples include seasonal altitudinal movements such as those up and down mountains by birds in both tropical areas and temperate zones (e.g., California spotted owl *[Strix occidentalis occidentalis]*), nomadic movements made in response to temporal and spatial variability of important resources (e.g., food for wide-ranging rainforest pigeons; Price et al. 1999), and large shifts in distribution in response to climate change (Parmesan 1996; McCarty 2001). Climate-related range shifts have typically been slow in the past (Keast 1981), but more rapid and extreme changes are expected in response to recent human-enhanced global climate change (Peters and Lovejoy 1992; Hughes 2000; Thomas et al. 2004). Habitat isolation may substantially inhibit such large-scale movements (Soulé et al. 2004).

Box 6.2. Habitat Isolation for Amphibians

Throughout Part II of this book, we emphasize the importance of considering that different species may respond uniquely to landscape modification, and that a given landscape pattern may pose a severe threat to some species but no threat to others. Amphibians are a useful group to consider in this context because their habitat requirements tend to be very different from those of birds and mammals—the two groups that have received by far the most attention by landscape ecologists and conservation biologists.

Joly et al. (2001) studied patterns of pond occupancy by three species of newts in southeastern France. For all three species, their work demonstrated a negative effect of the width of cultivated fields between ponds, indicating that such fields contributed to habitat isolation for each of these species. Notably, linear vegetation features that may be thought to contribute to landscape connectivity from a human perspective (see Chapter 12) were unlikely to mitigate habitat isolation for these species. Newt presence at a given pond was positively related to the number of ponds in the surrounding 50 ha area, suggesting that newts occurred as metapopulations (Joly et al. 2001).

In another study, Guerry and Hunter (2002) investigated the relationship of amphibian distributions with landscape pattern in Maine (USA). This study demonstrated that habitat isolation was a highly species-specific concept, with five species responding positively to the amount of forest cover surrounding a pond, and two species responding negatively to forest cover. Moreover, the spatial juxtaposition of forest in relation to ponds was also significantly related to the distribution of amphibians. Two species were more likely to occur in ponds that were directly adjacent to forest. Another species was more likely to occupy ponds adjacent to forests in areas with little overall forest cover but was less likely to occupy ponds adjacent to forests in areas with high overall forest cover. Occupancy of ponds, and habitat isolation therefore depended on a range of different factors for different species.

These case studies illustrate that (1) suitable habitat for some species may not be directly related to human-defined patches of native vegetation, (2) habitat isolation for some species may be related to factors other than human-defined landscape connectivity (see Chapter 12), and (3) a given landscape pattern may correspond to habitat isolation for some species but not for others.

What Represents Isolation?

The degree of habitat isolation experienced by individuals of a given species depends on many factors. For example, some authors have argued that above a particular level of habitat loss the physical distances between habitat patches increase exponentially (Gardner et al. 1987; Gustafson and Parker 1992). For many species, rates of movement between patches of suitable habitat can be reduced as a result (Powell and Powell 1987; Sarre et al. 1995; Desrochers and Hannon 1997).

Spatial scale, mobility, and mode of movement (e.g., flying versus crawling) are key issues associated with considerations of the impacts of habitat subdivision and habitat isolation. The spatial scales over which a species moves and over which it perceives its environment will strongly influence the extent to which a given modified landscape is, or is not, negatively subdivided or isolated for that taxon (Manning et al. 2004a). For example, for some small mammal and flightless insect species, a forest road may effectively subdivide and isolate the populations on either side of it (e.g., Mader 1984; Goosem 2000), whereas such a road would have very limited or no impact on more mobile species, such as a fast-flying parrot that forages over large areas (Forman et al. 2002). Although habitat subdivision can have detrimental consequences for many taxa, some species can effectively compensate for a certain level of habitat isolation by altering their behavior. For example, habitat subdivision may stimulate some species to move further to gather resources from more patches scattered over a larger area, as forecast for bears in northeastern North America (Boone and Hunter 1996) and shown empirically in a study of the tawny owl *(Strix aluco)* in England (Redpath 1995).

Many of the caveats associated with the themes of habitat loss and habitat degradation (see Chapters 4 and 5) are also relevant to considerations of habitat subdivision and habitat isolation. This is because, like habitat loss, what constitutes habitat subdivision and habitat isolation will be species-specific. Thus, for example, the isolation of vegetation patches defined from a human perspective may not lead to habitat isolation from the perspective of some species. Even in a landscape that is extensively modified by humans, the matrix may be highly permeable for some species. Hence actual levels of habitat isolation are relatively low for these taxa (Boone and Hunter 1996) and recolonization rates of patches can be high (Andrén 1999). For other species, the same matrix may be so hostile that neighboring patches may be close together but very isolated (Taylor et al. 1993)—as has been observed for highly patch-restricted geckos in the heavily cleared wheatbelt of Western Australia (Sarre et al. 1995; see earlier discussion).

In summary, habitat isolation is species-specific (Rosenberg et al. 1998), an outcome of dispersal behavior, mode of movement (e.g., whether the main form of movement is flying or crawling), and how these interact with landscape patterns (Box 6.2; also see Chapter 8).

Confounding of Habitat Loss and Habitat Isolation

Studying the effects of habitat subdivision and habitat isolation on species can be difficult because these processes are often confounded with habitat loss. Fahrig (2003) argued that the total amount of habitat tended to be more important in its effects on the persistence of biota than the spatial pattern of that habitat. In studies where habitat subdivision was examined and habitat quantity was held constant, the effects of habitat subdivision on biota were equivocal (Fahrig 2003). In some cases, subdivision had positive effects (e.g., due to factors such as increased opportunities for predators, their prey, and competitors to coexist in more subdivided landscapes) (Fahrig 2003). Negative effects were due to edge effects (Chapter 11) and the nonviability of populations occupying small fragments (Chapter 4). A potential problem when interpreting Fahrig's (2003) review is that she equated habitat loss with vegetation loss and did not explicitly distinguish between different species' unique habitat requirements. This is a common problem in many studies in human-modified landscapes (Chapter 4).

Despite inherent difficulties of disentangling the effects of habitat loss and habitat subdivision, determining the relative importance of these and other processes associated with landscape alteration can be important. If habitat loss is the primary factor influencing the persistence of species, then the thrust of landscape management may be to maximize the area of suitable habitat for a given species. Conversely, if habitat isolation is the key problem, then enhancing habitat connectivity might be an appropriate strategy (Harris and Scheck 1991; Cooper and Walters 2002)—although increasing the amount of available habitat can also contribute to habitat connectivity and, vice versa, increasing habitat connectivity can increase available habitat. Practical methods for mitigating habitat loss, habitat isolation, and other negative effects of landscape change on selected species and species assemblages are examined in Part V of this book.

Summary

Landscape alteration can lead to the subdivision and isolation of remaining habitat patches. Individuals and populations of a given species can respond negatively to the smaller size of the patches remaining after the subdivision of a formerly larger patch, and the increased distances between remaining patches. Habitat isolation can negatively impact on movements of individuals at various scales, including day to day movements within home ranges, dispersal from the natal territory, and larger-scale movements such as nomadic movements, seasonal migration, and long-term range shifts in response to climate change. The patchiness of habitat cover, often created by landscape modification, can result in the formation of metapopulations, although not all patchy populations have the key attributes of true metapopulations. Although habitat isolation is an important threatening process for many species in its own right, it is often extremely difficult to separate its impacts on species and populations from those of habitat loss. This is because these and other threatening processes usually co-occur in modified landscapes. Nevertheless, identifying which threatening processes are the most important ones for a given species can be essential for developing effective mitigation strategies.

Links to Other Chapters

The process of habitat subdivision often arises from spatially discontinuous land cover following from landscape modification. Landscape modification typically results in smaller habitat patches for individual species (Chapter 4), and in smaller patches of native vegetation (Chapter 9). The topics of habitat isolation and habitat connectivity are species-specific and may not necessarily be related to landscape connectivity as defined from a human perspective; the concept of landscape connectivity is discussed in Chapter 12. Spatial subdivision of patterns of vegetation cover (as opposed to habitat patches) leads to altered perimeter:area ratios in remaining areas of vegetation in modified landscapes. Edge effects that can manifest at these boundaries and their impacts on assemblages are discussed in Chapter 11.

Further Reading

Fahrig (1997, 2003) discusses why it is important to consider habitat subdivision as a distinct process from habitat loss. Examples of the many studies on movement patterns of individuals in relation to habitat subdivision include those of Guevara and Laborde (1993) and Fischer and Lindenmayer (2002b). Metapopulation biology is reviewed by Hanski and Simberloff (1997), Hanski (1999a, b), Chesson (2001), and Beissinger and McCullough (2002). Finally, useful empirical studies of the consequences of habitat isolation, and novel ways of identifying it as a key threatening process, are outlined by Cooper and Walters (2002) and Brooker and Brooker (2001, 2002).

CHAPTER 7

Changes in the Behavior, Biology, and Interactions of Species

Landscape change can have direct negative impacts on species via the interconnected processes of habitat loss, habitat degradation, and habitat subdivision (Chapters 4–6). However, the extinction of a given species is only the end point of a prolonged process of change (Simberloff 1988; Clark et al. 1990) and often follows subtle changes to the behavior and biology of that particular species. In addition, no species exists by itself, but interacts with others; for example, through competition, predation, and mutualisms. Landscape change can alter the habitat of many species, thus leading to changed species interactions. In this chapter, we provide several examples to highlight the range of ways in which landscape change may alter the behavior and biology of species. We also discuss some changes to immediate interactions between species, such as predator–prey relationships, interspecific aggression, and changed mutualisms. Species interactions are discussed primarily in the context of their effects on individual species. More complex changes to ecosystems involving many species are discussed in Chapter 15 in Part III.

Altered Behavior and Biology

Landscape modification can lead to many changes in the behavior and biology of individual species (Kozakiewicz and Kanopka 1991). Such changes can have major impacts on population persistence (Rolstad and Wegge 1987; McCarthy et al. 1994; McCarthy 1997). Some examples are illustrated in the following section.

Breeding Patterns and Social Systems

Landscape modification can alter the breeding biology of species (e.g., Huhta et al. 1999; Zanette and Jenkins 2000). For example, several studies of birds have found that, in comparison with unmodified areas, animals in remnant vegetation patches may (1) have a shorter breeding season (Zanette et al. 2000), (2) lay fewer eggs that are lighter in weight (Hinsley et al. 1999; Zanette et al. 2000), and (3) rear fewer smaller nestlings (Hinsley et al. 1999). Zanette et al. (2000) concluded that the reasons for such changes in the eastern yellow robin, an Australian passerine, were associated with per capita food limitations, which in turn explained why the species was less abundant in small habitat patches (see Box 4.3 in Chapter 4). Saunders (1980) studied Carnaby's cockatoo *(Calyptorhyncus funerus latirostris)* in the heavily altered wheatbelt of Western Australia and found that adult incubation behavior was impaired, leading to slower growth rates of young birds, reduced fledging weights, and overall lower levels of breeding success. Similarly, pairing success in the ovenbird (*Seiurus aurocapillus)* in North America can be significantly reduced in small forest patches where there is a lack of female birds (Villard et al. 1993).

For some animals, breeding success is closely related to particular social systems. Social systems are sometimes disrupted or heavily altered by landscape modification (Peacock and Smith 1997). As an example, Ims et al. (1993) documented how the social system of the capercaillie grouse *(Tetrao urogallus)* in southeastern Norway changed from a lekking system to one of solitary displaying males with increasing habitat loss and habitat subdivision. Altered social behavior can, in turn, influence breeding biology. This was demonstrated by Banks et al. (2004), who found that the amount of multiple paternity of litters in the small marsupial carnivore, the agile antechinus *(Antechinus agilis)*, was reduced in populations inhabiting small habitat patches in southeastern Australia.

Dispersal

Dispersal is one of the most significant biological processes shaping the distribution and abundance of all animals, plants, and other organisms on Earth (Chepko-Sade and Halpin 1987; see Chapter 6). It is fundamental to many attributes of populations such as recolonization dynamics (Hanski 1994a), the ability to reverse local extinctions and reinforce declining populations (Brown and Kodric-Brown 1977), the risk of extinction in isolated habitat patches (Brooker and Brooker 2002), and patterns of genetic variability (Saccheri et al. 1998; Young and Clarke 2000). Dispersal has three phases—emigration, travel, and immigration (Stenseth and Lidicker 1992a). Landscape change, habitat loss, and habitat subdivision can have a wide range of effects on all three phases of the dispersal behavior of individual species (Wiggett and Boag 1989). For example, landscape conditions may severely limit the number of dispersal events (Banks et al. 2004; Stow and Sannucks 2004) or the distances over which animals disperse (Matthysen et al. 1995). This, in turn, can have negative flow-on effects on breeding behavior such as the size of social neighborhoods (Cale 2003); the ability of animals to form breeding pairs (Gibbs and Faaborg 1990; Villard et al. 1993); elevated relatedness among animals in the same remnant patches (Banks et al. 2004); impaired inbreeding avoidance, which can decrease offspring fitness (Madsen et al. 1999); or the avoidance of mating altogether (Stow and Sannucks 2004).

For some species restricted to remnants of native vegetation, conditions in the matrix can be so hostile that individuals may be reluctant to disperse away from habitat fragments. This has been termed a "fence effect" (Wolff et al. 1997). Fence effects may increase the population density in habitat patches relative to unmodified areas or those patches where the surrounding matrix is less hostile. Fence effects are not often reported, but they have been documented for small mammals in the Northern Hemisphere (e.g., Krebs et al. 1969; Krohne 1997; Bayne and Hobson 1998). Fence effects demand some plasticity in life history parameters, such as increased population density, increased home range overlap, and reduced core home range size (Krebs 1992; Pope et al. 2004; see next section).

Altered Use of Space

The size and spatial configuration of habitat patches in modified landscapes can have profound effects on their use by the individuals of a given species. In some modified landscapes, animals may move greater distances to find food or mates than conspecifics inhabiting unmodified areas (Ricketts 2001; Fraser and Stutchbury 2004). Extraterritorial movements between habitat patches to find mates have been observed in some bird species such as the hooded warbler *(Wilsonia citrina)* in North America (Norris and Stutchbury 2001). Similarly, some bird and mammal species may need to move over larger distances in subdivided landscapes to find sufficient food (Boone and Hunter 1996). Recher et al. (1987) suggested that within linear strips of eucalypt forest surrounded by an extensive radiata pine *(Pinus radiata)* plantation in southeastern Australia, some species of birds and mammals had to move over large distances because they relied on dispersed food resources.

Home ranges and movement patterns of animals also may be altered within habitat patches as a result of habitat subdivision (Arnold et al. 1993). Pope et al. (2004) found that the home range size of the greater glider *(Petauroides volans)*, an Australian arboreal marsupial, decreased significantly with decreasing patch size and increased patch population density. Small patches had more animals per unit area, leading to smaller home ranges and greater home range overlap. The number of different den trees used by individual greater gliders also was lower in small patches than in large patches (Pope 2003; Lindenmayer et al. 2004a). In a similar North American example, Barbour and Litvaitis (1993) found that landscape modification in New Hampshire altered the use of remnants by New England cottontail rabbits *(Sylvilagus transitionalis)*. Animals in small areas of remnant habitat foraged in a greater range of areas with different understory cover patterns and ventured further from cover than animals in larger remnants (Barbour and Litvaitis 1993). Other differences characterized populations in small and large remnants; animals in small patches lived at higher densities and were more susceptible to predation. In addition, populations in smaller patches were highly male biased (Barbour and Litvaitis 1993).

Body Symmetry

The morphology and body symmetry of species can be altered in response to landscape change (Barbour and Litviatus 1993; see Box 7.1). For example, in the United Kingdom, Hill et al. (1999) found that the flight morphology of the rare butterfly *Hesperia comma* had changed as a result of landscape alteration with corresponding impacts on its dispersal ability and recolonization dynamics. Weishampel et al. (1997) studied the centipede *Rhysida nuda* in rainforest remnants in northeastern Australia. They made a wide range of morphological measurements and found that centipede populations in the smallest vegetation remnants had the highest levels of body asymmetry (also see Box 7.1). In addition, Weishampel et al. (1997) believed there was selection pressure for larger body sizes in centipedes in smaller remnants because of microclimatic edge effects leading to greater potential for desiccation in smaller animals (see Chapter 11).

Box 7.1. Landscape Alteration and Morphology: Two Case Studies on Australian Lizards

Two instructive examples of how landscape change can affect the morphology of species deal with Australian lizards. Sarre (1996) studied two gecko species (*Oedura reticulata* and *Gehyra variegata*) in the highly modified wheatbelt in Western Australia. The region used to be covered by native *Eucalyptus* woodlands but has been extensively cleared, especially since the early 1900s (Saunders et al. 1993b). Sarre (1996) investigated geckos in 12 woodland remnants and compared these with geckos inhabiting large, continuous woodland reserves. Sarre (1996) measured several morphological variables of both species, including head length, snout–vent length, and several other measures. In addition, he counted the number of scales directly above and below each gecko's mouth (called "labial" scales), separately for the left and right side of the head. Sarre (1996) found no differences in the overall morphology of geckos in patches versus continuous woodland areas. However, on average, individual geckos of both species exhibited a higher difference in the number of labial scales on their left and right body sides. The degree of random difference between left and right sides of bilateral body structures is referred to as "fluctuating body asymmetry." Increased body asymmetry is considered to be a sign of developmental instability in populations, typically caused by physiological stress (Sarre and Dearn 1991; Sarre et al. 1994). Sarre (1996) concluded that increased fluctuating body asymmetry in geckos in the Western Australian wheatbelt may be the result of environmental stress or the loss of genetic diversity due to landscape modification.

More obvious morphological changes were observed by Sumner et al. (1999) in a study on the prickly forest skink *(Gnypetoscincus queenslandiae)* in Queensland. In this study, lizard density was lower in rainforest fragments than in continuous areas of rainforest, and adult skinks in fragments were significantly smaller than skinks in areas of continuous forest.

Box 7.2. Landscape Change and Altered Acoustic Environments

Vocalization is a critical part of the biology of all birds (Catchpole and Slater 1995), and it is being studied intensively in New Zealand where major landscape change coupled with the impacts of introduced mammals has had massive impacts on the endemic avifauna. Offshore islands support some of the most complete assemblages of native bird species that remain in New Zealand. These assemblages are important because the song environment they create can have profound impacts on the development of the song repertoire of young birds. This, in turn, has significant effects on individual species such as the tui (*Prosthemadera novaeseelandiae*; Figure 7.1), which is a mimic and copies and incorporates sounds made by other species into its song repertoire. Mainland tui have simple repertoires, incorporating song elements from invasive avian species, compared to tui raised on islands where the acoustic environment is more complex and includes songs from predominantly native birds. These varying repertoires, in turn, have contributed to the development of geographic dialects and appear to influence the mate attraction and mating success of translocated birds such as those relocated to the mainland to help recover declining populations there. Therefore, landscape change has altered the acoustic environment of mainland New Zealand; novel approaches need to be employed to tackle the problems this has created for individual species and, in turn, increase the effectiveness of efforts to recover threatened endemic bird taxa (Dianne Brunton, pers. comm.).

Other Forms of Altered Behavior and Biology

Many other forms of behavior can be altered as a result of landscape change, habitat loss, and habitat subdivision. For example, Slabbekoorn and Peet (2003) demonstrated that great tits *(Parus major)* in the city of Leiden in the Netherlands vocalize at a higher frequency than elsewhere to prevent their songs from being masked by largely low-frequency urban noise. Similar changes to bird vocalization were reported by Lindenmayer et al. (2004b), who found different patterns of vocalization in habitat fragments than in continuous areas, and higher rates of vocalization in large patches than in small ones. This may be associated with what some authors have termed "conspecific attraction" (e.g., Reed and Dobson 1993). Dispersing birds may be more likely to settle in habitat patches where there are already conspecifics vocalizing (Smith and Peacock 1990; Eens 1994) because the amount of vocal activity may serve as a cue indicating habitat suitability (Alatalo et al. 1982). Therefore, increased activity in larger patches could be related to birds attempting to attract mates to colonize these areas (see also Box 7.2). These examples illustrate that acoustic changes to the environment may affect wildlife in complex and not yet fully appreciated ways (Slabbekoorn and Peet 2003; Katti and Warren 2004).

Group behaviors in birds that increase foraging efficiency and predator detection are other examples of the impacts of landscape change on behavior. With a reduction in group size, the benefits of group behavior can decline to the point where individuals and groups cannot persist (Gardner 2004; Box 7.3). The reliance on a sufficiently large number of individuals to function effectively is sometimes termed an "Allee effect." The effect is named after Warder Allee, who observed that some species, such as the common starling *(Sturnus vulgaris)* and red deer *(Cervus elaphus)*, survived only in groups, where group behaviors are needed to stimulate breeding, to avoid predators, and for group defense (Allee 1931; Allee et al. 1949). The disruption of Allee effects and a subsequent decline in breeding success are thought to have contributed to the extinction of the passenger pigeon *(Ectopistes simigratorius)* in North America (Schorger 1973).

Allee effects are also observed among plants in highly modified landscapes. For example, if plant population abundance and density are extremely low (e.g., as a result of clearing or intensive logging of tropical forests), there may be too few individuals to attract appropriate pollinators, and pollen transfer between individuals can become limited or halted (Bawa 1990;

Figure 7.1. New Zealand tui (photo by Dianne H. Brunton).

Groom 1998). Similarly, before their demise as a result of chestnut blight, it was believed that stands of the American chestnut *(Castanea dentata)* needed to be above a certain size to produce sufficient seed to overcome pressure from seed predators (P. Rutter in Rosenzweig 1995).

Altered Species Interactions

Many species are intimately associated and cannot survive in isolation from one another. Landscape modification can alter several kinds of species interactions (Taylor and Merriam 1995). Four types of immediate interactions between species are discussed in the following sections: competition, predation and parasitism, and mutualisms.

Competition

Landscape alteration can have profound impacts on competition for resources, or aggressive behavior between taxa. The noisy miner *(Manorina melanocephala)* prefers to live at the edges of temperate woodland remnants in eastern Australia (Piper et al. 2002), where it interacts aggressively with other species of birds. Populations of the noisy miner have increased considerably since European settlement, in part because woodland edges are now common in the remaining temperate woodlands. Landscapes once dominated by woodlands are now largely cleared (Benson 1999), and the remaining woodland patches are very small—most are 1 ha or smaller (Gibbons and Boak 2002). Noisy miners scare away other insect- and nectar-feeding birds that are smaller than them, thus leading to significantly re-

Box 7.3. Landscape Change and Allee Effects

The speckled warbler *(Sericornis sagittatus)* is a bird of shrubby woodland understories of temperate woodlands in southeastern Australia. These ecosystems have been heavily modified in the past 200 years (Prober and Thiele 1995; Lindenmayer et al. 2005b), and the speckled warbler is among a raft of bird taxa known to be declining in Australia's temperate woodland ecosystems (Reid 1999). Allee effects appear to be among the factors contributing to population declines of the speckled warbler (Gardner 2004). Individuals group into small foraging flocks to increase the efficiency of finding food and make it easier to detect predators. This strategy is extremely important for survival during harsh climatic conditions. The size of territories used by flocks of speckled warblers has been estimated to be 30 ha or larger. Group dynamics of warblers may be impaired in highly modified woodland landscapes where the remnants are too small. Reduced group size, in turn, can lead to significantly lower survival and recruitment of young birds (Gardner 2004).

duced bird species richness in many woodland remnants (Grey et al. 1997).

Issues associated with landscape change and interspecific competition can be particularly pronounced when exotic species are introduced to a human-modified environment. There are well-documented examples from many parts of the world of the highly detrimental effects of introduced plants on native species (Pimm 1992; Low 1999). Similar impacts are known for animals. Hadfield (1986) described the case of introduction of the African land snail *(Achatina fulica)* to a number of islands in the Pacific Ocean. The species was originally a food source but became a serious pest in highly modified landscapes where food crops had been established. Another snail, *Euglandia rosea*, was then introduced to control the African land snail but had limited effect on the target species. However, it rapidly extinguished substantial numbers of native snails, particularly those belonging to the genus *Partula*. Hawaii alone has lost nearly two-thirds of its original several hundred species of native tree snails (Hadfield et al. 1980; Hadfield 1986).

Predation and Parasitism

Predator–prey interactions and parasite–host interactions can be altered as a consequence of human landscape modification. Two of the most extensively researched species interactions in modified landscapes are nest predation and nest parasitism. Increased predation and parasitism can be prevalent as a result of landscape modification, particularly at the boundaries of vegetation remnants (Paton 1994; Robinson et al. 1995; Lahti 2001). These kinds of biotic edge effects (Chapter 11) can have substantial negative impacts on populations of individual species as well as entire assemblages of species (Murcia 1995). For example, in many parts of North America, the brown-headed cowbird *(Molothrus ater)* occupies vegetation edges and lays its eggs in the nests of a range of other bird species. This parasitic behavior can lead to significant losses in the number of offspring produced by the host (Brittingham and Temple 1983; Temple and Cary 1988), particularly reduced fledging success (Gates and Gysel 1978). In addition, recent work has suggested that cowbird parasitism may skew the sex ratios of the host's offspring (Zanette et al. 2005).

In response to landscape alteration, key changes may also occur to predation regimes, of both native and introduced predators. For example, in southeastern Australia, road development associated with landscape modification may provide conduits for the movement of introduced predators (Seabrook and Dettmann 1996), such as the red fox *(Vulpes vulpes)* and feral cat *(Felis catus)* (May and Norton 1996). This may allow predators to gain access to areas from which they were previously absent, thus increasing predation pressure on native animals (Robertshaw and Harden 1989; May 2001).

In another example of the altered predation effects in human-modified landscapes, New England cottontail rabbits were found to be more vulnerable to being captured by the red fox *(Vulpes vulpes)* and the coyote *(Canis latrans)* in small habitat remnants than in larger ones (Barbour and Litvaitis 1993).

Mutualisms

Landscape alteration, habitat loss, and habitat subdivision can disrupt mutualisms and other critical interrelationships between species (Table 7.1). The loss of mutualisms can lead to severe population declines or even extinctions (Box 7.4).

One of the most fundamental types of mutualism is the interaction between plants and their pollinators (Kearns and Inouye 1997). Many flowering plants are dependent on particular animals for pollination, and they decline in the absence of these pollinators. An example comes from the island of Mauritius in the Indian Ocean, which has been subject to massive human modification resulting in many extinctions and leaving numerous other taxa highly endangered (Safford 1997). Many plants are threatened because their pollinators have been lost or have declined to levels well below their functional effectiveness. In an elegant set of experiments and observational studies, Denis Hansen et al. (unpublished data) showed that in parts of Mauritius, endemic *Pandanus* spp. plants provide shelter for endemic *Phelsuma* spp. geckos that are also known to pollinate several other species of plants (Figure 7.2). These interactions, in turn, are strongly influenced by the spatial juxtaposition of stands of *Pandanus* spp. with other plant taxa, and plants growing close to areas of *Pandanus* spp. have a significantly higher fruitset than plants growing further away. Many *Pandanus* spp. are still declining on Mauritius, and loss of the dense patches they provide could lead to the decline of lizard populations and a subsequent decline of the plant species they pollinate, as well as alterations to the spatial distribution of plant assemblages and alterations to spatial juxtaposition of different plant species (Denis Hansen et al. unpublished data; see also Olesen and Valido 2003).

Kearns et al. (1998) argued that many ecosystems are currently facing a "pollination crisis." Anthropogenic landscape change and modern land use practices have resulted in the decline of introduced honeybees *(Apis mellifera)* as well as many native bee species in landscapes throughout the world. Birds, too, may be important pollinators and can be adversely affected by landscape change (Kearns et al. 1998). For example, in the Mount Lofty Ranges in South

Box 7.4. Disruption of the Mutualism between Frugivorous Birds and an Endemic Tree Species in Tanzania

Few empirical studies have directly demonstrated a link between landscape modification and the disruption of a mutualist relationship. A noteworthy exception is the study by Cordeiro and Howe (2003) in the East Usambara Mountains in Tanzania, which occur within one of Africa's biodiversity hotspots (Brooks et al. 2002). These authors investigated the mutualism between the endemic tree *Leptonychia usambarensis* and frugivorous birds that feed on the tree's fruit and upon which the tree is dependent to effectively disperse its seeds. Frugivorous birds were recorded within eight forest fragments ranging in size from 2 to 520 ha, and at three sites within continuous forest. Six bird species were restricted to continuous forest, and three species were more abundant in continuous forest. Seven other potential dispersing birds were equally abundant in forest fragments and continuous forest, and only one bird species was more abundant in fragments. The negative effects on *Leptonychia usambarensis* were clearly evident. The decline and loss of several of the most effective dispersal agents in forest fragments were not offset by the remaining frugivorous birds. Compared to continuous forest, in fragments, fewer seeds were removed from trees and seedling densities were higher directly beneath parent trees but lower at distances greater than 10 m from parent trees. The decline of dispersal agents thus appeared to be directly related to reduced recruitment in an endemic tree species.

Table 7.1. Examples of mutualist relationships that may be disrupted as a result of landscape change

Example	*Reference*
Pollination of trees on Pacific islands is disrupted by landscape alteration and overharvesting of flying foxes (*Pteropus* spp.)	Cox et al. 1991; Cox and Elmquist 2000
Ectomycorrhizal fungi can be lost in agricultural landscapes in Australia with negative effects on woodland restoration	Tommerup and Bougher 1999
Pollination of Proteaceae trees in southern Chile by hummingbirds is disrupted in modified landscapes because of altered patterns of territorial defense	Ramirez-Smith and Armesto 2003
Environmental and demographic stochasticity may threaten mutualist relationships between ants and plants in Amazonian rainforest fragments	Bruna et al. 2005
Northern flying squirrel *(Glaucomys volans)* is a key dispersal agent for many species of underground fruiting fungi that form mycorrhizal associations with tree and other plant species in the forests of northwestern North America	Carey et al. 1999a
Long-nosed potoroo *(Potoroo tridactylus)* is a key dispersal agent for many species of underground fruiting fungi that form mycorrhizal associations with tree and other plant species in the forests of eastern Australia	Bennett and Baxter 1989; Claridge 1993
Southern cassowary *(Casuarius casuarius)* is the only disperser of more than 100 plant species with large fruits in the rainforests of North Queensland; without the southern cassowary, the dynamics of rainforest communities would be changed markedly and the status of many plants (and the animals that are associated with them) would change; populations of the southern cassowary are negatively impacted in modified landscapes through problems such as collisions with motor vehicles	Crome 1976, 1994; Crome and Moore 1990; Crome and Bentrupperbaumer 1993
Crossbill birds can be very important consumers and dispersers of seeds of conifer trees in western North America, and most species in this species-rich assemblage are adapted specifically to forage on a single species or variety of conifer; populations of crossbills are negatively affected in modified forest landscapes, especially those in which old-growth conifer stands are reduced in extent	Benkman 1993

Australia, many honeyeaters leave the area in summer and autumn because vegetation clearing has led to a lack of available food during this time. This seasonal decline in bird populations, in turn, has adversely affected pollination for some bird-pollinated plants (Paton 2000).

Other disrupted mutualisms are more complex. A well-known example is the case of the large blue butterfly *(Maculina arion)* in Britain (Thomas 1984a, b). This species is strongly associated with open habitats, which have become severely reduced as a result of a modification in land use practices, particularly a reduction in grazing. Problems of habitat loss and habitat isolation compounded when populations of rabbits *(Oryctolagus cuniculus)*—which helped maintain open habitats for butterflies—were greatly reduced by the disease myxomatosis. Caterpillars of the large blue butterfly develop inside the nests of an ant species *(Myrmica sabuleti)*, but this species, too, was unable to survive in overgrown but formerly open habitats. Ant populations subsequently declined and butterfly populations went extinct (Ratcliffe 1979). The basis for decline and extinction was carefully established from an extensive understanding of the biology of the butterfly, and interrelationships with its ant host in relation to successional dynamics of vegetation cover (Elmes and Thomas 1992). The reintroduction of butterflies from northern Europe, together with habitat manipulation by careful burning

Figure 7.2. Native plants being pollinated by lizards in Mauritius (photo by Dennis Hansen).

and appropriate grazing regimes, eventually facilitated the recovery of the species in the United Kingdom (Caughley and Gunn 1996).

Strongly Interacting Species

Some species interact particularly strongly with other species, and therefore play a disproportionately important role in maintaining ecosystem function. Such species are sometimes referred to as keystone species or strongly interacting species (Paine 1969; Power et al. 1996; Soulé et al. 2005). The loss of keystone species can lead to major changes in ecosystem functioning. As an example, prior to the arrival of Polynesian people in New Zealand in about AD 950, that country's avifauna had probably twice as many species as presently. Vegetation clearing coupled with hunting and the introduction of exotic animals and plants extensively transformed New Zealand and led to numerous species' declines and extinction. The loss of the giant flightless moas is one of the most famous sets of bird extinctions in New Zealand (Holdaway and Jacomb 2000). Many species associations were disrupted as a result of the loss of the moas. Characteristic grazing intensities and dispersal of native plants were also heavily impacted (Kirkpatrick 1994). In addition, species that preyed on flightless moas, such as the giant eagle *(Harpagornis moorei)*, were lost as their prey went extinct (Milberg and Tyrberg 1993). A more extensive discussion of keystone species is presented in Chapter 15 in the context of cascading effects of landscape change.

Summary

Chapters 4 through 6 have outlined key processes that can have negative effects on species in modified landscapes. These processes may directly or indirectly impact on both the behavior and the biology of individual species. Biological changes include altered morphology and altered breeding success. Changes in behavior in modified landscapes include altered home range sizes, mating systems, patterns of dispersal, and vocalization. Because a given species does not exist in isolation from other elements of the biota, human modifications to landscapes can also lead to profound changes in species interactions such as competition, predator–prey relationships, and mutualisms. These changes in behavior, biology, and interspecies interactions can contribute substantially to the decline and eventual extinction of species in human-modified landscapes.

Links to Other Chapters

One major way in which landscape change can alter interspecific processes is via altered predator–prey or host–parasite relationships that often occur at the edges of patches of remnant vegetation. Such kinds of biotic edge effects are explored in detail in Chapter 11. Disrupted interactions between species can lead to many cascading effects of landscape alteration, some of which are discussed in Chapter 15. Changes in the biology and behavior of species interact with and may exacerbate the threatening processes of habitat loss (Chapter 4), habitat degradation (Chapter 5), and habitat isolation (Chapter 6). Such interactions and their effects on the extinction proneness of individual species are discussed in Chapter 8.

Further Reading

There are many hundreds of detailed autoecological studies that assess various changes in the biology and behavior of species, as well as species interactions, in response to landscape alteration. Interesting examples include Saunders (1980) and Zanette et al. (2000) for breeding success; Ims et al. (1993) for social systems; Brooker and Brooker (2002) for dispersal; Norris and Stutchbury (2001) for altered use of space; Sumner et al. (1999) for morphological changes; Grey et al. (1997) for competition; Barbour and Litvaitis (1993) for predation; and Cordeiro and Howe (2003) for disrupted mutualisms.

CHAPTER 8

Synthesis: Threatening Processes for Species in Modified Landscapes

In Chapters 4 through 7, we discussed key processes that are often associated with landscape change and which may pose a threat to the persistence of individual species. In this chapter, we provide a summary and synthesis of these processes. We outline how these threatening processes may lead to population declines and thus make species more susceptible to extinction as a result of stochastic (i.e., chance) events. Finally, we examine which biological attributes of a given species can make it either more prone or more resilient to extinction in modified landscapes.

Threatening Processes and Their Interactions

There are two broad forms of processes that are important in the dynamics of populations and can cause population declines (Gilpin and Soulé 1986). These are deterministic and stochastic events (Simberloff 1988; Figure 8.1).

Deterministic Threatening Processes

Gilpin and Soulé (1986) defined deterministic threatening processes as those "which result from some inexorable change or force from which there is no hope of escape." Chapters 4, 5, and 6 have outlined four major deterministic processes threatening individual species in modified landscapes—habitat loss, habitat degradation, habitat subdivision, and habitat isolation. Two additional groups of deterministic effects resulting from landscape modification were discussed in Chapter 7: disruptions in the biology of a species and changed interspecific interactions. The processes discussed in Chapters 4 through 6 (habitat loss, degradation, subdivision, and isolation) may be classified as exogenous because they influence a given species' population size but are not directly part of the species' inherent biology. In contrast, the changes in a given species' biology and behavior discussed in Chapter 7 are endogenous; that is, they are internal to a given species' life cycle (e.g., social systems and dispersal). Changed interspecific interactions like predation, competition, and mutualisms are partly exogenous and partly endogenous; they are partly external and partly internal to a given species' biology (see Figure 8.1).

Stochastic Threatening Processes

Stochastic threatening processes result from random fluctuations or perturbations (Gilpin and Soulé 1986). For small populations in particular, stochastic threats can exacerbate the risk from deterministic threatening processes. Four broad types of stochastic events are outlined in the following section.

EXOGENOUS STOCHASTIC PROCESSES

Environmental stochastic processes correspond to year to year changes in birth and death rates within populations that result from temporal variation in factors that are external to a population. Such factors include climate (e.g., rainfall and temperature), changes in habitat parameters, and populations of prey, competitors, predators, parasites, and diseases. For example, the availability of invertebrate prey for the Florida snail kite *(Rostrhamus sociabilis)*

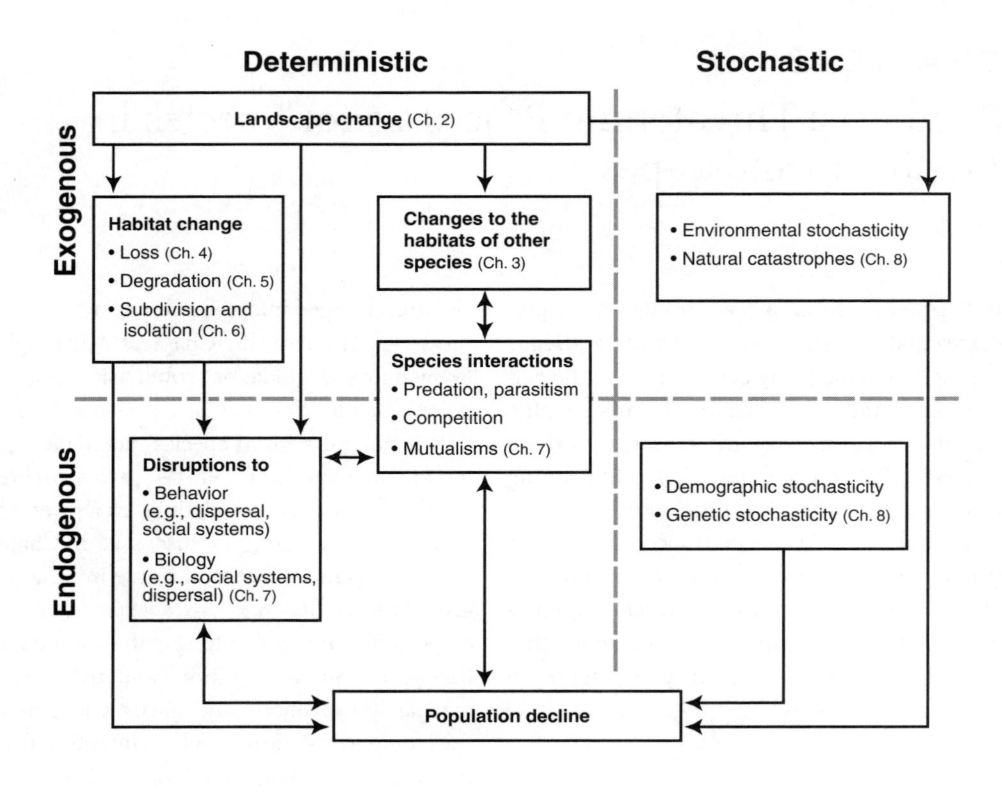

Figure 8.1. Effects of anthropogenic landscape change from the perspective of a declining species. The figure classifies the threatening processes discussed in various chapters as deterministic, stochastic, exogenous, and endogenous. Deterministic threats predictably lead to declines, whereas stochastic threats are driven by chance events. Exogenous threatening processes are external to a species' biology, whereas endogenous threats arise as part of a species' biology (see text for details).

was shown to be correlated with rainfall (Beissinger 1986). Environmental stochastic processes may influence the demographic properties of a population (Simberloff 1988).

Natural catastrophes are an extreme form of environmental stochasticity and include events such as floods, fires, and droughts, which occur at random intervals through time (Simberloff 1988). Catastrophes are an extreme form of perturbation and may have a dramatic impact on populations (Gilpin 1987; Menges 1990). For example, a single hurricane was directly linked to the loss of a significant proportion of the remaining wild population of the Puerto Rican parrot (*Amazona vittata;* Lacy and Clark 1990).

Endogenous Stochastic Processes

Endogenous stochastic processes occur as part of a species' life cycle. Demographic stochastic processes correspond to random variation in demographic parameters such as sex ratio and the intrinsic survival, mortality, and reproductive success of individuals (May 1973; Roughgarden 1975; Clark et al. 1990). An example of a stochastic demographic event is the birth of just one sex in a small population in any given year (Simberloff 1988). Demographic stochasticity is intrinsic to populations of organisms because life-history events are probabilistic. The influence of such effects will be most pronounced in small populations (Murphy et al. 1990; McCarthy et al. 1994, 2004).

Genetic stochastic processes result in changes in the frequencies of genes within a population. Small populations are inherently more prone than larger ones to detrimental genetic processes (e.g., genetic drift and founder effect; Ralls et al. 1986; Lacy 1993). In general, larger populations have more genetic diversity than small ones (Billington 1991; Frankham 1996; Madsen et al. 1999). Genetic diversity protects species against unpredictable environmental change and contributes to the reproductive effectiveness of individuals and the viability of their offspring. The loss of individuals and populations invariably results in a loss of genetic variation. Elevated homozygosity, in turn, can lead to reduced survival or fecundity, a process that is termed inbreeding depression (Hedrick and Kalinowski 2000; Keller and Waller 2002). This can, in turn, lead to increased extinction risk (Olsson et al. 1996; Keller 1998; Kruuk et al. 2002).

Interactions of Threatening Processes

Typically, species declining as a result of landscape modification respond to a range of deterministic threats (Gilpin and Soulé 1986), as well as their complex interactions with stochastic processes (Falk 1990; Caughley and Gunn 1996; see Figure 8.1). In addition, different processes can be important at different times for different species and at different stages in a species' decline (Andrén 1994; Zuidema et al. 1996; see Box 8.1). Often, exogenous threatening processes lead to the initial decline of species. The resulting smaller populations, in turn, are more susceptible to endogenous threats that reinforce the decline of the species (Clark et al. 1990).

A classic example is the heath hen (*Tympanuchus cupido cupido*) in North America. Human-induced habitat degradation coupled with hunting dramatically reduced the population. Numbers further declined as a result of fires, windstorms, and unfavorable seasonal conditions. Later, a combination of disease, inbreeding depression, reduced sexual vigor, and an imbalanced sex ratio led to the extinction of the heath hen (Simberloff 1986, 1988). This example illustrates the complex interactions of exogenous and endogenous, as well as deterministic and stochastic, threatening processes (see Figure 8.1).

Although there are no generic rules for predetermining which of the threatening processes associated with landscape modification already listed is most important for a particular species, habitat loss is widely regarded as the most important and influential threat on the persistence of individual species (Fahrig 1999, 2003). This is intuitive because access to suitable habitat is the most fundamental of all preconditions necessary for a species' survival (see Chapter 4).

Box 8.1. Interactions between Multiple Threatening Processes—Habitat Loss and Habitat Subdivision

Habitat loss, habitat subdivision, and the resulting isolation of remnant habitat patches typically occur simultaneously as a result of landscape change. The negative effects of habitat isolation are sometimes particularly pronounced when the total amount of remaining habitat for a given species is low. An interaction between habitat loss and habitat isolation has been predicted from several simulation studies (e.g., With and Crist 1995; Fahrig 1997). Andrén (1994) reviewed empirical work on population declines of birds and mammals in landscapes with different amounts of remaining habitat cover. Consistent with theoretical predictions, Andrén (1994) suggested that populations initially tended to decline proportionally with a decline in habitat cover. However, population declines appeared to accelerate when habitat cover was less than ~30% in a given landscape, presumably because habitat isolation at this stage compounded the negative effects of habitat loss (see also Fahrig 2003). Homan et al. (2004) demonstrated such compounding of threatening processes in a case study on the pool-breeding spotted salamander (*Ambystoma maculatum*; Figure 8.2) and wood frog (*Rana sylvatica*) in Massachusetts (USA). Suitable habitat for the spotted salamander was upland forest. The spotted salamander was less likely to occupy ponds where the amount of upland forest within 500 m was less than ~40%. Similarly, upland and wetland forests were suitable habitat for the wood frog, and this species was less likely to occupy ponds where the amount of suitable forest habitat within 1000 m was less than ~45% (Homan et al. 2004). These findings provide empirical support for theoretical predictions that multiple threatening processes (e.g., habitat loss and habitat isolation) may interact, thereby posing a particularly severe threat to some species.

Which Species Are Extinction Prone?

Landscape modification can pose a range of threats to some species (see earlier discussion). However, not all species will respond in the same way to landscape modification.

Figure 8.2. Spotted salamander (photo by J.D. Willson).

Some may become extinct, some may decline, and others may even thrive as a result of new habitat being created for them (see Box 4.2 in Chapter 4 on the galah). The effects of landscape change on species' populations can take a long time to manifest (the "extinction debt"; Tilman et al. 1994; see Chapter 9 for more details). For this reason, it is not always easy to ascertain which species are most at risk from landscape change. This can be problematic because knowledge about species' susceptibilities to decline is important for targeted conservation planning. Indeed, certain kinds of species such as primates, parrots, and pheasants appear to be disproportionately threatened relative to other groups (Groombridge and Jenkins 2002; Gaston and Spicer 2004). An important problem in this context, then, is to determine if there are generic traits of species that make them more resilient to landscape change or, conversely, more prone to extinction.

This section examines a range of topics associated with attempts to determine the most extinction-prone taxa in landscapes subject to human modification. First, we discuss some of the life history and biological attributes of species susceptible to decline or at risk of extinction in modified landscapes and relate these biological attributes to the threatening processes already described. We then consider the extinction proneness of species in response to landscape modification, particularly with respect to other threatening processes that are not directly related to landscape change.

Linking Threatening Processes and Extinction Proneness

Many factors have been related to a species' extinction proneness arising from human landscape modification (Table 8.1). Some biological attributes of species may have direct implications for their susceptibility to a given threatening process; for example, a given species' low mobility directly increases its susceptibility to habitat isolation and habitat subdivision (other things being equal). Other biological attributes of species are more indirectly linked with their extinction proneness. For example, a species' body size is broadly correlated with its position in the food chain, its home range size, its likelihood of being hunted by humans, and its population density (Reynolds 2003). A range of examples of species' attributes that are directly or indirectly related to their extinction proneness are summarized in Table 8.1. A more general list of biological attributes that may relate to extinction proneness, and their relationship with threatening processes, is proposed in Table 8.2. The explicit link between threatening processes, biological attributes, and extinction proneness suggests that, in general: a given species' ability to withstand human landscape modification is related to the extent to which landscape change causes habitat loss and isolation, and the disruption of biological and interspecific processes for this species; small population size (natural or human induced) will further exacerbate a species' risk of extinction due to stochastic events (Figure 8.3; also see Table 8.2).

Box 8.2. Why Accurate Predictions of Extinction Proneness Are Difficult

Although landscape modification can be a major contributor to the extinction of many species, there can be significant problems in making accurate predictions of the species at greatest risk of extinction. These difficulties arise because studies of taxa that have been lost show that almost all extinctions result from multiple processes, which often act in unexpected ways (Simberloff 1988; Caughley and Gunn 1996). A useful illustrative case is the loss of the red-fronted parakeet *(Cyanoramphus novaezelandiae)* from the sub-Antarctic Macquarie Island as a result of the interacting impacts of introduced cats *(Felis catus)* and rabbits (*Oryctolagus cuniculus*; Taylor 1979). The red-fronted parakeet coexisted with cats for 80 years on the island until rabbits were introduced in 1879. Rabbits increased in abundance as did numbers of cats that preyed upon them. Altered vegetation cover through rabbit grazing and increased numbers of cats subsequently led to the loss of the red-fronted parakeet. Given the interactions between various threats, it would have been very difficult to accurately forecast the parakeet's extinction proneness.

Empirical Evidence: Predictions of Extinction Proneness

Many studies have attempted to predict the extinction proneness of a species or a suite of species (see Table 8.1). Not surprisingly, different threatening processes are important for different species, and these different processes interact in complex ways (see earlier discussion). Hence it is unlikely that a single generic life-history trait will determine the extinction proneness for a wide range of species in a wide range of landscapes (Box 8.2). As a result, relatively few specific predictions about extinction proneness have been reliable to date.

Habitat Specificity and Matrix Persistence

There is widespread empirical support that species with broad habitat requirements are more likely to persist in heavily modified landscapes (e.g., Mac Nally and Bennett 1997). For example, Koh et al. (2004) examined ecological correlates of extinction proneness for butterflies in Singapore. Among other biological attributes, habitat specialization was a good predictor of extinction proneness for this group of species.

Consistent with this, many studies have demonstrated that those species least susceptible to

Table 8.1. Overview of biological attributes that have been related to species' extinction proneness in various case studies*

Biological attribute	*Example*
Habitat specialization	• Florida scrub jay *(Aphelocoma coerulescens)*, southeastern United States (Breininger et al. 2006) • Butterflies, southeast Asia (Koh et al. 2004) • Northern spotted owl *(Strix occidentalis caurina)*, northwestern North America (Lamberson et al. 1994) • Reptiles, western Australia (Sarre et al. 1995)
Niche specialization	• Marine limpet specializing on seagrass, east coast North America (Carlton et al. 1991) • Frugivorous primates (Johns and Skorupa 1987)
Complex behavioral patterns (e.g., seasonal aggregations)	• General review (McKinney 1997)
Mating system	• Mammals, Ghana, Africa (Brashares 2003)
Home range size	• Mammals, Ghana, Africa (Brashares 2003) • Large carnivores (Woodroffe and Ginsberg (1998)
Dispersal ability	• Arboreal marsupials, Australia (Lindenmayer and Possingham 1996) • Fish, North America (Angermeier 1995)
Mobility	• Mammals, United Kingdom (Bright 1993) • Beetles, Australia (Driscoll and Weir 2005) • Butterflies, Finland (Kotiaho et al. 2005) • General review—flightlessness of birds on islands (Whittaker 1998)
Extent of geographic distribution	• Butterflies, southeast Asia (Koh et al. 2004) • Mollusks, United States (Jablonski and Valentine 1990)
Island biota (see Figure 8.3)	• General review (Simberloff 1995, 2000; Frankham 1998; Whittaker 1998)
Population density, population size, rarity	• General review (Gaston 1994; O'Grady et al. 2004) • Butterflies, southeast Asia (Koh et al. 2004) • Spiders, Bahamian islands (Schoener and Spiller 1992) • Invertebrates, Europe (Thomas 1990; Tscharntke 1992)
Edge sensitivity	• Herpetofauna, Madagascar (Lehtinen et al. 2003)
Dependence on particular mutualist or prey species	• Mites dependent on hummingbirds (Colwell 1973) • Giant eagle *(Harpagornis moorei)* preying on flightless moas in New Zealand (Holdaway and Jacomb 2000)
Inability to use matrix	• Rainforest mammals, North Queensland, Australia (Laurance 1991) • Forest mammals, Brazil (Viveiros de Castro and Fernandez 2004) • Range of rainforest vertebrate groups, Brazil (Gascon et al. 1999) • Invertebrates, southeastern Australia (Davies et al. 2000)
Genetic variability	• Macropods on Western Australian islands (Eldridge et al. 1999) • Butterflies on Finnish islands (Saccheri et al. 1998)
Disturbance tolerance	• General review (Diamond 1989; Balmford 1996)
Body weight/body size	• Mammals, arid and semiarid Australia (Cardillo and Bromham 2001; Brook and Bowman 2004) • Large-bodied, long-lived primates (Johns and Skorupa 1987)
Carnivorous diet	• General review (Terborgh 1974) • Mammals, Ghana, Africa (Brashares 2003)
Persecution by humans	• Gray wolf *(Canis lupus)*, North America (Beier 1993)

*Attributes that are directly linked with key threatening processes are listed near the top of the table; more indirectly linked attributes are listed toward the bottom of the table.

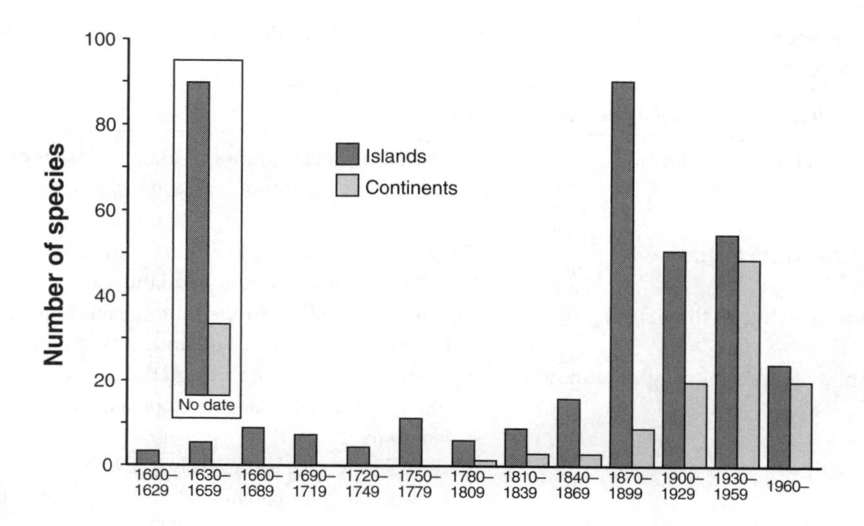

Figure 8.3. Comparisons of extinctions of island (dark gray) and mainland (light gray) species of mollusks, birds and mammals since AD 1600 (redrawn from Groombridge, B., ed. 1992: Fig. 16.5. Time series of extinctions of species of molluscs, birds and mammals from islands and continents since about AD 1600. In: *Global Biodiversity: Status of the Earth's Living Resources* [a report compiled by the World Conservation Monitoring Centre], Chapman and Hall, London with kind permission of Springer Science and Business Media). The skew for islands is pronounced for some groups such as birds—approximately 85% of the world's bird extinctions are from islands (Groombridge and Jenkins 2002).

landscape change are typically those for which the areas surrounding original vegetation remnants represent suitable or partially suitable habitat (Szaro and Jakle 1985; Renjifo 2001). For such species, a heavily modified landscape is not necessarily unsuitable (Davies and Margules 1998; Ås 1999) because it may be used for foraging (Yahner 1983; McAlpine et al. 1999), extends home ranges, or supports breeding populations (McCarthy et al. 2000). Reduced sensitivity to landscape alteration among matrix-using species has been demonstrated for a range of taxonomic groups, including:

- Birds (Blake 1983; Howe 1984; Diamond et al. 1987; Stouffer and Bierregaard 1995, Bierregaard and Stouffer 1997; Warburton 1997; Renjifo 2001; Sekercioglu et al. 2002)
- Mammals (Laurance 1990, 1991)
- Frogs (Tocher et al. 1997)
- Insects (Ricketts 2001)

A seminal study illustrating the importance of matrix persistence for resilience to extinction was completed by Laurance (1991), who investigated extinction proneness among 16 mammal species in a heavily modified former rainforest-dominated landscape in Queensland, northern Australia. The mammal species represented three guilds: arboreal folivores, predators, and omnivorous rodents. Long-lived species with low fecundity, as well as dietary specialists, were the most extinction prone (see also Johns and Skorupa 1987). Species that were rare or absent in disturbed environments were most likely to have been lost or to have declined substantially in the rainforest remnants (e.g., brown Antechinus *[Antechinus stuartii]*, tiger quoll *[Dasyurus maculatus]*, and Atherton Antechinus *[Antechinus godmani]*). Conversely, those species that persisted in the matrix were also likely to occur in the remnants.

Several factors may explain why matrix-using species in modified landscapes are those likely to be least extinction prone. Laurance (1991) concluded that those species which persist in areas of remnant vegetation are there because they can tolerate edge effects (Chapter 11) and use modified habitats in the surrounding landscape to feed and disperse between vegetation remnants. In addition, individuals dispersing through the matrix may "rescue" declining populations in the remnants (Laurance 1997). Conversely, animals lost from the matrix are more likely to become extinct in remnant patches and hence may also become extinct at a more regional scale.

Table 8.2. Proposed relationship between key threatening processes associated with landscape modification and biological attributes that ameliorate extinction proneness

Threatening process	*Ameliorating biological attribute*	*Explanation*
Habitat loss and habitat degradation (Chapters 4, 5)	Low habitat specialization	Specialized species are more likely to lose their habitats as a result of landscape change
	Disturbance tolerance	Disturbance-tolerant species are more likely to find suitable habitat in modified landscapes
	Ability to live in the matrix	Species that can live in the matrix experience no habitat loss as a result of landscape modification
Habitat isolation and subdivision (Chapter 6)	Ability to move through the matrix	Species that can move through the matrix are less likely to suffer the negative consequences of habitat isolation
	Mobility	Mobile species are more likely to be able to move between habitat patches for resources, or between subpopulations
	Dispersal ability	Strong dispersers are more likely to maintain viable metapopulations
Disrupted species interactions (Chapter 7)	Limited dependence on particular prey or mutualist species	Species that can switch prey or mutualists are more likely to withstand landscape change
	Competitive ability	Species that are strong competitors are less likely to be outcompeted by species whose habitat expands as a result of landscape change
Disrupted biology (Chapter 7)	Low biological and behavioral complexity	Species with a complex biology (e.g., social or breeding systems) are more likely to have their biological processes disrupted as a result of landscape change than species with simpler biological systems
Stochastic events (Chapter 8)	Population density	High-density populations contain many individuals even in small areas, and hence are more resilient to stochastic events

Mobility

There is empirical support demonstrating that habitat isolation can lead to the decline of some species (Chapter 6). Consistent with this, several authors have argued that the extinction proneness of species should be related to their dispersal abilities (Bright 1993; Mac Nally and Bennett 1997). For example, Kotiaho et al. (2005) demonstrated that, among other factors, poor dispersal ability was a key biological attribute contributing to the extinction proneness of butterflies in Finland (Box 8.3). Similarly, Driscoll and Weir (2005) studied beetles in an agricultural landscape in Central New South Wales, Australia. The beetle species most vulnerable to decline in this landscape were species living underground and flightless species. Although it is likely that a range of different factors contributed to the extinction proneness of different species (Driscoll and Weir 2005), taxa that were unable to fly between remnants of unmodified habitat were particularly threatened by habitat isolation.

Population Density and Rarity

Because they are less prone to stochastic threats, common and numerically abundant species have a greater chance of persisting in modified landscapes than rare ones (Lacy 1987; Fagan et al. 2002). Thus the overall regional abundance of species can play a significant role in influencing their occurrence in vegetation remnants (Lynch and Whigham 1984; Freemark and Collins 1992). Indeed, a number of studies have shown that the persistence of species in modified landscapes is strongly related to their abundance at larger spatial scales (Table 8.3).

Landscape Modification in Context

A range of factors can influence whether species go extinct (see Figure 8.1 and Table 8.2). Hence, although the three factors just discussed in detail (i.e., habitat specificity, mobility, rarity) are often important, they cannot fully explain the extinction proneness of all species in all landscapes. An instructive example was the work by Mac Nally and Bennett (1997). These authors were interested in the extinction proneness of 43 species of woodland birds in Eucalyptus woodlands in northern Victoria, Australia. They developed a predictive model of extinction proneness based on each species' habitat specificity, mobility, and population density (Figure 8.4). Later, Mac Nally et al. (2000) tested this model against field data and found it had virtually no predictive power. They suggested that the effects of habitat specificity, mobility, and population density were clouded by natural variability in habitat suitability and interspecific interactions in the study area, including competition with the aggressive noisy miner (*Manorina melanocephala;* see Chapter 7). This example underlines a key theme of this chapter. That is, threatening processes typically interact, and predictions of extinction proneness may fail unless they simultaneously consider all of the most relevant threatening processes (see Figure 8.1 and Table 8.2).

Relationships between extinction proneness and biological attributes that are more indirectly linked with threatening processes may similarly depend on the specific situation at hand. In a global review of empirical studies on extinction proneness, Reynolds (2003) showed that body size was a reasonable predictor of extinction proneness in many cases. A likely reason for this trend was that large body size tended to be correlated with a number of more directly relevant ecological attributes (e.g., area requirements, position in food chain). In addition, Reynolds (2003) suggested that large-bodied animals were more likely to be hunted by humans in many parts of the world (e.g., as a food source; Box 8.4). Notably, hunting by humans is not directly related to landscape modification—rather, it is a flow-on effect (albeit a common one in some regions; Redford 1992). This supports another important consideration we alluded to in Chapter 1. That is, the direct effects of human landscape modification are not the sole reason why species occur where they do, or why they may be threatened or prone to extinction.

Different relationships between extinction proneness and body sizes have been found

Box 8.3. Extinction Proneness of Finnish Butterflies

Kotiaho et al. (2005) were interested in predicting the risk of extinction for 95 butterfly species in Finland based on their ecological characteristics. Twenty-three of these species were classified as threatened, and 72 were classified as nonthreatened. For each species, Kotiaho et al. (2005) quantified six ecological characteristics: larval specificity, resource distribution, dispersal ability, adult habitat specialization, length of flight period, and body size. The work demonstrated that species recognized as threatened tended to have high larval specificity, restricted resource distribution, poor dispersal ability, and a relatively short flight period. Importantly, these correlations allowed Kotiaho et al. (2005) to identify several species that were not currently listed as threatened but that shared many ecological characteristics with threatened species. Given their similar ecological characteristics, Kotiaho et al. (2005) considered that these species were likely to be extinction prone and that their conservation status therefore should be reconsidered. This case study illustrates that the identification of species potentially at risk on the basis of their biological attributes can be useful to target conservation efforts more directly at these species.

Table 8.3. Examples where the persistence of species in modified landscapes was related to their abundance at larger spatial scales

Example	*References*
Birds, Oregon Coast Range, United States	McGarigal and McComb 1995
Birds, Connecticut, United States	Askins et al. 1987; Askins and Philbrick 1987
Birds, Maryland, New York, Pennsylvania	Boulinier et al. 2001
Birds, southwestern Western Australia	Arnold and Weeldenburg 1998
Birds, Alberta, Canada	Schmiegelow et al. 1997

elsewhere. In Australia, Burbidge and McKenzie (1989) undertook a pioneering study on extinction proneness of mammals. Unlike Reynolds (2003), they found that almost all the species that had declined or had been lost were nonflying taxa weighing between 35 and 5500 grams—a category Burbidge and McKenzie (1989) named the critical weight range (e.g., Figure 8.5). Factors that correlated with the demise of species in this weight range included their diet, habitat requirements, and regional patterns in rainfall. Despite the intuitive appeal of this finding, a more recent study has suggested the critical weight range is actually an artifact because most Australian mammals fall within this range (Cardillo and Bromham 2001). Instead, Cardillo and Bromham (2001) argued that the smallest Australian mammals were those most resistant to extinction (Figure 8.6).

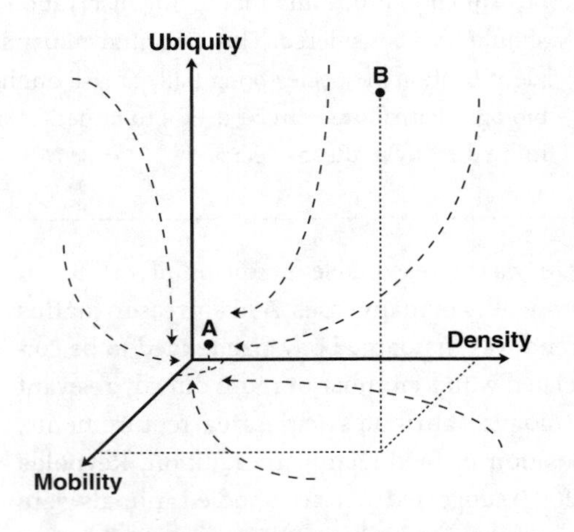

Figure 8.4. Mac Nally and Bennett's (1997) extinction proneness model based on population density, mobility, and habitat specificity. The arrows show paths of increasing habitat specificity in which species B would not be at risk but species A would be highly extinction prone (redrawn from Mac Nally, R. and Bennett, A.F. 1997: *Biological Conservation*, **82**, 147–155, Species-specific predictions of the impact of habitat fragmentation: local extinction of birds in the box–ironbark forests of central Victoria, Australia, with permission from Elsevier).

Finally, in some cases, the causes of extinction, and hence which species are likely to go extinct first, are simply not known. Amphibians represent the most prominent example of this situation. Stuart et al. (2004) reviewed the status and trends of amphibian declines and extinctions throughout the world. Approximately every third amphibian species worldwide is currently listed as threatened, and amphibians are being lost at a rate greater than both mammals and birds. Similarly to mammals and birds, landscape modification is linked to the decline of many species. However, almost every second threatened amphibian species is declining for reasons that are simply not known. A wide range of factors has been implicated in the decline of amphibians. A fungal disease, chytridiomycosis, has been identified as one of the factors contributing to the decline of a number of frog populations in Australia and Central America (Daszak et al. 2000). Introduced fish and other exotic species (including other amphibians; e.g., the cane toad *[Bufo marinus]*) have also been implicated in the decline of some amphibian species (Gillespie 1995; Williamson 1999; Lever 2001; Hamer et al. 2002). Thus, as with the declines of almost all species and species assemblages, there are multiple contributing factors, and the magnitude of their relative effects varies depending on the species, the location, and other taxa in the assemblages. The lack of precise knowledge on threatening processes and their interactions, in turn, makes accurate predictions of which species are most extinction prone very difficult and hinders the development of targeted conservation strategies (Stuart et al. 2004).

Box 8.4. Hunting Pressure in African Forests

Human hunting pressure on mammals in African forests highlights its importance for affecting animal population sizes. Mammal populations in tropical forests have low annual production rates, and Fa et al. (2002) estimated that, while harvesting rates for mammals in the Congo Basin approximate 5 million tonnes per year, production rates were less than half that. Given demand for bush meat from urban populations, hunting pressure on African forest mammals is clearly unsustainable, and its effects outweigh a range of other threatening processes.

Summary

Threatening processes discussed in previous chapters included habitat loss, habitat degradation, habitat isolation, changes to a species' biology, and changed interactions between species. These threatening processes typically interact and lead to the decline of populations. Small populations, in turn, are more at risk from stochastic (i.e., chance) events, including environmental, demographic, and genetic stochasticity. Because multiple threatening processes typically interact, and different threatening processes dominate at different stages of a species' decline, predictions of which species are at most risk from landscape change are difficult. In general, a given species' ability to withstand human landscape modification is related to the extent to which landscape change

Figure 8.5. The numbat *(Myrmecobius fasciatus)*, an Australian mammal falling within the "critical weight range" (sensu Burbidge and McKenzie 1989; photo by Tony Friend).

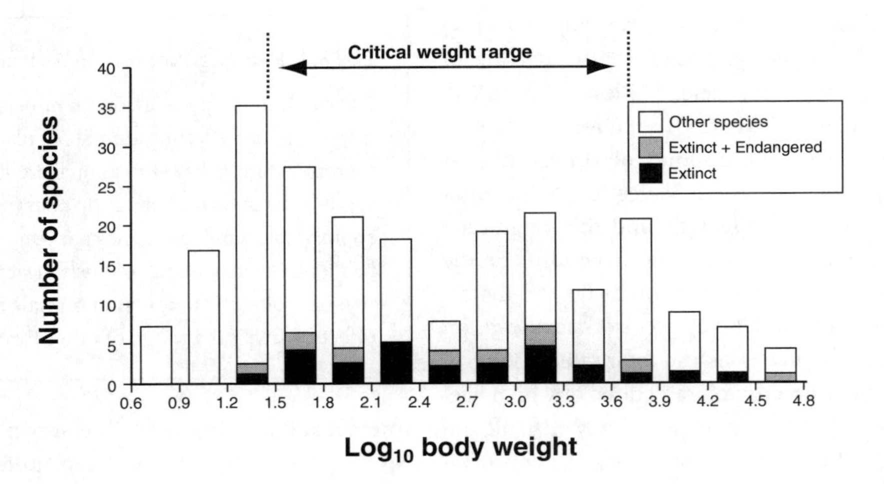

Figure 8.6. Frequency distribution of extinct, endangered, and other species of Australian terrestrial mammals (redrawn from Cardillo, M. and Bromham, L. 2001: *Conservation Biology*, **15**, 1435–1440, Body size and risk of extinction in Australian mammals, with permission from Blackwell Publishing).

causes habitat loss and isolation and the disruption of biological and interspecific processes for this species; small population size (natural or human induced) will further exacerbate a species' risk of extinction due to stochastic events. Key biological attributes linked to a species' extinction proneness are directly related to key threatening processes. For example, high habitat specificity increases the risks from habitat loss; low mobility increases the threat from habitat isolation; and highly complex biological processes and species interactions increase the likelihood of these processes being disrupted by landscape change.

Links to Other Chapters

This chapter is a synthesis of Part II of this book, which is devoted to single species and the processes threatening their persistence in modified landscapes. Key threatening processes are habitat loss (Chapter 4), habitat degradation (Chapter 5), habitat subdivision and habitat isolation (Chapter 6), and changes to a species' biology and interactions with other species (Chapter 7). The next part of the book, Part III, discusses landscape patterns that are broadly related to the threatening processes discussed in Part II.

Further Reading

Zuidema et al. (1996) and Hobbs and Yates (2003) give valuable overviews of the various kinds of factors that can influence populations in human-modified landscapes; including interactions between threatening processes. McKinney (1997) provides a wide-ranging and thought-provoking review of the factors that contribute to extinction proneness. Primack (2001) is a very readable account of extinction processes. Pimm (1995), May et al. (1995), and Primack (2001) review knowledge about extinction rates. Laurance (1991) contains an excellent example of the use of matrix environments and species persistence in human-modified landscapes. Particularly interesting case studies on extinction proneness include Mac Nally and Bennett (1997), Mac Nally et al. (2000), Lehtinen et al. (2003), Koh et al. (2004), and Kotiaho et al. (2005). Adams (2004) provides a useful compendium of information on extinction.

PART III

The Human Perspective: Landscape Patterns and Species Assemblages

In Part II, we summarized the processes associated with landscape change that may pose a particular threat to individual species. Studying these processes in detail can be useful to understand and mitigate the effects of landscape change on single species. The investigation of individual species and the processes threatening their persistence corresponds to a suite of "process-oriented" approaches to developing a better understanding of how landscape change affects species. This suite of approaches aims to develop a detailed understanding of as many species and ecological processes as possible. One of the strengths of process-oriented approaches is that they deal directly with the ecological processes that affect the distribution patterns of species. A limitation of process-oriented approaches is that it can be very difficult to study enough species in enough detail to make reasonable management decisions. In many cases, it is inappropriate to assume that detailed management prescriptions suitable for one species will also suit other species.

In Part III of the book, we discuss alternative approaches to develop a better understanding of how landscape change affects biota. The approaches outlined in this part may be broadly classified as pattern-oriented approaches be-

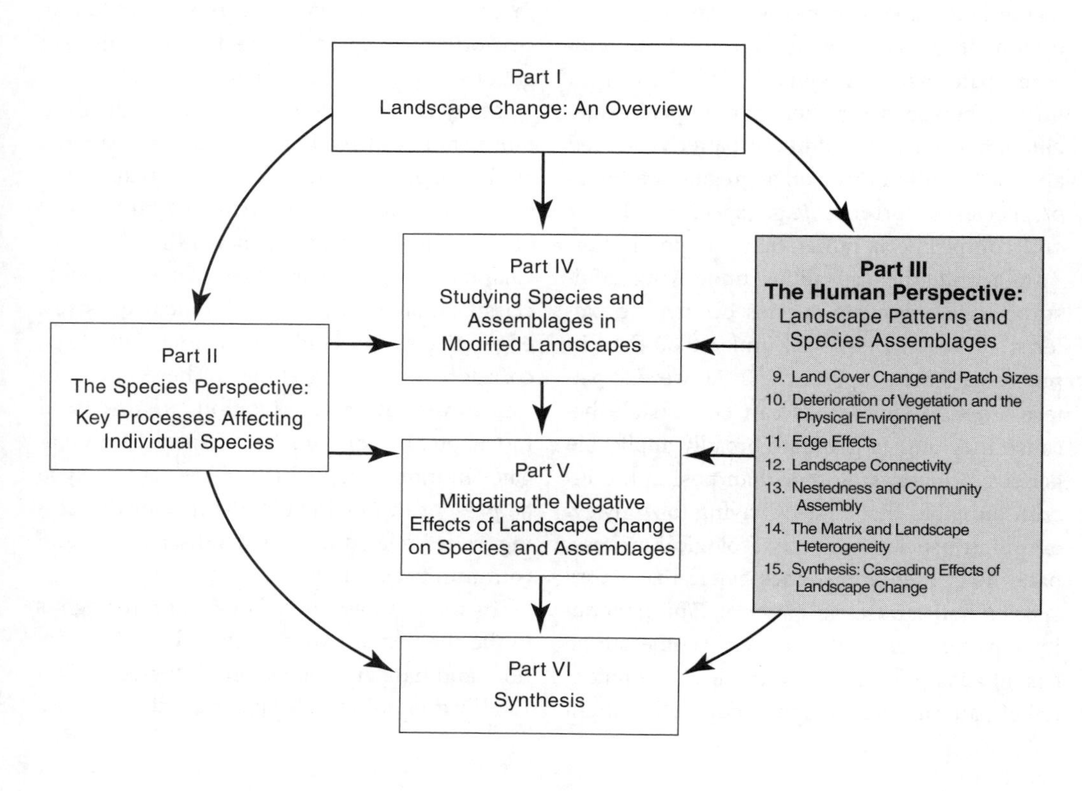

Figure III.1. Topic interaction diagram showing links between the different parts of the book, and the topics covered in the chapters of Part III.

Mixed agricultural landscape in Costa Rica (photo by David Lindenmayer).

cause they relate easily identifiable patterns in the landscape to patterns of species distribution. In pattern-oriented approaches, landscape patterns are frequently defined from a human perspective rather than a species-specific perspective. In addition, pattern-oriented approaches often focus on aggregate measures of species occurrence (e.g., species richness and composition) rather than single species. As a result, the causalities underlying landscape-scale aggregate species occurrence patterns are usually less well understood than in process-oriented approaches. However, pattern-oriented approaches can be valuable because they aim to generate broadly applicable general principles. Rather than posing the insurmountable challenge of having to study every single species and every ecological process, pattern-oriented approaches generalize across species and landscape patterns. The generality of pattern-oriented approaches comes at the cost of a loss of detail. Nevertheless, the potential of pattern-oriented approaches to highlight general trends relevant to management means that they are an important complement to process-oriented approaches and detailed investigations of single species (Figure III.1).

Many concepts describing species distribution patterns in relation to landscape patterns are based on a classification of landscapes into patches and thus are related to the island model or patch–matrix–corridor model introduced in Chapter 3. Important topics associated with this classification of landscapes include patch sizes, edge effects, and landscape connectivity (e.g., corridors or stepping stones). These topics, as well as vegetation deterioration, are discussed in Chapters 9 through 12. Chapter 13 investigates in more detail which species are likely to be present in modified landscapes by discussing issues related to nested subset theory and community assembly.

In many ways, these topics are analogous to the themes of habitat loss, habitat degradation, and habitat subdivision discussed in Part II. However, they tackle landscape change from the "top" (i.e., they focus on landscape patterns), rather than beginning at the "bottom" (i.e., with a focus on the detailed requirements of individual species). Recognizing the complementarity of process-oriented and pattern-oriented approaches acknowledges that landscape change is associated with both a multitude of ecological processes and changes to landscape pattern. Although there is some overlap between the topics discussed in Part II and this part of the book, we do not repeat earlier information in this part. Rather, we reiterate the importance of carefully distinguishing between approaches that recognize different species' unique habitat requirements (Part II) and approaches that take a broader perspective and generalize across species (Part III). Confusion of these two approaches has been the norm in the fragmentation literature to date. However, we believe that the distinction between Parts II and III is key to tackling Bunnell's (1999a) fragmentation panchreston; that is, to break the catch-all term "fragmentation" into its subcomponents and develop consistent and coherent terminology (see Chapter 1).

We explicitly recognize that a view of landscapes as mosaics of patches and corridors within a relatively inhospitable matrix is not always appropriate to understand species distribution patterns or promote conservation management. In Chapter 14, we discuss alternative and complementary approaches to studying or managing modified landscapes. In particular, we highlight the importance of managing entire landscapes rather than only predefined patches of largely undisturbed vegetation. Finally, Chapter 15 synthesizes information presented in Part III and discusses cascading and interacting changes to landscape pattern, vegetation structure, and species composition, and their effects on ecosystems.

In combination, Part II and Part III represent different but complementary ways of understanding the effects of landscape change on biota (Figure III.1). This understanding is the basis for later parts of the book, which are devoted to studying the effects of landscape change on species assemblages (Part IV), and mitigating the negative effects of landscape change on species and assemblages (Part V).

CHAPTER 9
Land Cover Change and Patch Sizes

In Chapter 4, we discussed the negative effects of habitat loss on individual species. At the landscape scale, the clearing of native vegetation often leads to the loss of habitat for many individual species (see introduction to Part III). Thus a common outcome of the loss of native vegetation in modified landscapes is reduced species richness (Box 9.1). At the local scale, the shrinkage (sensu Forman 1995; see Chapter 2) of patches of native vegetation can also have negative effects on groups of species and even entire assemblages of native species. In this chapter, we briefly summarize the negative impacts of vegetation loss on species richness. Our discussion covers well-known relationships between the size of patches of remnant vegetation and species richness, and we highlight key mechanisms that tend to reduce species richness in smaller patches. Although species–area relationships clearly indicate that large patches of native vegetation are important in modified landscapes, we also discuss several caveats associated with a focus on species–area relationships in a land management context. We conclude this chapter by arguing that the maintenance of large patches of native vegetation should be concomitant with other, complementary conservation approaches such as matrix management and the protection of medium-sized and small vegetation patches (Turner 1996).

Landscape Modification, Loss of Native Vegetation Cover, and Species Loss

Landscape alteration (e.g., land clearing, urbanization, agricultural intensification) typically results in the modification of native vegetation cover (Chapter 2). The loss of native vegetation can, in turn, lead to the loss of habitat for many species (Fahrig 2003; Box 9.2), although the amount of suitable habitat for some individual species may increase (such as generalists that can use the heavily modified areas; see Chapter 3). Despite such exceptions, a myriad of studies have shown that overall levels of species richness tend to be reduced in landscapes subject to extensive human modification (reviewed by McGarigal and Cushman 2002; Table 9.1).

Box 9.1. Vegetation Loss and Extinction—the Case of Easter Island

Extinction rates on many islands around the world have been substantial, particularly where losses of native vegetation cover have been pervasive (Whittaker 1998; Groombridge and Jenkins 2002; Brook et al. 2003). Easter Island is one of many illustrative examples (Figure 9.1). The island, the most isolated inhabited land on Earth, was covered in native forest prior to the arrival of Polynesian people about AD 400. Virtually the entire island was cleared of native vegetation, and areas of exotic eucalypt stands are now the only forest cover. Landscape modification has been associated with the extinction of an endemic land snail fauna, the loss of all native species of land birds (including two species of parrots, two species of rails, and a heron and owl species), and the extinction of 15 of 22 seabirds, of which the remaining 7 are restricted to a handful of tiny offshore islands (Steadman 1997a, b). Similar situations occur around the world, ranging from Singapore (Brook et al. 2003) to Mauritius (Groombridge and Jenkins 2002). Indeed, from the 800-plus islands in Oceania, Steadman (1995) has estimated that 2000 species of rails alone may have been lost. Thus some authors believe that the world's bird diversity might be 20% richer than it is now if humans had not colonized the islands of the Pacific (Whittaker 1998).

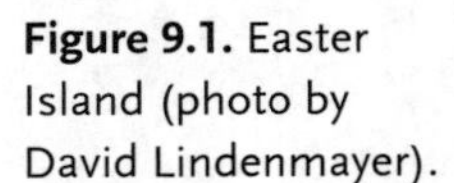

Figure 9.1. Easter Island (photo by David Lindenmayer).

Vegetation Loss, Remaining Patch Size, and Species Richness

The species–area relationship (Figure 9.2) is one of the primary factors used to account for reduced species richness in landscapes where native vegetation cover is reduced. The well-known relationships in Figure 9.2 can be found at many spatial scales (continents, regions, landscapes, islands). They are intimately related to the widespread observation in many modified landscapes that larger patches of remnant vegetation typically support more species than smaller ones (e.g., Kitchener et al. 1980a, b, 1982; Kitchener and How 1982; Smith et al. 1996; Edenius and Sjöberg 1997; Kjoss and Litvaitis 2001). A detailed review of this topic, presented by Hanski (1994a), demonstrates the strong relationships that have been quantified between patch size and patch occupancy by many species.

The following sections in this chapter outline the mechanisms that underpin species–area relationships, quantification of species–area relationships, and some of the difficult problems in using such relationships to predict species loss.

Mechanisms Underlying Species–Area Relationships

Strong positive relationships between the size of an area and species richness have long been known (e.g., Arrhenius 1921; Preston 1962; reviewed by Rosenzweig 1995). Species–area relationships can occur for many reasons (Scheiner 2003). Key mechanisms include local immigration and extinction rates, disturbance, habitat diversity, and random placement (Connor and McCoy 1979; McGuinness 1984; Newton 1995). These mechanisms can be closely related. For example, larger patches may have lower extinction rates for populations of individual species (Chapter 4), but they may also be less disturbed (Ambuel and Temple 1983) (e.g., from edge effects; see Chapter 11) and capture more environmental variability, thus providing habitat for a larger number of species. Because of the inherent relatedness of mechanisms potentially underlying the species–area relationship, many studies have simply described an increase in species richness with patch size, rather than analyzing the relative importance of potential causal mechanisms underlying the relationship.

Immigration and Extinction Rates

The theory of island biogeography (MacArthur and Wilson 1963, 1967) was developed to explain species–area phenomena for assemblages on islands. The theory considers aggregate species richness on islands of varying size and isolation in relation to a mainland source of colonists (Shafer 1990). According to island biogeography theory, the number of species on an island arises from the dynamic equilibrium between the local extinction and local immigration of species (Whittaker 1998). Larger islands have larger population sizes (Haila 1983; Haila et al. 1983) and that means fewer species should suffer extinction on larger islands (Rosenzweig 1995). The same reasoning as for oceanic islands has frequently been applied to patches of native vegetation in modified landscapes (e.g., Kitchener et al. 1982).

Disturbance

Differences in susceptibility to disturbance between small and large patches may contribute to species–area relationships. That is, larger areas are likely to contain at least some area that remains unaffected by disturbance (Pickett and Thompson 1978; Baker 1992) and which will be important refugia for some species (Mackey et al. 2002). Larger areas are likely to support "interior zones" away from edges that can have negative impacts on some species (Harris 1984; see Chapter 11). In addition, disturbance regimes in larger areas may result in the creation of a greater range of niches and types of habitat that, in turn, enable more species to persist or coexist (Shiel and Burslem 2003).

Habitat Diversity and Random Placement

Two other mechanisms that may give rise to observed species–area relationships are habitat diversity and random placement. Habitat diversity is considered to be important because larger patches cover more space and are therefore more likely to support more types of

Box 9.2. Large-Scale Human Landscape Modification and Threats to Species Persistence

The extent of anthropogenic landscape modification is not uniform across the world. Some types of ecosystems are under severe human pressure, and only limited areas of relatively unmodified vegetation remain. Hoekstra et al. (2005) analyzed the extent to which different biogeographic regions were either modified by human activities or reserved in protected areas. The analysis highlighted that it was important to complement conservation activities in species-rich "hotspots" (such as Brazil's Atlantic Forest; see Myers et al. 2000; Brooks et al. 2002) with increased conservation efforts in some of the world's most severely modified but less species-rich biogeographic regions. Hoekstra et al. (2005) defined "crisis ecoregions" as areas where the vast majority of natural vegetation was severely modified or received little formal protection; examples include temperate grasslands and Mediterranean biomes. Large-scale modification of native vegetation associations and ecosystems in these regions poses a major threat to the survival of many species. A study by Kerr and Deguise (2004) illustrates this relationship for Canada. Even though Canada still contains extensive areas of largely unmodified wilderness, over 50% of the ranges of threatened species overlap with highly modified agricultural or urban areas. This overlap is particularly high in the prairie ecozone in central Canada and the mixed-woods ecozone of southeastern Canada. Notably, these areas are also highly suitable for agricultural production (Kerr and Deguise 2004). These examples reinforce the key point that throughout the world, the removal of native vegetation poses a major threat to species persistence.

Table 9.1. Examples of reduced species richness in landscapes subject to human modification

Group and location	*Notes*	*References*
Primarily birds, southwest western Australia	Compilations of studies of the decline in species richness of many groups	Saunders et al. 1987, 1993a; Saunders and Ingram 1987
Birds, Atlantic rainforest, Brazil	Range of datasets and field sampling	Ribon et al. 2003
Invertebrates, United Kingdom	Compilations of studies	Thomas and Morris 1995
Mammals, Eastern North America	Data on mammal species lists, minimum area requirements for mammal species, and national park sizes	Gurd et al. 2001
Birds, Canada and Finland	Long-term bird count data from both countries	Schmiegelow and Mönkköen 2002
Many groups (e.g., butterflies, freshwater fish, birds, mammals), Singapore	Various species lists linked with various data analyses	Brook et al. 2003
Reptiles, southeastern Australia	Systematic sampling of sites within selected landscapes	Driscoll 2004

habitats that will enable a greater number of species to persist within them (e.g. Fox 1983). Harner and Harper (1976) examined relationships between plant species richness and habitat diversity in the pinyon–juniper woodlands in Utah and New Mexico, USA. Although area and habitat diversity (represented by soil types) were strongly correlated, when sample sizes were standardized by area, habitat diversity was the best explanatory variable for species richness (Figure 9.3).

The concept of random placement (sometimes called random sampling) describes the fact that, other things being equal, larger areas are more likely to capture a greater number of patchily distributed species than small areas (Connor and McCoy 1979; Haila et al. 1993).

Equations for Species–Area Relationships and Forecasting Species Loss

Several equations have been proposed to calculate species richness *(S)* for a given area (e.g., Connor and McCoy 1979; McGuinness 1984). A widely used expression is:

$$S = cA^z$$

where A is the area, and c and z are fitted constants (Preston 1962; Connor and McCoy 1979; Dial 1995). Often, the logarithm of A is used in this equation, and normally, the value of z is taken to be about 0.25 (Preston 1962), with a range of about $0.15 \leq z \leq 0.40$ (Connor and McCoy 1979; Dial 1995). The value of z can vary according to (1) whether areas are part of continents or islands, (2) the latitude of the area in question, and (3) the range in size of the areas in question (Palmer and White 1994; Rosenzweig 1995; Crawley and Harral 2001; Gaston and Spicer 2004). Assuming that most species are restricted to patches of native vegetation, this expression can give an approximation of the expected species loss as a result of vegetation clearing. The proportional change in the number of species, ΔS, is given by (Preston 1962; Koopowitz et al. 1994; Dial 1995):

$$\Delta S = 1 - \left(\frac{S_1}{S_0}\right) = 1 - \left(\frac{A_1}{A_0}\right)^z$$

where S_0 and A_0 are the number of species and the area, respectively, before landscape alteration, and S_1 and A_1 are the number of species and the area after landscape alteration. The

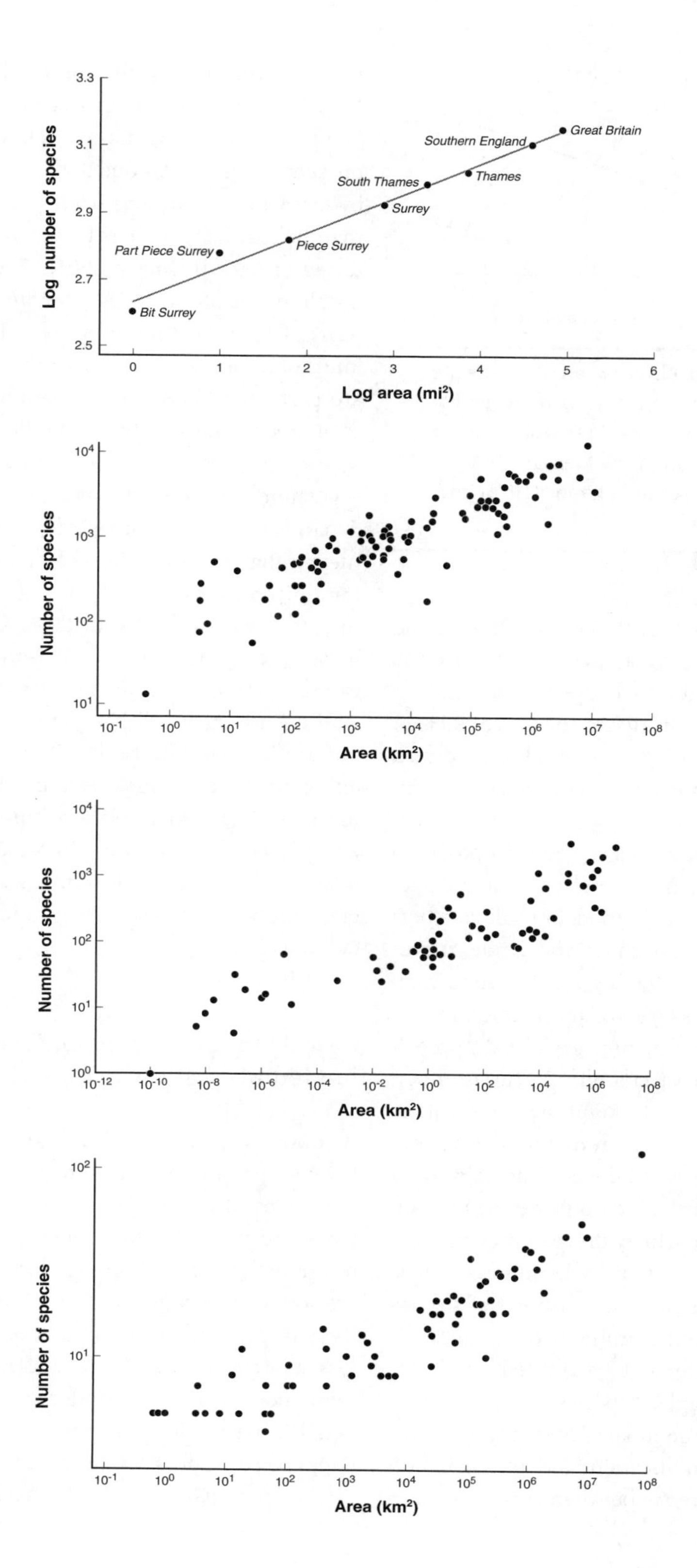

Figure 9.2. Examples of species–area curves. Top: Plants in Great Britain showing values for subareas starting with the county of Surrey (redrawn from Rosenzweig, M.L. 1995: *Species Diversity in Space and Time,* Cambridge University Press, Cambridge). Second from top: Native plant species from around the world (redrawn from Gaston, K.J. and Spicer, J.I. 2004: *Biodiversity: An Introduction,* 2nd edition, with permission from Blackwell Publishing). Second from bottom: Benthic macrofauna in the Arctic Ocean (redrawn from Gaston, K.J. and Spicer, J.I. 2004: *Biodiversity: An Introduction,* 2nd edition, with permission from Blackwell Publishing). Bottom: Land snails on the Aegean Islands (redrawn from Gaston, K.J. and Spicer, J.I. 2004: *Biodiversity: An Introduction,* 2nd edition, with permission from Blackwell Publishing).

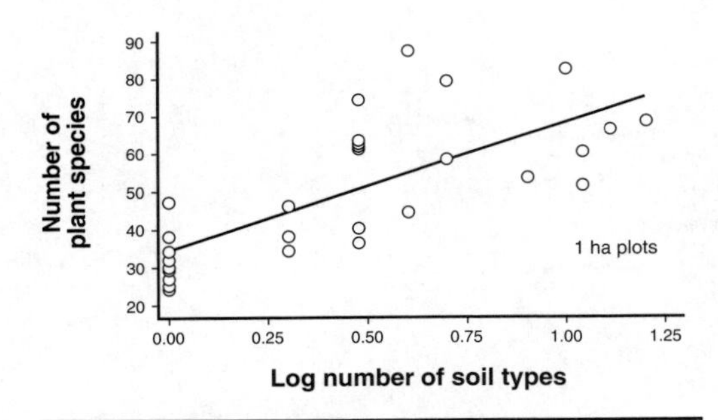

Figure 9.3. Relationship between plant species diversity and habitat diversity (as represented by the number of soil types) in Pinyon–Juniper woodlands in Utah and New Mexico, USA. Redrawn from Harner and Harper (1976) and Rosenzweig (1995).

equation predicts that the loss of 90% of the native vegetation in an area will result in a loss of between 30 and 60% of species (depending on the value of z). Similarly, the loss of 99% of native vegetation in an area would be predicted to lead to approximately 75% of species becoming extinct.

Other expressions have been proposed to predict the relationship between area and species richness. No single model fits all data best (Simberloff 1992; Burgman and Lindenmayer 1998). Connor and McCoy (1979) found that, of the 100 species–area relationships they examined, a power function explained approximately 50% of the variation in species number; habitat variation was thought likely to contribute significantly to the remaining variation. Not surprisingly, confidence intervals were typically wide, particularly outside the domain of the data from which the parameters were estimated. Hence attempts to forecast levels of species loss and thus the number of species expected that result from landscape alteration are often characterized by limited predictive ability. For example, "species richness relaxation" is the decline in species richness following landscape modification, and an "extinction debt" is the difference between current species richness and the number of species expected (Box 9.3). However, forecasts of the number of species following faunal relaxation based on species–area relationships are complicated by extinctions resulting from processes other than habitat loss, such as the impacts of introduced species (Primack 2001). There are also problems resulting from (1) nonrandom patterns of landscape modification where certain kinds of places (such as more productive areas) are particularly prone to alteration (Seabloom et al. 2002; Chapter 2); (2) the fact that some species may persist outside patches of native vegetation (Daily 2001); and (3) successional dynamics within native vegetation, which means that the suitability of areas as habitat for a given species or sets of taxa change over time (Odum 1971; Attiwill 1994). As a result of these (and other) factors, there is no theory that can accurately predict the rate at which the new equilibrium state will be approached (Simberloff 1992; Doak and Mills 1994). Importantly, although there is much evidence for the existence of species–area relationships (see Rosenzweig 1995; Gaston and Spicer 2004), there is almost no evidence for an equilibrium species number in terrestrial vegetation remnants (Whittaker 1998).

Caveats for Species–Area Relationships in Modified Landscapes

As outlined earlier, although species–area relationships are a widespread phenomenon and many studies have confirmed their existence, there can be difficulties in using them to predict the rate of species loss. Thus there are some important caveats that need to guide their use in examining reductions in species loss associated with landscape modification. First, not all parts of landscapes are created equal in terms of the number of species they support. Some places act as small-scale biological hotspots (Gentry 1989; Lindenmayer and

Franklin 2002) and their loss can have disproportionately high impacts on the loss of species—even though they may encompass a relatively small part of a landscape. For example, parts of South American tropical forest landscapes are subject to persistent fog and can support very high bird species richness (Laurance et al. 1997). Other examples include cliffs, caves, mound-springs, and other features that can be small but support many unique species (e.g. Culver et al. 2000). Similarly, many plant species are endemic to special types of soils and rock formations with restricted distributions such as serpentine soils (Whittaker 1954a, b; Harrison et al. 2000). The corollary to the hotspot issue is that unmodified areas in many altered landscapes are often those of low productivity (e.g., with poor soils or on steep, rocky, or icy terrain) that are unsuitable for many elements of the biota. Hence relatively large areas may remain, but species diversity may be low (see Chapter 2).

A second caveat is that, although strong relationships occur between species richness and area, species richness is also a function of much more than simply area (reviewed by Gaston and Spicer 2004). It is related to (among others): latitude (Armesto et al. 1998), elevation (MacArthur 1972), spatial juxtaposition with other areas (e.g., islands; MacArthur and Wilson 1967), and disturbance regimes (Connell 1978; Shiel and Burslem 2003). Thus species loss and species number will be a function of many more factors than the total area remaining (Mac Nally et al. 2000).

A third caveat is that many critically important habitat attributes like vegetation structure are often correlated with remnant area. Thus species richness may be a response to them rather than to area per se (Robbins 1980; Lynch and Whigham 1984; Dobkin and Wilcox 1986; Haila 1986; Jellinek et al. 2004). Management may require careful attention to the maintenance of habitat quality (for a particular species or set of taxa) rather than simply reserving large areas (Berglund and Jonsson 2003).

Box 9.3. Extinction Debts and Species Richness Relaxation

The effects of landscape history can sometimes be related to concepts such as extinction debts and species richness relaxation, which are well known in the theoretical conservation biology literature (MacArthur and Wilson 1967; McCarthy et al. 1997; Brooks et al. 1999). These concepts suggest that populations go extinct well after anthropogenic landscape change, such as vegetation clearing, has taken place (Loyn 1987; Tilman et al. 1994; Robinson 1999), and that some species will respond to circumstances that are no longer directly visible to researchers. Under such circumstances, present relationships between patch size and species richness may be complicated by the existence of small, nonviable, isolated populations in remaining vegetation remnants (Berglund and Jonsson 2005). For example, Barro Colorado Island is a former hilltop that became an island between 1911 and 1914 when its surrounds were flooded to create the Panama Canal. The island is approximately 1600 ha in size. Since its isolation, approximately 65 bird species have gone extinct on the island (Robinson 1999). However, these extinctions occurred over a prolonged period of time, and species were lost well into the 1990s (Robinson 1999; Figure 9.4). Similarly, although less than 7% of the original cover of Brazilian Atlantic forest now remains, no bird species has yet suffered extinction at the regional scale. However, the large number of species that are highly threatened is approximately equivalent to the number of taxa predicted to be lost as a result of deforestation (Brooks and Balmford 1996).

In summary, species–area functions highlight the fundamental importance of limiting the amount of native vegetation loss from landscapes and maintaining large patches of remnant vegetation for biological conservation. However, the caveats listed above need consideration as part of any analysis of the effects of landscape modification on biota, and they provide part of the context for the responses of various species and assemblages of taxa. An overly simplistic focus on species–area functions may mask the underlying reasons for such relationships and divert attention away from key problems that may require management. Moreover, species–area relationships focus, by definition,

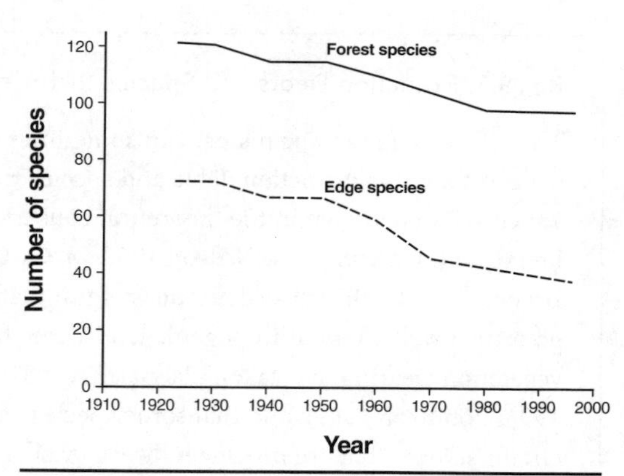

Figure 9.4. Loss of bird species on Barro Colorado Island since the Panama Canal was created between 1911 and 1914 (redrawn from Robinson, W.D. 1999: *Conservation Biology*, **13**, 85–97, Long-term changes in the avi-fauna of Barro Colorado Island, Panama, a tropical forest isolate, with permission from Blackwell Publishing).

on the number of species. Although this can be a valuable measure of biodiversity, it is often the composition of biotic communities that is more important (Murphy 1989). This is exemplified by work on mammals in heavily modified landscapes in the Narigal region of southwestern Victoria, Australia. At the time of European settlement there were 33 species of native mammals in the Naringal region. Of these, six became regionally extinct. However, six introduced taxa invaded. Thus, although the total number of mammal species was unchanged, identity of taxa had changed dramatically (Bennett 1990b). Similar results have been recorded in other studies, such as in tropical ecosystems in Brazil, where species richness patterns were strongly shaped not only by the loss of species from vegetation remnants but also by the influx of new taxa capable of utilizing heavily altered surrounding areas (e.g., Tocher et al. 1997; Gascon et al. 1999). These findings highlight two key points. First, a focus on species richness may not always be particularly informative for management and conservation (Noss and Cooperrider 1994). Second, the matrix is rarely nonhabitat for all species (Chapter 14), and landscape alteration will disadvantage some species but benefit others (Chapter 3)—with corresponding effects on species composition (Chapter 13).

Large Patches and Other Conservation Strategies

Larger areas support more species than small ones; this is one of the reasons why the protection of large ecological reserves and large patches of remnant vegetation in modified landscapes is a critical element of sound strategies for biological conservation (Lindenmayer and Franklin 2002). However, an overly simplistic application of species–area relationships can divert attention from the need for concomitant conservation strategies, such as the retention of small patches and improved management of the matrix (Chapter 14). For example, while large patches are considered to be critical, this can sometimes reduce the perception of importance of small patches (Simberloff 1988; Table 9.2). Many examples from around the world highlight the importance of small patches, and sometimes they are the only remaining options for conservation (Semlitsch and Bodie 1998; Lindenmayer et al. 1999b). The forest, open-woodland, and tallgrass prarie remnants in the heavily altered Midwest of the United States are important examples (Schwartz 1997). This point is further illustrated by a study of temperate woodland remnants scattered throughout farming landscapes in southeastern Australia (Freudenberger 2001). Birds, plants, and ants were surveyed in 35 different patches ranging from <0.5 to >300 ha in size. Although larger patches supported the largest number of species of each group, every patch supported a different mix of species. In the case of the 106 species of birds that were recorded, it took almost 35% of the patches to record each species at least once.

Table 9.2. Examples highlighting the conservation significance of small patches of remnant vegetation

Group	*Location*	*References*
Plants	Tropical rainforests, Asia	Turner 1996
Plants	Tasmania, Australia	Kirkpatrick and Gilfedder 1995
Plants	NSW, Australia	Prober and Thiele 1995
Forest biota	Sweden	Angelstam and Pettersson 1997, Fries et al. 1997
Vertebrates	Florida, United States	McCoy and Mushinsky 1999
Birds	NSW, Australia	Fischer and Lindenmayer 2002c
Bats	NSW, Australia	Law et al. 1999
Invertebrates	California	Rubinoff and Powell 2004
Invertebrates	Western Australia	Abensperg-Traun and Smith 2000

For ants, almost 80% of the patches needed to be surveyed before each species was recorded at least once (Freudenberger 2001).

Summary

Human landscape modification often involves the partial removal of native vegetation cover. Shrinkage (sensu Forman 1995) of remaining patches of native vegetation cover can also occur. Reduced species richness is typically associated with both the overall reduction in landscape levels of native vegetation cover and the smaller sizes of individual patches of remnant native vegetation at a more localized scale. Species–area functions are often used to describe the widely observed relationships between species richness and the amount of native vegetation cover or the size of individual vegetation patches. Mechanisms that underpin species–area relationships include extinction-recolonization dynamics, increased disturbance in smaller patches, greater habitat diversity in larger areas, and the random placement effect in which larger areas capture more species with patchy distributions simply by chance. Although species–area relationships are widespread and well established in the literature, predictions of the numbers of species likely to be lost and the rate of species loss as a function of vegetation loss remain highly problematic—forecasts of species extinctions using such relationships have limited predictive accuracy. Finally, in a land management context, while species–area relationships emphasize the critical role for conserving large patches of remnant vegetation, other complementary strategies such as matrix management and the retention of small vegetation patches can also make an important contribution to biological conservation in human-modified landscapes.

Links to Other Chapters

Chapter 3 discusses the island model of landscape cover, which is closely related to concepts associated with the species–area relationship. Chapter 4 focuses on habitat loss for individual species and is the parallel chapter to this one on vegetation loss and species richness. The influence on species richness of the matrix and associated patterns of landscape heterogeneity are discussed in Chapter 14. Chapter 18 in Part V explores ways in which vegetation loss might be curtailed and reversed to mitigate species loss.

Further Reading

Rosenzweig (1995) and Gaston and Spicer (2004) are useful texts that contain considerable valuable material on patterns of species richness at a range of spatial scales. Gaston (1996) discusses problems with estimating the number of species as do Lindenmayer and Burgman (2005) and Gotelli and Colwell (2001). The series of books by Saunders et al. (1987, 1993a) and Saunders and Hobbs 1991) gives an excellent set of case studies of species loss associated with landscape change. Rosenzweig (1995) probably provides the most comprehensive review of the wealth of material on species–area relationships. Connor and McCoy (1979) and McGuinness (1984) explore species–area relationships and the form of these functions. Simberloff (1992) gives a good outline of why applying species–area relationships to forecast the number and rate of extinctions is highly problematic. Many studies have found significant positive relationships between patch size and species richness. Hanski (1994a) reviews this topic and provides many useful examples. The edited volume by Schwartz (1997) contains many examples illustrating the conservation significance of small native vegetation remnants—particularly in the heavily altered landscapes of midwestern United States.

CHAPTER 10
Deterioration of Vegetation and the Physical Environment

This chapter continues the overarching theme of Part III of exploring the impacts of landscape modification on assemblages, and examines the deterioration of the structural complexity of vegetation and the physical environment. "Structural complexity" can be broadly defined as the range and variability in vertical attributes of a given local area, including aspects of ground and vegetation cover. Thus structural complexity is a locally scaled attribute, in contrast to landscape heterogeneity, which encompasses horizontal and hence spatial changes in land cover and environmental gradients (Franklin and van Pelt 2004; see Chapter 14). This chapter focuses on changes in structural complexity that can occur as a consequence of human land-use practices. We illustrate such impacts on assemblages by briefly discussing three common land uses that often result in vegetation deterioration and loss of structural complexity: (1) domestic livestock grazing, (2) forestry, and (3) firewood collection. In many ways this chapter parallels Chapter 5 in Part II, which discusses the process of habitat degradation and its impacts on individual species. However, here we focus on landscape patterns from a human perspective, and their effects on sets of species rather than on the particular habitat requirements of individual species. In contrast to the landscape-scale focus of this chapter, the more localized phenomenon of edge effects is discussed in Chapter 11.

Deterioration of the Physical Environment due to Grazing and Pastoralism

Severe landscape modification afflicts nearly 20% of Earth's vegetated land (World Resources Institute 1992; Ghassemi et al. 1995; UNEP 1999) and is predominantly associated with agricultural development and extensive grazing by domestic livestock (Peterken 1996; Wang and Hacker 1997; Hobbs 2001). Accelerated deterioration of the physical environment associated with human activities includes wind and water erosion, dryland salinity, irrigation salinity and water logging, soil compaction, vegetation loss, invasion of exotic species such as weeds, mass movement of soil, chemical contamination of soil and water, and soil acidification. These processes can, in turn, have significant negative impacts on vegetation structure and result in habitat degradation for many individual species (Mercer 1995; Recher 1999; Dale et al. 2000).

Vegetation clearing to promote grass growth for grazing removes key plant species and modifies or destroys habitat for the array of organisms dependent on that vegetation. As grazing pressure increases, native plant diversity is reduced and exotic plant species diversity and cover increase (Burrows 1999; Yates et al. 2000). There tends to be limited natural regeneration of native plants (Spooner et al. 2002). Many types of native vegetation and their associated biota around the world have declined dramatically as a direct or indirect consequence of grazing (Noss and Cooperrider 1994; Hobbs and Yates 2000; Groombridge and Jenkins 2002; Box 10.1).

Box 10.1. The Global Grazing Industry

The impacts of domestic livestock grazing on landscape change and potential species losses can be gauged, in part, from the size of the area subject to pasture conversion, the size of the global livestock herd, and annual global rates of meat production. It has been estimated that landscape modification has resulted in 11% of the world's land surface now being covered by cropland and 23% dominated by pastureland. The United Nations Development Programme (2000) estimated that for the period between 1996 and 1998 the global average meat production from livestock was 215 million tonnes per annum. This was based on a herd size approaching 3.4 billion animals consisting of 1.3 billon cattle, 1.8 billion sheep and goats, 0.1 billion horses and donkeys, and 0.2 billion buffaloes and camels (United Nations Development Programme 2000). Notably, while problems associated with pasture quality and levels of ground cover can manifest as a result of domestic livestock grazing (e.g., Wang and Hacker 1997), further complications arise when additional pressure on vegetation occurs from nondomesticated animals (Freudenberger 1995).

Grazing by domestic livestock can also cause soil compaction (Bezkorowajynj et al. 1993) (and therefore reduce water infiltration into the soil), damage remnant native trees (e.g., by rubbing; McIntyre et al. 2002), and add nutrients to the soil, which can, in turn, promote the invasion and growth of nonnative plants (Janzen 1983; Carr et al. 1992; Primack 2002). Livestock often concentrate in more productive parts of pastures and near watercourses (Jansen and Robertson 2001; see Chapter 2). Their impacts in these areas often lead to the impoverishment of mammal, bird, reptile, frog, insect, and other biotic communities. For example, Bock et al. (1990) studied the lizard *Sceloporus scalaris* in southeastern Arizona. For years, it had been thought that the species was naturally restricted to montane areas. However, Bock et al. (1990) argued that the lizard's higher abundance in montane areas was primarily related to the absence of grazing pressure in these areas. In contrast, the more productive valleys in the study area were heavily grazed. A possible causal explanation for the lizard's negative response to grazing was that the species sought refuge from predators in tall grass, and tall grass was sparse in the grazed valleys (Bock et al. 1990). Similar negative effects of grazing on lizard populations have been reported in a range of other studies elsewhere in the United States (Busack and Bury 1974) and in Australia (Brown 2001; Box 10.2). Other practices commonly associated with grazing such as the use of pesticides (Hall and Henry 1992), ploughing (Dorrough and Ash 1999), the removal of woody debris (Mac Nally et al. 2001), and a range of other management activities can also negatively affect species (McIntyre and Lavorel 1994; Bromham et al. 1999; Hobbs 2001). Fencing, which is a key part of pastoral enterprises, can be a barrier to the movement of some species (van der Ree 1999), and in parts of North America native predator eradication has had significant impacts on many species, including those lower in the food chain (Berger et al. 2001; Ripple and Beschta 2005).

Although livestock grazing in temperate ecosystems often leads directly to a loss of vegetation cover, in many of the world's semiarid savannahs it has been linked to encroachment of woody vegetation (Table 10.1). Naturally, the spatial pattern defining savannahs of widely scattered trees surrounded by grassland and some shrubs is maintained by the interplay of fire, the grazing of native herbivores, and rainfall (Mistry 2000). However, domestic livestock grazing can change this spatial pattern, especially in combination with reduced fire frequency (Mistry 2000), and can facilitate the encroachment of dense shrubs (see Table 10.1). In African savannahs, a possible causal link between livestock grazing and excessive encroachment of native shrubs lies in the fact that domestic livestock browse on some shrub species with fleshy fruit, which are particularly common under scattered trees. Later, cattle often defecate the seeds of shrubs in the more open grassland areas of the savannah, thereby facilitating the long-distance dispersal of shrubs (Jeltsch et al. 1996). The resulting shrub encroachment can degrade the habitats of many native species (see Table 10.1), including birds (see Box 5.1 in Chapter 5) and lizards (Meik et al. 2002). Counteracting shrub encroachment by managing livestock grazing and fire regimes is widely considered to be important for biological conservation in semiarid savannahs throughout the world (e.g., Milton et al. 1994; Dean et al. 1999).

Deterioration of the Physical Environment due to Forestry

Native forests provide multiple values for human societies—timber, pulp, firewood, water, honey, wildlife, and recreation. Over 4 billion cubic meters of timber and pulpwood are harvested annually (FAO 2001). Forestry has been one of the most controversial forms of land use around the world for much of the past 30 years (Yaffee 1994; Angelstam 1996; Lindenmayer and Franklin 2003). Many forms of forestry result in the maintenance of vegetation cover, but forest suitability for many species and assemblages can be significantly reduced (Franklin et al. 1981). As will be outlined here, vegetation deterioration can manifest in several ways.

Box 10.2. Pastoralism and Species Loss in Australia

Grazing by sheep *(Ovis ovis)* or cattle *(Bos taurus)* on natural or seminatural pastures is the major land use for 4.5 million km^2 (60%) of Australia's land surface (Commonwealth of Australia 2002). Products from pastoralism contribute about 3% of the nation's annual average gross domestic product, particularly through exports (Australian Agriculture, Fisheries and Forestry 2003). However, pastoralism has had many negative impacts on vegetation structure (including weed invasion; Burrows 1999) and has led to habitat degradation for many taxa, leading to their decline (Dickman et al. 1993). A key problem in grazed Australian ecosystems is that hard-hoofed animals did not inhabit the continent prior to European settlement, and many ecosystems are not well adapted to soil compaction and degradation that are commonly associated with large numbers of domestic livestock (Hobbs and Hopkins 1990).

A study of pastoral use of the open woodlands of northern Queensland found that many pastures are now suffering the effects of overgrazing, including loss of herbaceous cover, invasion of unpalatable shrubs, and soil erosion (Gardener et al. 1990). In arid and semiarid areas, domestic livestock congregate near permanent water, especially during drought (Landsberg et al. 1997; James et al. 1999). They eat and trample native vegetation, compact soil, alter soil structure, and disrupt the soil crust (Hobbs and Hopkins 1990). Landsberg et al. (1997) and James et al. (1999) measured vegetation in several Australian rangelands along 10 km gradients between water points and areas remote from water. Grazing intensity increased closer to water. They found about 30% of the native species declined near heavily grazed areas around artificial water points. Spatial analysis of the distribution of water points in the rangelands showed that potential habitat had become highly degraded for species adversely affected by grazing.

Logging and the Structural Complexity of the Forest

Stand structural complexity of forests includes a wide variety of structural features such as:

- Trees from multiple age cohorts within a stand
- Large living trees and standing dead trees
- Large-diameter logs on the forest floor
- Vertical structural complexity created by multiple or continuous canopy layers
- Mixtures of tree species and other plant taxa
- Canopy gaps and antigaps (i.e., areas with very dense canopy coverage under which understory development can be limited; Halpern and Franklin 1990; Franklin et al. 2002)

These elements of stand structural complexity provide critical habitat components for a vast array of forest-dependent taxa (Hunter

Table 10.1. Conceptual model of the steps involved in the degradation of semiarid rangelands*

Step	Description	Symptoms
0	Biomass and composition of vegetation varies with climatic cycles and stochastic events	Perennial vegetation varies with weather
1	Herbivory reduces recruitment of palatable plants, allowing populations of unpalatable species to grow	Demography of plant populations changes
2	Plant species that fail to recruit are lost, as are their specialized predators and symbionts	Plant and animal losses
3	Biomass and productivity of vegetation fluctuate as ephemerals benefit from loss of perennial cover	Perennial biomass reduced (short-lived plants and instability increase), resident birds decrease, nomads increase
4	Denudation and desertification involve changes in soil function and detritivore activity	Bare ground, erosion, aridification

*Modified from Milton et al. (1994). Note that herbivory in Step 1 often involves domestic livestock grazing, which is a key degrading process in many parts of the world.

Box 10.3. Vegetation Deterioration, Deterioration of the Physical Environment and Weeds

There are many potential environmental effects of weeds that can contribute significantly to vegetation deterioration and deterioration of the physical environment (Williams and West 2000). Examples of the effects of weeds include (1) competition with indigenous plants for space, light, nutrients, and other resources (sometimes leading to the complete replacement of indigenous plant communities; (2) the restriction of natural regeneration of native plants; (3) altered movement of water through the soil and in watercourses and ultimately altered water quality; (4) the addition of toxins to the soil and water; (5) altered microtopography of landscapes; and (6) altered fire regimes. The direct and indirect economic costs of weeds are enormous. Invasive species (both weeds and exotic animals) cost the economy of the United States US$137 billion annually (Pimentel et al. 2000). The overall total economic impact of weeds in Australia in terms of lost production and costs of control is more than $US2 billion (McNeeley et al. 2003).

1990, 1999; Lindenmayer and Franklin 2002). In contrast, logging operations and regeneration aimed at maximizing timber production can lead to long-term changes in forest composition and structure (Kellas and Hateley 1991; Halpern and Spies 1995; Mueck et al. 1996; Ough 2001), including the invasion of weed species (e.g., Brown and Gurevitch 2004; Box 10.3). In addition, particular tree species can be severely reduced in abundance; for example, red cedar (*Toona australis*) in eastern Australia and mahogany (*Swietenia* spp.) in South America (Primack 2002).

Changes in stand structural complexity can negatively affect taxa dependent on particular structural attributes such as hollow-bearing trees, fallen logs, and decaying wood (Harmon et al. 1986; Gibbons and Lindenmayer 2002), including many formerly abundant generalist species (Niemelä et al. 1993). For example, the loss of structural complexity can impair the suitability of foraging habitat for vertebrates, such as birds and bats (Brown et al. 1997; Woinarski et al. 1997; Brokaw and Lent 1999; Box 10.4; Figure 10.1). Old-growth forest–dependent taxa can be at particular risk from the reduction or loss of structural features typical of old-growth stands (Franklin et al. 1981; Scotts 1991; Marcot 1997).

Other Ways in which Logging Can Deteriorate the Physical Environment

In addition to reducing the structural complexity of vegetation, there are other ways in which logging can lead to habitat degradation for many species in an assemblage. Logging operations can modify the environment directly through (1) the construction of roads and tracks; (2) burning to suppress wildfires, encourage regeneration, or control pest species; (3) the inadvertent introduction of weeds; (4) soil compaction; (5) the application of fertilizers; and (6) an increase in silting streams from roads and cutting areas (RAC 1991; Lindenmayer and Franklin 2002).

Forest roads can have particularly severe effects on the physical environment. For example, relative to the area they occupy, roads exert disproportionate, persistent, and intense impacts on aquatic ecosystems (Forman 1998; Forman et al. 2002), and several studies have shown that road density has a negative effect on aquatic fauna (Vos and Chardon 1998). Many hydrological changes arising from road networks are permanent because subsurface flows and patterns are interrupted and water flows are rerouted into extensive constructed surface channels (e.g., road ditches and culverts) (Jones and Grant 1996; Walker 1999).

In addition, roads and railways constructed for harvesting, even in well-managed forests, provide access to previously remote regions, thereby encouraging the expansion of human populations and associated agricultural practices (Harcourt 1992; Putz et al. 2000), including domestic livestock grazing, which can have cumulative effects with logging operations (Smith et al. 1992). They may also facilitate the movement of native predators (Taylor et al. 1985) and feral predators (May and Norton 1996; May 2001). Large native carnivores can be highly vulnerable to the access provided by roads because they may be eliminated by legal or illegal hunting and trapping (Trombulak and Frissel 2000).

Logging can significantly alter patterns of landscape heterogeneity (Franklin and Forman 1987). The resulting changes in the spatial pattern of forest age classes may affect a wide range of species, such as large forest owls (Milledge et al. 1991; Lamberson et al. 1994), wide-ranging herbivores (Thompson and Angelstam 1999), and some wide-ranging species of arboreal marsupials (Lindenmayer et al. 1999a). Inappropriate rates or patterns of forest harvest can also negatively affect hydrological and geomorphological processes as well as aquatic biota (Doeg and Koehn 1990; Naiman 1992; Trayler and Davis 1998). Harvesting schedules resulting in extensive areas of recently clearcut forest can increase flood flows (Jones and Grant 1996), with massive effects on aquatic ecosystems (e.g., Silsbee and Larson 1983; Graynoth 1989).

Box 10.4. Vegetation Deterioration and the Loss of Stand Structural Complexity in European Forests.

The problem of vegetation deterioration and loss of structural complexity of forests has been recognized for a long time, especially in Europe (Figure 10.2). Almost 120 years ago, Gayer (1886) expressed concerns about the simplification of German forests. In Sweden, a century of intensive management in a 123,000 ha area of boreal forest transformed stand structure from one dominated by widely spaced, large-diameter trees to young, densely stocked forests. The number and volume of large living and dead standing trees were reduced by 90% and the extent of old stands by 99% (Linder and Östlund 1998; Figure 10.1). It has been recognized that large, dead trees are particularly valuable for many Scandinavian forest taxa (Samuelsson et al. 1994). For example, Berg et al. (1994) calculated that almost 50% of the threatened species in Sweden were dependent on dead standing trees or logs. Similarly, mature deciduous trees have also become a rare element of managed stands in Scandinavia (Esseen et al. 1997), although they are a key component of forest composition for a wide range of animal and plant groups (Enoksson et al. 1995).

Deterioration of the Physical Environment due to Firewood and Fuelwood Collection

Firewood harvesting and fuelwood collection are forms of land use that can have profound impacts on vegetation structure and, in turn, can lead to significant habitat degradation for

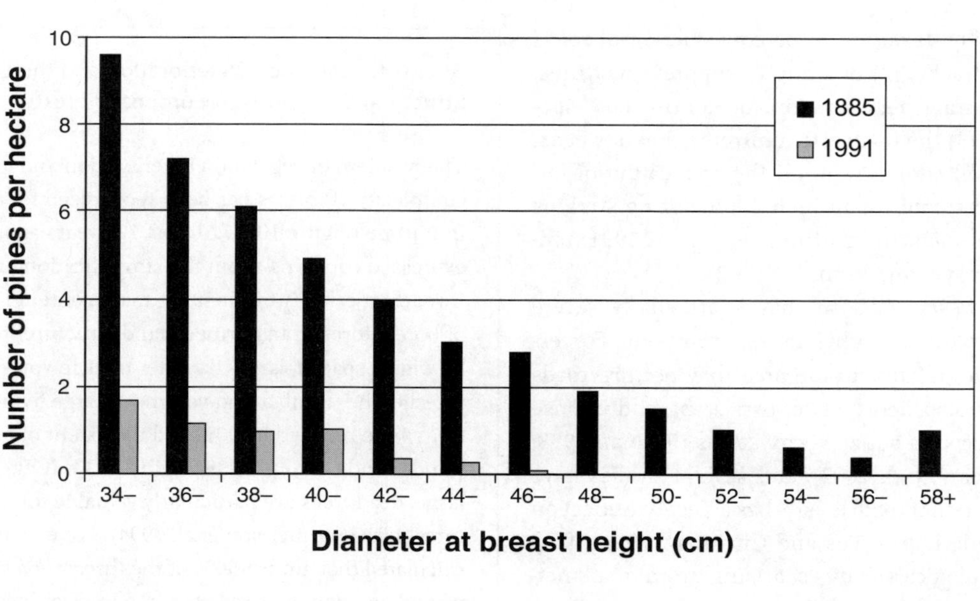

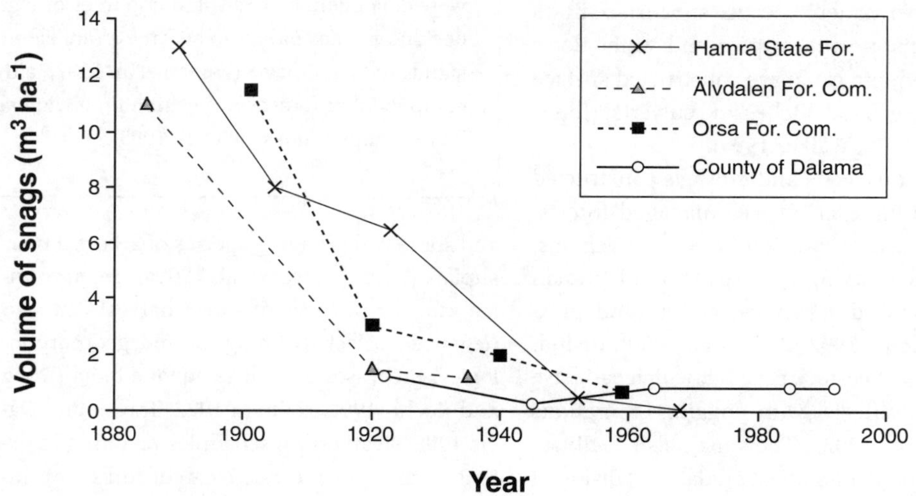

Figure 10.1. Altered structural attributes in four districts of Swedish boreal forest subject to a century of intensive harvesting. Data are shown for the number of pine trees in different diameter classes and the volume of dead standing trees ("snags") (redrawn from Linder, P. and Östlund, L. 1998: *Biological Conservation,* **85**, 9–19, Structural changes in three midboreal Swedish forest landscapes, 1885–1996, with permission from Elsevier).

many elements of the biota, including entire assemblages of taxa dependent on cavities and fallen-logs (Driscoll et al. 2000; Manning et al. 2004b). Firewood and fuelwood collection can target both dead wood and living trees.

Firewood harvesting is known to have significant impacts in forest and woodland ecosystems in many parts of the world (United Nations Development Programme 2000). In many developing countries, severe deterioration of vegetation and the physical environment occurs as a result of firewood collection for domestic and industrial use, particularly in savannah woodlands, scrub, and, increasingly, tropical forests (Harcourt 1992; FAO 2001). Approximately one-third of the world's population depend on biomass fuels for energy; hence demands for firewood are substantial in many nations. For

Figure 10.2. This area of forest near Copenhagen in Denmark has been harvested many times in the past centuries and is characterized by trees of similar diameter and uniform spacing as well as an almost complete lack of coarse woody debris and understory plants (photo by David Lindenmayer).

example, in the late 1980s, it was estimated that firewood (together with animal dung) provided more than 90% of the primary energy needs of countries such as Nepal, Tanzania, and Malawi (McNeely et al. 1990). Similarly, at least 90% of Africans depend on firewood (and other biomass) for their energy needs (UNEP 1999). The consumption of firewood rose from 250 to 502 million m^3 between 1970 and 1994 on that continent and is predicted to rise by a further 5% by 2010 (FAO 1997). Impacts of firewood harvesting on mammals, birds, reptiles, and other biota have been substantial (UNEP 1999; Groombridge and Jenkins 2002).

In China, several species are threatened, at least in part, by fuelwood collection; for example, the giant panda *(Ailuropoda melanoleuca)* in the Wolong Nature Reserve in Sichuan Province (Liu et al. 1999; Box 4.4), and the white-headed langur *(Trachypithecus leucocephalus)* in Guanxi Province (Huang et al. 2002). Similarly, in India, Chettri et al. (2001, 2002) studied patterns of firewood collection and bird community composition along the Yuksam-Dzongri trekking corridor of Sikkim Himalaya. Their work showed that firewood harvesting could lead to severe forest degradation, especially near settlements and in easily accessible areas along walking tracks. Firewood in the Sikkim area was used both by local people and to maintain the tourism industry. Chettri et al. (2002) suggested that a key solution to halting vegetation deterioration in this area lay in collaborative investigations with local people into alternative fuel sources.

The impacts of firewood harvesting on vegetation deterioration and associated forest and woodland biota are not confined to developing nations. In the United States there are substantial issues associated with educating people that dead trees are important wildlife habitat and should not be removed for firewood

(Styskel 1983; James and M'Closkey 2003). In Australia, the State of the Environment Report (2001) noted that firewood was the third largest source of energy in Australia after gas and coal. Between 4.5 and 7.1 million tonnes of fuelwood are cut annually for domestic consumption in Australia—it is difficult to quantify firewood consumption in Australia because most harvesting takes place outside any regulatory framework (Wall 1999; Driscoll et al. 2000; Pears 2000). The Australian firewood industry is based on cutting dead standing and fallen trees that often contain hollows (Gibbons and Lindenmayer 2002). These trees and logs have significant value for many elements of the biota—including some rare and threatened species such as the superb parrot (*Polytelis swainsonii;* Webster and Ahern 1992; Manning 2004). Over 50 species of Australian vertebrates are threatened by firewood harvesting (Driscoll et al. 2000; Garnett and Crowley 2000).

Interactions with Other Processes Associated with Landscape Alteration

Like most processes associated with landscape change, large-scale deterioration of the physical environment usually co-occurs with other processes such as vegetation clearing and edge effects. Key processes often interact in complex ways that can have significant impacts on assemblages (Chapter 15). Many studies have shown that when vegetation is lost, there can be flow-on effects on the condition of remaining vegetation (Ambuel and Temple 1983). Edge effects that result from contrasts between remaining vegetation remnants and the condition of the surrounding matrix can strongly affect the type and rate of deterioration of the physical environment across entire landscapes (Dale et al. 2000; Chapter 11). Similarly, McIntyre et al. (2000) noted that the level of vegetation deterioration caused by tree dieback in Australian temperate woodland remnants was strongly related to the amount of vegetation clearing in the surrounding landscape. Dieback levels were significantly higher where less than 30% of original levels of vegetation cover remained (McIntyre et al. 2000; see Chapter 15). In the Brazilian Cerrado (see Box 2.1), where land clearing to establish crops (e.g., soybeans) is extensive, deterioration of remaining areas of native vegetation can be substantial through other processes linked with landscape alteration such as increased frequency of fire and highly elevated rates of soil erosion (Klink and Machado 2005). Similarly, mining, fire, and logging can be key processes contributing to the deterioration of vegetation associated with karst outcrops in Malaysia, with corresponding negative impacts on a wide range of biotic groups, including endemic species–rich invertebrate assemblages (Schilthuizen et al. 2005).

The fact that vegetation deterioration co-occurs with other processes in the landscape, particularly vegetation clearing, has been highlighted in models (e.g., Ambuel and Temple 1983), and is also apparent from empirical studies on small mammals (Dunstan and Fox 1996; Knight and Fox 2000) and birds (Mac Nally et al. 2000). In these cases, species composition can be influenced as much by conditions in the matrix and levels of vegetation deterioration within patches of remnant vegetation as by patch size per se (Chapter 9).

Summary

Human land-use practices can lead to substantial vegetation modification and alteration of the physical environment. This can result in widespread losses of biota. Short case studies of three significant and widespread forms of human land use feature in this chapter—domestic livestock grazing and pastoralism, forestry, and fuelwood harvesting. Vegetation deterioration can be a longer-term and more incremental process than direct vegetation loss, and carefully developed monitoring programs will often be required to assess its impacts on vegetation structure and associated effects on biota. Importantly, the deterioration of vegetation structure and aspects of the physical environment usually co-occurs with other landscape-scale changes such as clearing of native vegetation (Chapter 9) and the loss of landscape connectivity (Chapter 12). Often, it is the interaction and compounding of these various changes to landscape pattern that can negatively affect assemblages and ecosystem function (Chapter 15).

Links to Other Chapters

Chapter 5 in Part II explores the threatening process of habitat degradation from the perspective of individual species. Habitat degradation for individual species is often closely linked to the overall deterioration of vegetation and the physical environment. However, we reiterate that different species respond in different ways to landscape change (Chapter 2), which is why we make an explicit distinction between habitat degradation and landscape-scale deterioration of the physical environment. Some ways of tackling vegetation deterioration are examined in Part V of this book. Appropriate strategies will vary depending on the ecosystem in question, the nature of the degrading processes, and a range of other factors.

Further Reading

The impact of grazing on biota is a complex topic, and several authors provide valuable introductions to this area, including Noss and Cooperrider (1994), Landsberg et al. (1997), and James et al. (1999). An excellent review in a rangeland context is given by Milton et al. (1994). There are many useful texts and reviews on the impacts of forestry on biota, including those by Hunter (1990, 1999) and Lindenmayer and Franklin (2002, 2003). FAO (1997, 2001) and UNEP (1999) provide background information on the range of negative effects that can result from firewood collection. Driscoll et al. (2000) and Wall (1999) discuss the extent of the problem in an Australian context. Weed invasion is often a key problem associated with the deterioration of native vegetation, and there is a vast literature on exotic plant invasions. Interesting books and papers on weeds include (among numerous others) those by Richardson et al. (1994), Cronk and Fuller (1995), Low (1999), Lonsdale (1999), Pimentel et al. (2000), Myers et al. (2000), McNeely et al. (2003), and Simberloff et al. (2005). Hobbs (2001) provides a valuable case study of the combined sets of processes giving rise to the deterioration of vegetation structure and the physical environment in the temperate woodlands of the wheatbelt in Western Australia. Bennett (2003) outlines ways in which vegetation deterioration might be tackled in landscapes subject to extensive human modification.

CHAPTER 11
Edge Effects

Chapter 2 discussed changes in landscape spatial patterns typically associated with human landscape modification. A common change in landscape pattern resulting from landscape alteration is the increase in the length of boundaries or edges between remaining patches of original vegetation and the surrounding matrix. For example, in the Cadiz township in Green County (Wisconsin, USA; see Chapter 2), the length of edge increased from 61 to 208 m per hectare of forest between 1882 and 1950, concurrent with major deforestation during that period (Burgess and Sharpe 1981). Edge effects can occur at such boundaries, penetrating deeply into vegetation remnants in some heavily modified landscapes (Laurance 2000; Siitonen et al. 2005). Such altered spatial patterns in modified landscapes can have profound impacts on assemblages of taxa. This chapter discusses abiotic and biotic edge effects, some factors that are related to the variability of edge effects, and the relationship between edge effects and species extinctions.

Types of Edges and Edge Effects

The edge of a vegetation patch can be broadly defined as a marginal zone of altered microclimatic and ecological conditions that contrasts with its interior (Matlack 1993). A range of edge types has been recognized. One way of classifying edges is by their origin—natural or human-derived (Luck et al. 1999a, b). Edges usually occur naturally at the interface of two ecological communities. Such natural boundaries are sometimes termed ecotones and can support distinctive plant and animal communities (Figure 11.1; Witham et al. 1991).

In modified landscapes, many edges are created by humans. Examples are the boundaries between recently clearcut forest and adjacent unlogged stands (Chen et al. 1990, 1992) or between cropped areas and remnant native vegetation (Sargent et al. 1998). Edges can be "soft," where the transition between different types of patches is gradual; or "hard," at boundaries with marked contrasts in vegetation structure and other features (e.g., clearfelled/unlogged forest; Forman 1995).

Edge effects refer to changes in biological and physical conditions that occur at an ecosystem boundary and within adjacent ecosystems (Wilcove 1985; Temple and Cary 1988; Kremsater and Bunnell 1999). The ecological edge that results from a disturbance is the result of interactions between the kind and intensity of the disturbance event and the ecological dynamics within the adjacent, undisturbed environment.

In a major review of edge effects in forest landscapes, Harper et al. (2005) classified edge effects as either primary responses that arise directly from edge creation or secondary responses that arise indirectly as a result of edge creation. Primary responses include structural damage to the vegetation, disruption of the forest floor and soil layer, altered nutrient cycling and decomposition, changed evaporation, and altered pollen and seed dispersal (Harper et al. 2005). Secondary or indirect responses (that flow from primary responses) include patterns of plant growth, regeneration, reproduction, and mortality and are manifested as altered patterns

Figure 11.1. Natural ecotone of gallery riparian rainforest and savanna at Rio Negro River, Brazil (photo by David Lindenmayer).

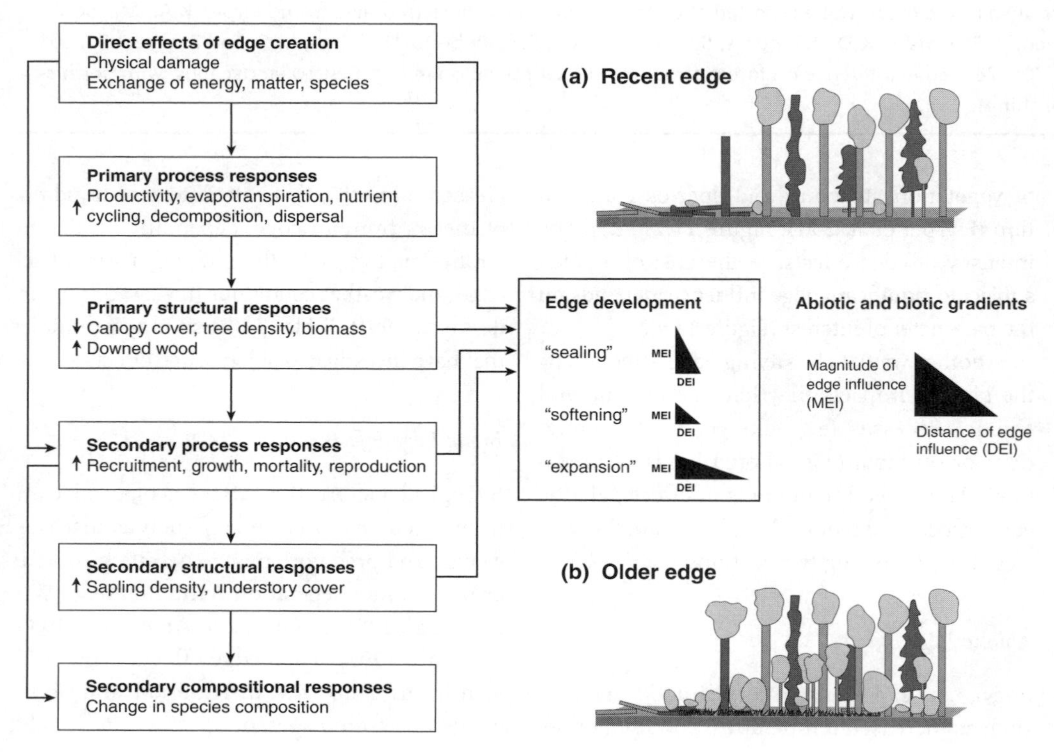

Figure 11.2. Processes and responses following edge creation in a forest landscape (redrawn from Harper, K.A., Macdonald, S.E., Burton, P.J., Chen, J., Brosofske, K.D., Saunders, S.C., Euskirchen, E.S., Roberts, D., Jaiteh, M.S. and Essen, P-E. 2005: *Conservation Biology*, **19**, 768–782, Edge influence on forest structure and composition in fragmented landscapes, with permission from Blackwell Publishing).

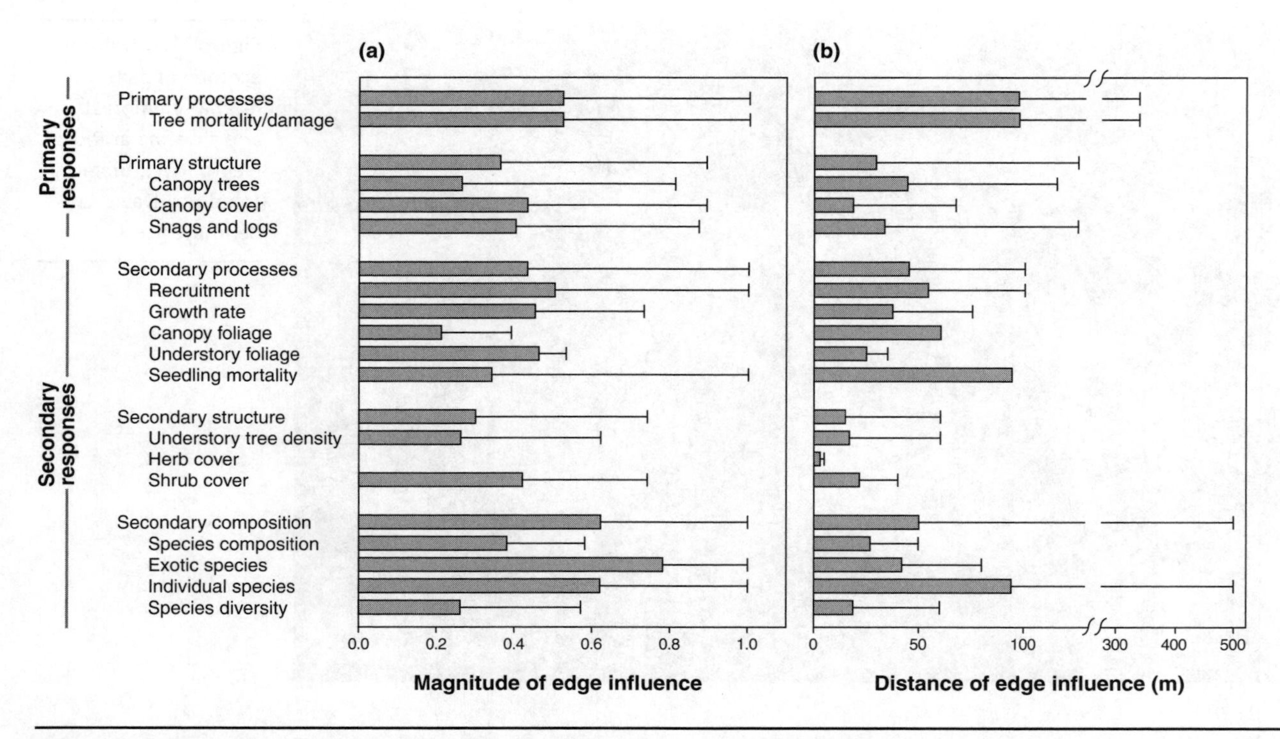

Figure 11.3. Penetration of primary and secondary edge effects in forests. The magnitude of a given edge effect was calculated by Harper et al. (2005) relative to the mean value reported at edges and patch interiors (redrawn from Harper, K.A., Macdonald, S.E., Burton, P.J., Chen, J., Brosofske, K.D., Saunders, S.C., Euskirchen, E.S., Roberts, D., Jaiteh, M.S. and Essen, P-E. 2005: *Conservation Biology*, **19**, 768–782, Edge influence on forest structure and composition in fragmented landscapes, with permission from Blackwell Publishing).

of vegetation structure and species composition (Harper et al. 2005; Figure 11.2, 11.3). The intensity of edge effects, or the area of a patch subject to significant edge influence, depends on the parameter of interest (Figure 11.4).

Another way of classifying edge effects is by the kind of impacts they have on climate and abiotic processes (e.g., altered wind penetration) or on biota (e.g., altered levels of predation). These two broad kinds of effects, abiotic edge effects and biotic edge effects, are the subject of the following two sections.

Abiotic Edge Effects

Edges may experience microclimatic changes, such as increased temperature and light or decreased humidity, that extend tens or hundreds of meters from an edge, depending on the environmental variable, the physical nature of the edge, and weather conditions (reviewed by Saunders et al. 1991). Table 11.1 lists a range of studies that have investigated abiotic edge effects.

Biotic Edge Effects

Biological factors that affect ecological communities across a boundary such as diseases, weeds, and predators may penetrate hundreds of meters into vegetation remnants (e.g., Wilcove et al. 1986; Andrén and Angelstam 1988; Laurance 1997). Thus edge effects can significantly influence the distribution and abundance of assemblages of species that inhabit

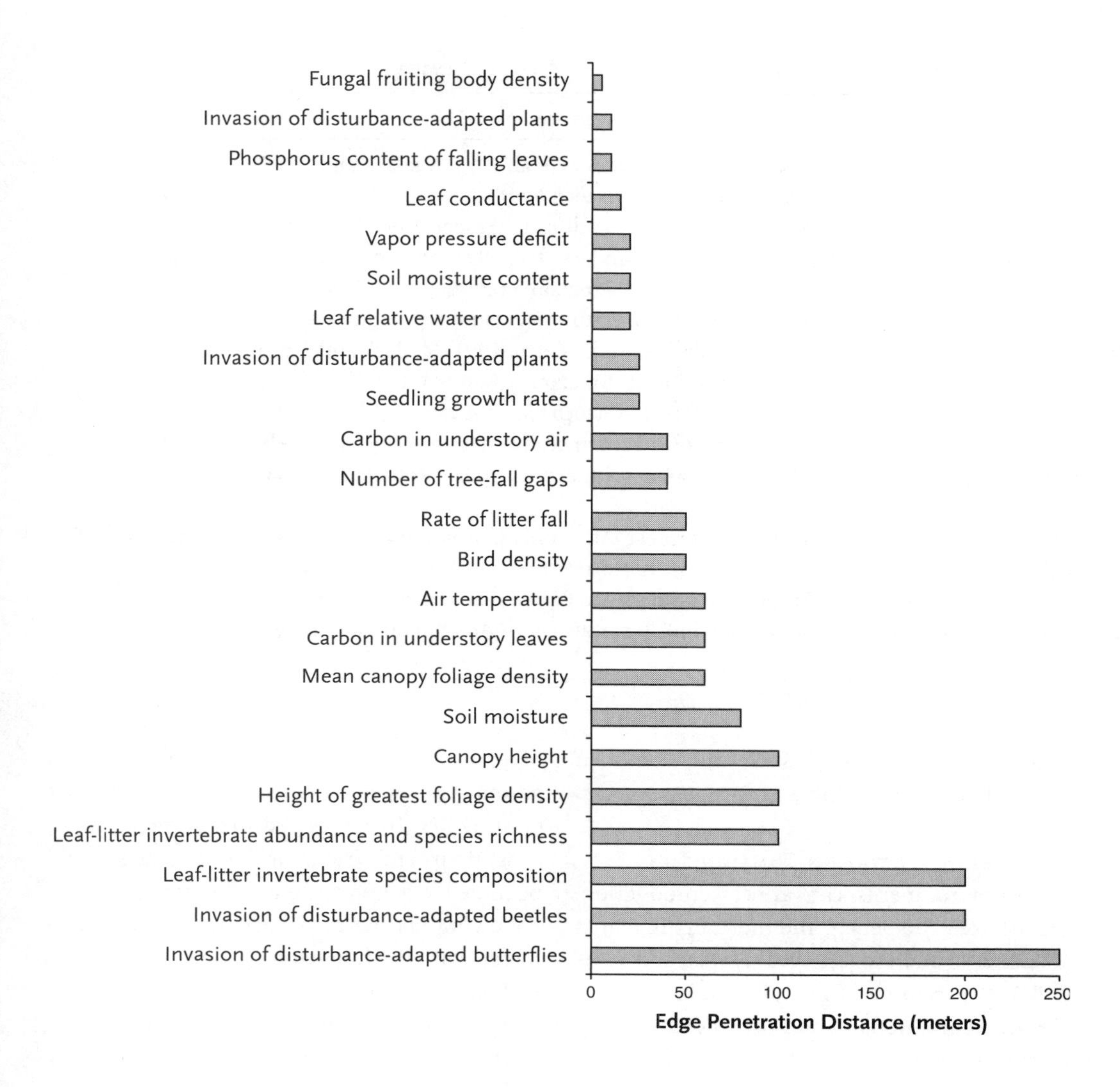

Figure 11.4. Variation in edge penetration for a range of measures in the Biological Dynamics of Forest Fragments Project in Brazil (modified and redrawn from Laurance, W.F., Bierregaard, R.O., Gascon, C., Didham, R.K., Smith, A.P., Lynam, A.J., Viana, V. M., Lovejoy, T.E., Sieving, K.E., Sites, J.W., Andersen, M., Tocher, M.D., Kramer, E.A., Restrepo, C. and Moritz, C. 1997: Fig 32.1: Penetration distances of various edge effects into forest remnants, measured as part of the Biological Dynamics of Forest Fragments Project in central Amazonia, p. 508, Tropical forest fragmentation: synthesis of a diverse and dynamic discipline. pp. 502–525. In: *Tropical Forest Remnants. Ecology, Management and Conservation of Fragmented Communities.* [W.F. Laurance and R.O. Bierregaard, Eds], published by, and reproduced with permission from, The University of Chicago).

Table 11.1. Examples of abiotic edge effects

Abiotic edge effect	*Ecosystem, location, reference*
Modification of wind speeds	Review, global (Saunders et al. 1991; Matlack and Litvaitis 1999)
	Temperate forest, western North America (Chen et al. 1990, 1992)
	Midboreal forest, Scandinavia (Esseen 1994; Zeng et al. 2004)
Modification of temperature regimes and humidity	Review, global (Saunders et al. 1991, Matlack and Litvaitis 1999)
	Tropical rainforest, Amazonia (Lovejoy et al. 1984)
	Temperate forests, southern Australia (Margules et al. 1994)
Modification of light fluxes	Review, global (Saunders et al. 1991, Matlack and Litviatis 1999)
	Tropical rainforest, Amazonia (Kapos 1989)
	Temperate forest, western North America (Chen et al. 1995)
	Temperate forest, southwestern United States (Meyer et al. 2001)
Changes in localized precipitation and frost intensity	Plantation forest, southern Australia (Roberts 1973)
Altered fire ignition probabilities	Tropical rainforest, Amazonia (Lovejoy et al. 1984; Kauffman and Uhl 1991)
Altered levels of nutrients, chemicals	Hardwood forest, North America (Yanai 1991; Weathers et al. 2001)
Accelerated levels of windthrow and disrupted landscape connectivity between vegetation remnants	Tropical rainforest, Amazonia (Lovejoy et al. 1986)
	Temperate forest, southeastern Australia (Lindenmayer et al. 1997)

vegetation remnants (Esseen and Renhorn 1998; Fletcher 2005; Table 11.2).

Changes in Vegetation Communities

Edge environments can affect reproduction, growth, seed dispersal, and mortality in plants (Cadenasso and Pickett 2001; Hobbs and Yates 2003). For example, Chen (1991) observed increased reproduction and growth of surviving mature trees in old-growth forests bordering recently clearfelled areas in the Pacific Northwest of the United States. Similarly, weed invasion is a major biotic edge effect in many heavily disturbed landscapes (Brothers and Spingarn 1992; Beer and Fox 1997). Altered microclimatic conditions at edges may make conditions particularly favorable for the growth of nonnative plants (Fox and Fox 1986; Carr et al. 1992; Honnay et al. 2002). Moreover, regenerating vegetation and patch edges often experience a seed rain of environmental weeds and other introduced plants that are frequently better adapted to exposed and disturbed environments (Janzen 1983; Yates et al. 2004).

In the case of forests, based on their model of edge influence that included primary and secondary responses to edge creation, Harper et al. (2005) hypothesized that some kinds of ecosystems and vegetation communities would be more likely to suffer edge effects than others. Following Harper et al. (2005), these included forests with the following characteristics:

- High mean air temperatures
- Low levels of cloud cover (particularly in the growing season)
- Frequent windy conditions
- Major stand-replacing disturbance events
- Shallow soils
- Tall dense canopies
- Abrupt edges
- Regional flora and fauna consisting of many pioneer, exotic, and/or invasive taxa
- Natural landscapes exhibiting low levels of heterogeneity in vegetation cover, topography, and soil type

Changes in Animal Communities

Several studies have demonstrated the existence of altered animal community composition at

Table 11.2. Examples of biotic edge effects

Biotic edge effect	*References*
Increased nest predation and brood parasitism in birds and reptiles	Paton 1994; Robinson et al. 1995; Lahti 2001; Kolbe and Janzen 2002; Huhta et al. 2004
Impaired breeding success in birds and other animals	Brittingham and Temple 1983; van Horn et al. 1995; Chalfoun et al. 2002; Zanette et al. 2005
Altered patterns of animal and plant dispersal	Ries and Debinski 2001; Schultz and Crone 2001; Ness 2004
Altered patterns of behavior such as nest building and food gathering	Rodriguez et al. 2001; Brotons et al. 2001; Doherty and Grubb 2003
Lowered rates of fledging success among birds	Zanette et al. 2005
Reduced habitat quality for individual species	Anderson and Burgin 2002; Watson et al. 2004; Wolf and Batzli 2004
Increased numbers of browsing and/or game animals	Johnson et al. 1995; Lidicker 1999; Wahungu et al. 1999; Cadenasso and Pickett 2001
Increased human hunting of large vertebrates	Revilla et al. 2001; Saj et al. 2001
Altered invertebrate community composition	Bellinger et al. 1989; Hill 1995; Magara 2002; Kitahara and Watanabe 2003
Altered levels of insect activity	Simandl 1992; Magara 2002
Increased levels of weed invasion	Brothers and Spingarn 1992; Burdon and Chilvers 1994; Lindenmayer and McCarthy 2001; Yates et al. 2004
Altered density, reproduction, growth, and mortality in native plants	Esseen and Renhorn 1998; Hobbs and Yates 2003; Lienert and Fischer 2003; Harper et al. 2005
Altered composition of soilborne bacterial and fungal populations	Jha et al. 1992; Bradshaw 1992

edges (e.g., Hansson 1998; Berry 2001; Laurance 2004) including assemblages of invertebrates (e.g., Bellinger et al. 1989; Hill 1995; Magara 2002). For instance, amphipods responded negatively to changes in wind, moisture, and temperature at the edges of forest fragments in southeastern Australia (Margules et al. 1994). Similarly, some vertebrates avoid forest edges and are classified as "interior forest" species (Gates and Gysel 1978; Robinson et al. 1995). Many forest bird species fall into this category (Terborgh 1989; Frumhoff 1995), in part because arthropod densities may be lower near edges (Burke and Nol 1998; Zanette and Jenkins 2000) or because foraging efficiency may be impaired by edge conditions (McCollin 1998; Huhta et al. 1999).

Another common reason for species avoiding edges is that higher abundances of predators and parasites at edges can have negative impacts on prey and host species near edges (Andrén 1992; Robinson et al. 1995; Chalfoun et al. 2002; Box 11.1). For example, bird populations inhabiting edge environments can experience lower rates of successful pairings and impaired breeding success compared with populations of the same taxa occupying other parts of a landscape (Brittingham and Temple 1983; van Horn et al. 1995; Kurki et al. 2000; Chalfoun et al. 2002). Frequently studied predators in this context include corvids, small mustelids, and other small mammals, such as the red squirrel *(Tamiasciurus hudsonicus)*, as well as larger carnivores, such as the foxes (*Vulpes* spp.) and badger *(Meles meles)* (e.g., Andrén and Angelstam 1988; Small and Hunter 1988; Telleria and Santos 1992; Huhta et al. 2004). By far the most frequently studied nest parasite is the brown-headed cowbird in North America *(Molothrus ater)* (Brittingham and Temple 1983; Zanette et al. 2005).

Box 11.1. Issues Associated with Nest Predation Studies

A widely implemented type of study in human-modified landscapes is to examine rates of nest predation at the boundaries between different types of patches. Nest predation experiments are common, particularly in agricultural landscapes in the Northern Hemisphere where there are large contrasts between vegetation remnants and the surrounding environment (e.g., Andrén 1992; Bayne and Hobson 1997; Hannon and Cotterill 1998).

However, a wide range of factors can limit the generality of the outcomes from these studies (reviewed by Major and Kendal 1996). For example, many studies are confined to a short period of time (e.g., 2 weeks) and use artificial nests that are not attended by adult birds (Piper et al. 2002; Roos 2002; Figure 11.5). Relatively few studies collect data on predation levels of natural nests, making it impossible to contrast results with rates observed for artificial ones (but see for example Zanette and Jenkins 2000; Stuart-Smith and Hayes 2003). Quail eggs are often used in nest predation studies, but it is possible that some predators are unable to break and eat them (Roper 1992). Moreover, the study period is often less than the typical egg incubation time of most birds, and when longer periods of exposure time are employed edge effects may no longer remain significant (Batary and Baldi 2004).

These potential problems highlight the fact that predation rates in experimental studies may not always be validly extrapolated to natural nests (Major and Kendal 1996; Batary and Baldi 2004; Fraser and Whitehead 2005).

In the Northern Hemisphere, some game species have strong positive preferences for edge environments where foraging and cover are close together (Patton 1974; Matlack and Litvaitis 1999). Large populations of white-tailed deer *(Odocoileus virginianus)* heavily graze food plants at forest edges (Johnson et al. 1995) including endangered taxa (Miller et al. 1992). This can limit stand regeneration (Tilghman 1989) and alter patterns of species diversity (McShea and Rappole 2000). In other regions, rare or threatened species exhibit strong positive associations with edges such as the boundaries between forests and more open vegetation types. Rare butterflies around Mount Fuji in Japan provide one of many examples (Kitahara and Watanabe 2003).

Variation in Edge Effects

Regional Variation

Strong spatial variation in edge effects has been recognized by a range of authors (Rudnicky and Hunter 1993; Matlack and Litvaitis 1999; Batary and Baldi 2004) and has even been used to hypothesize which ecosystems and vegetation communities are likely to be most prone to edge effects (Harper et al. 2005; see the earlier section on vegetation communities). As an example, elevated nest predation and rates of nest parasitism show strong patterns of variation between locations (Batary and Baldi 2004). Indeed, while such effects are commonly associated with edges, patterns are far from universal (Berg et al. 1992; Hanski et al. 1996, reviewed by Kremsater and Bunnell 1999). Forest edges on the boundary of clearcuts in Sweden have lower bird species diversity than continuous forest areas (Hansson 1983), but no such pattern has been detected at some clearcut forest edges in northeastern United States (Rudnicky and Hunter 1993). Similarly, patterns of brood parasitism characteristic of edge environments in many temperate Northern Hemisphere landscapes appear to be rare in Australia (Piper et al. 2002) and are less pronounced in central Europe than they are in northwestern Europe (Batary and Baldi 2004). They are also not prominent in many tropical landscapes (Stratford and Robinson 2005). Even within North America, increased nest predation and brood parasitism as observed on the eastern side of the continent are less common in the northwest (Kremsater and Bunnell 1999; Marzluff and Restani

Figure 11.5. An artificial nest with a quail egg used in a nest predation experiment at Tumut, southeastern Australia (photo by Matthew Pope).

Box 11.2. Edge Effects in Brazilian Rainforest

Edge effects have been studied in detail in the Biological Dynamics of Forest Fragments Project (BDFFP) in north-central Brazil. When edges were created at the boundary between primary rainforest and adjacent pasture, light penetrated the forest both vertically and horizontally, altering microclimatic conditions, such as temperature regimes and relative humidity (Kapos 1989; Sizer and Tanner 1999). Wind speeds also changed dramatically at forest–matrix edges (Lovejoy et al. 1986). The extent that edge effects penetrated varied substantially depending on the measured attributes (see Figure 11.4) such as leaf fall, tree mortality, seedling recruitment, or, ultimately, plant species composition (Laurance et al. 1997). Laurance et al. (2001) reported dramatic impacts on the liana community structure. Increased ecophysiological stresses on trees through altered light and nutrient regimes led to heavy infestations of lianas and contributed to significantly increased tree damage and mortality, especially near the edges of rainforest remnants. Edge effects in the BDFFP varied depending on whether the surrounding matrix was cleared pasture or regrowth forest recovering from clearing. For example, annual tree mortality in primary rainforest remnants was significantly higher if the surrounding vegetation was pasture compared with regrowth forest (Mesquita et al. 1999).

1999). The reasons underpinning regional differences in edge effects are not clear. Regional differences in vegetation types might account for some of the observed differences. For example, levels of canopy density and cover are lower in many kinds of Australian eucalyptus forest than they are in tropical forests or North American and European coniferous and deciduous forests, and this may mean that edge contrasts with surrounding areas such as cleared fields will be lower.

The Influence of the Matrix

As noted earlier for birds, the magnitude of many types of edge effects is related to the strength of the contrast between the matrix and other landscape units; where the contrast is strong there will often be more intense interactions and edge effects (Laurance et al. 1997; Mesquita et al. 1999; see Box 11.2). The extent of the area supporting high contrast conditions influences the magnitude of edge effects. For example, microclimate edge effects in forests may be greater where a large clearfelled area abuts a retained patch than where a cutover is small (Lindenmayer et al. 1997). Elevated nest predation is often observed in agricultural landscapes where there are strong contrasts between vegetation remnants and the surrounding environment (e.g., Andrén 1992; Bayne and Hobson 1997, 1998; Hannon and Cotterill 1998). Such effects may even be stronger in urban areas than in agricultural ones, where matrix contrasts are extremely pronounced (Wilcove 1985). Conversely, nest predation is less pronounced at edges in landscapes where contrasts are weaker (Schmiegelow et al. 1997) such as continuous eucalyptus forest dissected by minor bush tracks or native hardwood forest juxtaposed with softwood plantations (Sargent et al. 1998; Lindenmayer et al. 1999e).

Edge Sensitivity and Extinction Proneness

Edge effects vary markedly between species (Schlaepfer and Gavin 2001) and some authors have argued that edge-sensitive species may be among those at particular risk in heavily modified environments (Lehtinen et al. 2003). Laurance and Yensen (1991) developed a model (the core-area model) that attempted to calculate the amount of interior habitat within a vegetation remnant. From this, it might then be possible to identify edge-sensitive species at risk in altered landscapes. The underlying premise of this approach is that species which are edge-avoiders and/or respond negatively to edge effects will be more extinction prone (Laurance 1999). Although the ecological and conservation literatures are replete with numerous studies of edge effects, relationships between extinction proneness and edge responses have almost never been demonstrated empirically. A notable exception is the work of Lehtinen et al. (2003), who found that edge-avoiding species of reptiles and frogs in Madagascar were less likely to persist within forest remnants than taxa insensitive to boundaries with the surrounding modified landscape matrix. Conversely, species that frequently inhabited edges or were ubiquitously distributed were extinction resistant (Lehtinen et al. 2003; see Figure 11.6).

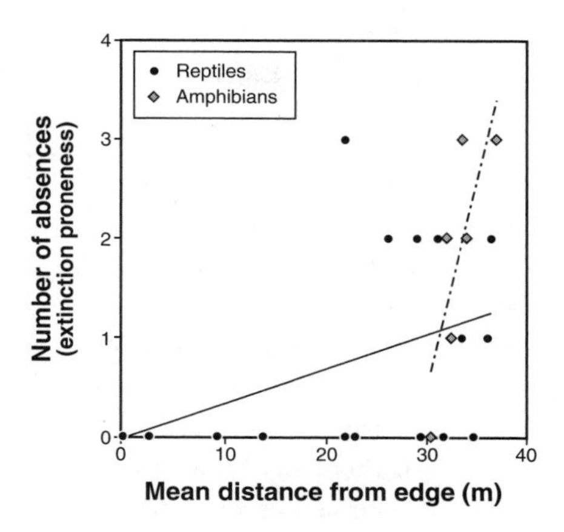

Figure 11.6. Relationships between extinction proneness of various species of reptiles and frogs (defined as a function of the number of absences from remnant patches) and the mean distance from the edge of forest fragments in Madagascar (redrawn from Lehtinen, R.M., Ramanamanjato, J-B. and Raveloarison, J.G. 2003: Fig. 2. The number of observed absences [local extinctions] of each species versus the mean distance from the edge [m] along 50 m transects in the dry season [1999], p. 1365, *Biodiversity and Conservation,* **12**, 1357–1370, Edge effects and extinction proneness in a herpetofauna from Madagascar, with kind permission of Springer Science and Business Media).

Predicting Edge Effects

To date it appears there is limited predictive ability in attempts to forecast accurately which species will be edge sensitive in landscapes subject to significant human modification (Murcia 1995). This appears to be a result of several factors. First, there are markedly different results from studies of edge effects from, for example, different ecosystem types (e.g., forests versus woodland), similar ecosystem types in different regions (e.g., central and western North America; Kremsater and Bunnell 1999), the same ecosystem but where the surrounding landscape matrix varies (Bayne and Hobson 1997), and different species in the same ecosystem (Margules et al. 1994; Lehtinen et al. 2003). In addition, even within the same local area, abiotic edge effects can vary between aspects (Hylander 2005) and seasons (Gascon 1993; Parry 1997; Schlaepfer and Gavin 2001), and the response of individual species can change markedly as a result (Gambold and Woinarski 1993; De Maynadier and Hunter 1998; Lehtinen et al. 2003). Similarly, the shape of vegetation patches as well as the distance from the edge may influence the intensity of edge effects (Box 11.3). Some authors have argued that ecologists are often lacking sufficient appreciation of the range of spatial and temporal complexities that characterize the type and magnitude of edge effects (e.g., Murcia 1995).

A second factor limiting predictive ability lies with the use of markedly different methods between different investigations. This leads to major problems in making rigorous meta-analytical comparisons across investigations (Murcia 1995). A third problem is that many factors influence the distribution and abundance of organisms, leading to high variability in datasets and resulting statistical relationships. For example, although Lehtinen et al. (2003) demonstrated significant extinction proneness for edge-sensitive Madagascan herpetofauna, there were several exceptions to their general results. Such variability can limit the ability to make accurate forecasts of extinction-prone species.

Box 11.3. Edge Effect and Patch Shape

There can be important relationships between the manifestation of edge effects and the shape of vegetation patches. Linear patches may be more prone to abiotic and biotic edge effects than elliptical or circular ones. On this basis, some authors have considered that circular reserves will be superior to linear reserves (Diamond 1975a; Blouin and Connor 1985). For example, Reading et al. (1996) considered that round reserves would be better reintroduction sites than linear reserves for the eastern barred bandicoot *(Parameles gnunii)* in the Australian state of Victoria. There may be other indirect kinds of problems related to edge influences and patch shape. Based on the theory of central place foraging, Recher et al. (1987) predicted that colonial species of birds that feed on widely dispersed food may be disadvantaged in narrow vegetation patches. Long, thin areas may not support the minimum home range requirements of a group or allow some types of food to be collected efficiently. Such predictions were consistent with empirical results for arboreal marsupials inhabiting linear strips of retained forest in southeastern Australia (Lindenmayer et al. 1993).

Although some authors have argued that there currently appears to be a limited ability to predict edge-sensitive ecosystems (e.g., Box 11.4), vegetation communities, and species, a recent review of the topic by Ries et al. (2004) proposed an alternative view. They introduced a conceptual model of edge effects that considered four mechanisms underlying ecological changes near edges: changed ecological flows, species accessing spatially separated resources, resource mapping, and species interactions. Using these four broad categories as a conceptual basis, Ries et al. (2004) argued that ecological responses to edges were far less idiosyncratic and unpredictable than often suggested. Rather, the correct prediction of edge effects depended on the careful definition of the types of processes under consideration (Ries et al. 2004). A similar logic is inherent in the edge influence model by Harper et al. (2005; see Figure 11.2) from which they developed hypotheses about which vegetation communities were likely to be most susceptible to edge effects (see earlier discussion). These hypotheses have yet to undergo rigorous testing, but they at least represent a formal mechanism to explore the generality of some kinds of edge effects.

Caveats

It is often forgotten in studies of edge effects that humans define patch boundaries. These "boundaries" may or may not be relevant to particular species or processes. What actually defines a patch boundary for a particular species will be defined by that species. For example, edge effects might mean that a 60 ha area of native forest contains only 30 ha of suitable habitat for a species that avoids edge environments. In other cases, individuals of a species will move well beyond a patch boundary into the adjacent matrix to gather additional resources—sometimes termed landscape supplementation or a "halo" effect (Figure 11.7). In southeastern Australia, birds are more likely to occur in stands of radiata pine *(Pinus radiata)* when they are adjacent to patches of native eucalyptus forest (Tubelis et al. 2004). Similarly, native mammals that are regarded as pests by forest managers often move from remnant native forest patches into adjacent planted stands and browse and defoliate trees (Bulinski 1999; McArthur 2000). In these cases, patches exert "reverse edge effects" onto the matrix. In many cases, the level of contrast between different adjacent patch types has a major effect on the ecological interactions at edges. Often, the boundary between two habitat types may be much more gradual than the sharp edges defined by mapping tools like geographic information systems (see Kremsater and Bunnell 1999; Dangerfield et al. 2003).

There are other important considerations in the study and interpretation of edge effects. First, because the nature and strength of edge effects are species-specific (Hansson 2002), not all elements of an assemblage will respond to boundaries in the same way. Some will be edge avoiders, others may be interior avoiders, and others will not respond to boundaries in a positive or negative way (e.g., Noss 1991; Anderson and Burgin 2002; Magara 2002). Moreover, many kinds of abiotic edge effects are likely to display marked temporal variability. As climate regimes vary throughout the year, including at boundaries (Parry 1997), biotic edge effects also may exhibit seasonal changes. For example, in Madagascar, Lehtinen et al. (2003) found that a suite of reptiles and amphibians that avoided edges during the dry season showed no response to boundaries in the wet season. Similarly, amphibians in northeastern North America are most sensitive to edge effects during the drier parts of the year (Gibbs 1998). Such seasonal variations in edge effects are rarely considered (Schlaepfer and Gavin 2001). Finally, edge effects are sometimes regarded as univariate entities, but in fact many factors can influence the magnitude and nature of edge effects, including the existence of multiple edges in the same local area (Fletcher 2005). As an example, in Australia, aggressive interactions at edges between small native honeyeaters and the noisy miner *(Manorina melanocephala)* were influenced by the body size of small honeyeaters, the remnant patch size, and the density of understory vegetation within areas of remnant native vegetation (Piper and Catterall 2003).

Box 11.4. Surprises and Exceptions

Much of this chapter has discussed general principles governing edge effects. Levels of contrast between remaining native vegetation patches and the adjacent matrix have been found to be important in many studies. However, as in many aspects of ecology there can be exceptions.

An example comes from the mid-elevation forests of Costa Rica, which abut pastures of exotic grasslands established for livestock grazing. There appear to be few major biotic edge effects on bird communities in the adjacent undisturbed forests, perhaps because the avian assemblages typical of rainforests and pastures are so markedly different they do not penetrate the abutting areas of vegetation, and interactions are limited. Conversely, partially cleared pastures with remnant large trees and a diverse understory result in lower patch–matrix contrasts (sensu Forman 1995; see Chapter 2), and bird assemblages can have more taxa in common. In these landscapes, pronounced biotic edge effects occur more commonly. Some abundant pasture species remain common more than 100 meters from the vegetation boundary into the adjacent rainforest (Sisk and Zook 1996; Tom Sisk, pers. comm.).

Summary

Human landscape modification often results in an increase in perimeter to area ratios of remaining patches of native vegetation. The boundaries created between remaining vegetation patches and the surrounding matrix can be subject to edge effects. Two broad types of edge effects are recognized—abiotic and biotic edge effects. Abiotic edge effects are primarily concerned with altered microclimatic conditions (e.g., modification of wind speeds, light fluxes, and temperature regimes). Biotic edge effects may refer to changes in ecological processes, community composition, or species interactions. Examples of biotic edge effects include changes in the rate of organic matter decomposition, increased nest predation and brood parasitism, and lowered rates of fledging success among birds whose territories are located at edges. Notably, not all species respond negatively to edges, and some taxa can be more common at patch boundaries than elsewhere in a landscape.

Land managers may deliberately manipulate vegetation cover to promote the creation of edges as part of conservation management for a given species or assemblage of taxa. In the United Kingdom and parts of Europe, open linear grassy areas are maintained within forests and woodlands to ensure suitable areas are provided

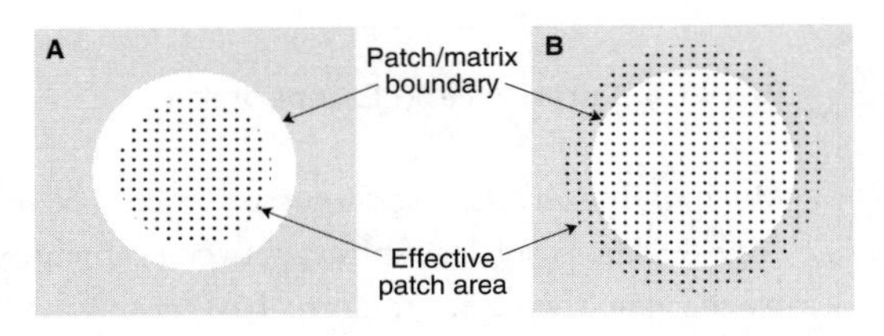

Figure 11.7. Negative and "reverse" edge effects. In Part A processes from the matrix reduce functional patch size for the species or assemblage of interest. In Part B suitable conditions in the matrix enable members of an assemblage to increase functional patch size.

for species such as plants (Peterken and Francis 1999), and reptiles (Jäggi and Baur 1999).

The magnitude of edge effects varies in response to a number of factors, such as the type of ecosystem and the level of contrast between remaining patches and the surrounding matrix. There is no generic edge depth, and the intensity of edge effects will vary with the processes and species in question. Edge effects also vary between locations—some types of edge effects, such as increased nest parasitism found in the midwestern United States and northern Europe, appear to be less important elsewhere (e.g., western North America, Australia, central Europe, and tropical environments).

It remains unclear if it is possible to predict accurately which ecosystems, vegetation communities, and individual species are most susceptible to edge effects. However, recent conceptual models that attempt to capture the range of processes, responses, and interspecies interactions that can occur across patch boundaries indicate that edge effects may be more predictable than many authors had earlier proposed.

Links to Other Chapters

Chapter 2 discusses issues associated with altered patterns of landscape cover that typically accompany human landscape modification. Edge effects are often particularly strong in small patches of remnant vegetation (Chapter 9). Not all species respond negatively to edges. Chapter 8 discusses some aspects related to a species' extinction proneness in response to landscape modification.

Further Reading

Recent in-depth reviews of edge effects are provided by Lidicker (1999), Ries et al. (2004), and Harper et al. (2005), and they provide interesting conceptual models for predicting the impacts of edge effects on vegetation communities and species. Saunders et al. (1991), Laurance et al. (1997), Kremsater and Bunnell (1999), Matlack and Litvaitis (1999), and Lindenmayer and Franklin (2002) give syntheses of edge effects. Rudnicky and Hunter (1993), Paton (1994), Lidicker (1999), Lahti (2001), Stephens et al. (2003), Batary and Baldi (2004), and Parker et al. (2005) provide detailed reviews of nest predation and parasitism and other kinds of biotic effects that manifest near edges.

CHAPTER 12
Landscape Connectivity

In Chapter 6 we discussed how landscape modification can produce negative habitat isolation effects on the individuals and populations of a given species. Especially in combination with other threatening processes such as habitat loss, increased habitat isolation can lead to population declines of some species and increase their proneness to extinction (Chapter 8). The vegetation cover patterns created by human landscape modification can also result in landscapes becoming subdivided for suites of taxa and even entire assemblages (Noss 1991). In this chapter, we discuss the implications of losses of landscape connectivity for suites of species in human-modified landscapes. First, we define landscape connectivity and contrast it with other related connectivity concepts. We then provide a brief overview of how landscape connectivity can be quantified, and discuss the negative effects of reduced landscape connectivity on assemblages and ecosystem processes. Finally, we outline three types of features that contribute to landscape connectivity: corridors, stepping stones, and a "soft" matrix.

Landscape Connectivity and Other Connectivity Concepts

"Connectivity" is a widely used term in the literature on landscape change. Although there is general agreement that maintaining or enhancing connectivity is important (Taylor et al. 1993), different aspects of connectivity are often confused. Along with the term "habitat" (Chapter 4), "connectivity" therefore is a central aspect of Bunnell's (1999a) fragmentation panchreston (Chapter 1). That is, it is a concept that is universally agreed on to be important but that is often conceived too broadly, thereby rendering it difficult to use in practice and sparking much academic debate (e.g., on the ecological value of wildlife corridors; see Simberloff et al. 1992; Beier and Noss 1998).

To clarify various themes associated with connectivity, we make an explicit distinction between three types of connectivity. First, we define "habitat connectivity" as the connectedness between patches of suitable habitat for an individual species; it is the opposite of habitat isolation (see Chapter 6). Second, we define "landscape connectivity" as a human perspective of the connectedness of patterns of vegetation cover in a given landscape (e.g., Figure 12.1). Third, we define "ecological connectivity" as the connectedness of ecological processes across multiple scales. As illustrated in Table 12.1, ecological connectivity includes processes related to trophic relationships, disturbance processes, and hydroecological flows (Soulé et al. 2004). Although these three connectivity concepts are interrelated (Figure 12.2), they are not synonymous with one another. Landscape connectivity may increase habitat connectivity for some species but not for others (Box 12.1). Similarly, low habitat connectivity for functionally redundant species (sensu Walker 1992; see Chapter 15) may have relatively little impact on the overall connectedness of ecological processes. However, for other species that fulfill irreplaceable ecological functions, the loss of habitat connectivity can have major impacts on ecological connectivity. For example, some bat and bird species in Central America are in-

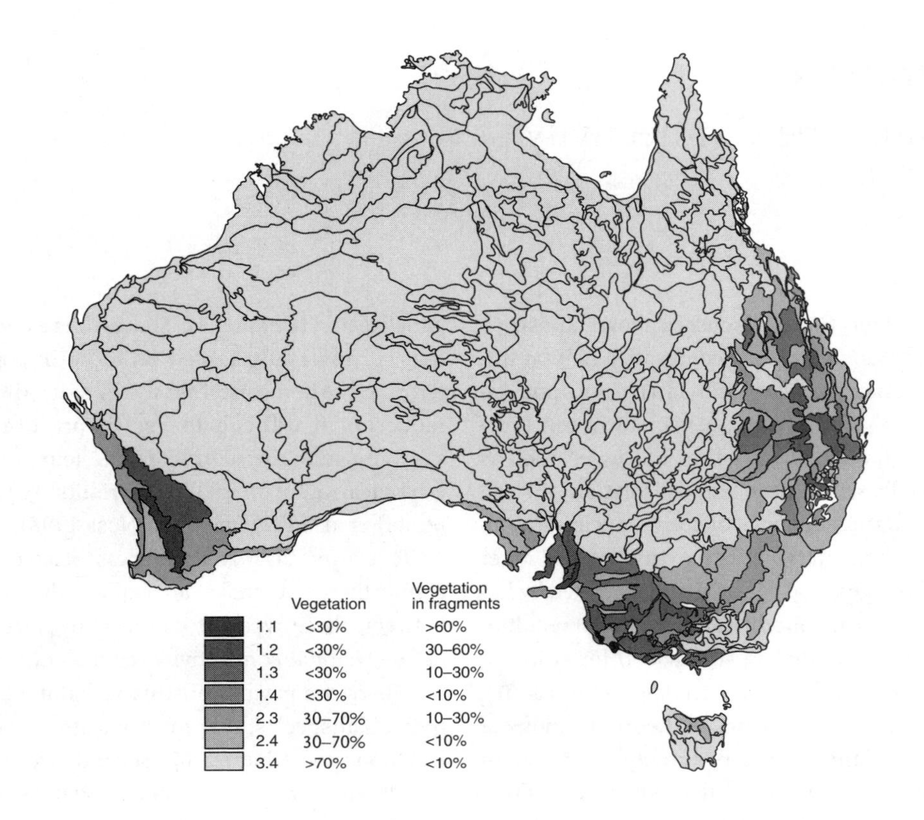

Figure 12.1. A map of Australia showing patterns of vegetation cover in different bioregions and a measure of landscape connectivity for vegetation in the intensive land use zone (IUZ) (redrawn from Commonwealth of Australia (2002). *Australia's natural resources 1997–2002 and beyond.* National Land and Water Resources Audit, Commonwealth of Australia, Canberra.

strumental in dispersing the seeds of rainforest plants across agricultural areas (Galindo-González et al. 2000), thereby contributing to the genetic viability of plant populations (Cascante et al. 2002). The loss of habitat connectivity for these vertebrate species would have severe implications for ecological connectivity because the key ecological process of seed dispersal would be lost, with likely negative consequences for numerous plant species and the animals that depend on them (also see Chapter 15 on cascading effects of landscape change).

Consistent with other chapters in Part III, this chapter considers a human perspective of ecosystems, with a primary focus on landscape connectivity. Where appropriate, we identify links with habitat connectivity and ecological connectivity.

Quantifying Landscape Connectivity

Landscape connectivity can be considered to be one of three key interrelated attributes that can be used to describe patterns of vegetation cover in a landscape—the other two are typically composition (or the identity and characteristics of the different types of patches), and configuration (or the spatial arrangement of the patches). There is a vast array of indices that can be used to characterize patterns of landscape cover, including ones that quantify landscape connectivity (Turner 1989; Wegner 1994; reviewed by Haines-Young and Chopping 1996; Gustafson 1998; Moilanen and Nieminen 2002). For instance, O'Neill et al. (1988) devised a "spatial contagion" index to measure the extent and adjacency of K landscape units or patches:

$$D_2 = 2K \, In(K-1) \sum_{i=1}^{K} \sum_{j=1} p_{ij} \, Inp_{ij}$$

where p_{ij} is the probability of land use type i being found adjacent to land use type j. Landscapes with smaller and more dispersed patches have lower values. Similarly, Knaapen et al. (1992) developed a connectivity measure to capture organisms' ability to traverse a landscape successfully. If vegetation patches are discrete, then Knaapen et al. (1992) considered that landscape connectivity could be quantified as a function of the number of connections between these patches, relative to the maximum number of potential connections. Thus, if there are direct links between all N patches in a landscape, there will be $N(N-1)/2$ such links. Landscape connectivity (C) of patches and links can then be calculated as follows:

$$C = \frac{2L}{N(N-1)}$$

where L is the number of links and N is the number of patches. A fully connected landscape with complete levels of landscape connectivity would have a value of 1, and a landscape that has no landscape connectivity would have a value of 0. Knaapen et al. (1992) noted

Box 12.1. Landscape Connectivity versus Habitat Connectivity

Landscape connectivity reflects human perceptions of the vegetation cover patterns of a landscape. A given landscape pattern can correspond to low habitat connectivity for some species but high habitat connectivity for other species, even within the same assemblage. In the Tumut Fragmentation Experiment in southeastern New South Wales, species such as the red wattlebird *(Anthochaera carunculata)* and golden whistler *(Pachycephala pectoralis)* were among a suite of taxa significantly less abundant in areas with low levels of landscape connectivity and many subdivided remnant vegetation patches, than in areas where remaining areas of native eucalypt forest were consolidated as a small number of contiguous stands (Lindenmayer et al. 2002b). In contrast, the common ringtail possum *(Pseudocheirus peregrinus)* and the crimson rosella *(Platycercus elegans)* showed the reverse response, possibly because they are edge-attracted species, and the longer boundaries created in more subdivided landscapes made these landscapes more suitable for them. This emphasizes the fact that higher levels of landscape connectivity as perceived by humans will not always directly correspond to higher levels of habitat connectivity for a given individual species, or vice versa. It also reinforces the rationale for the key distinction we make in this book between habitat connectivity for individual species and landscape connectivity as the human-defined connectedness of vegetation cover within a landscape.

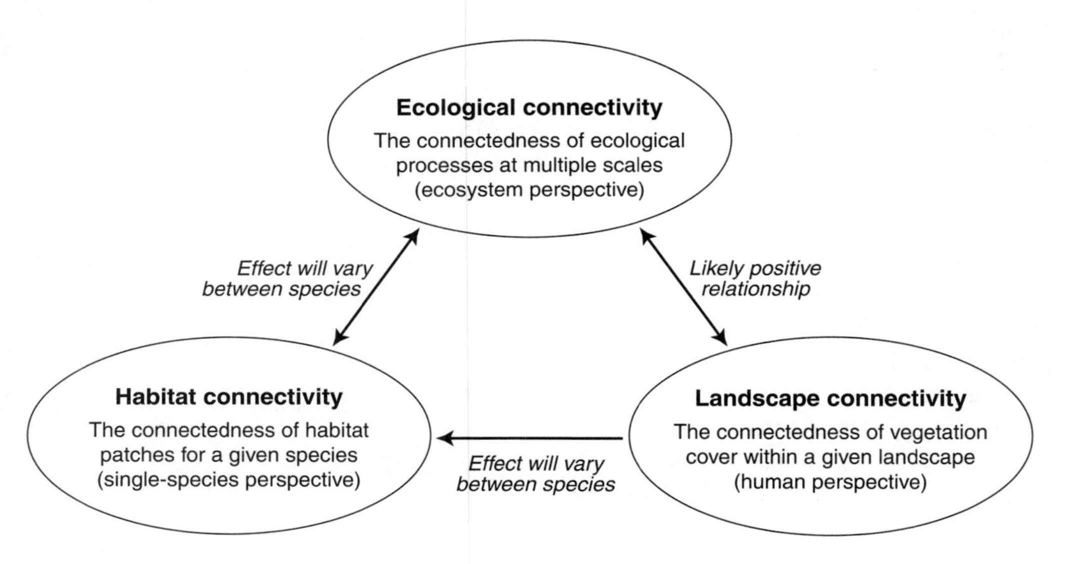

Figure 12.2. The interrelationships between three connectivity concepts.

Table 12.1. Aspects related to ecological connectivity*

Ecological process	*Explanation*
Critical species interactions	The dispersal and movement routes of highly interactive species (e.g., pollinators, large carnivores) are part of ecological connectivity
Disturbance at local and regional scales	Both natural and anthropogenic disturbance (e.g., fire, grazing, land clearing) can disrupt or maintain ecological connectivity
Long distance biological movement	Large-scale migration (seasonal or resource driven) of animals can be part of ecological connectivity
Hydroecological processes	Hydroecology relates to the role of vegetation in regulating hydrological flows, such as water infiltration, subsurface water flows, and runoff; the spatial continuity of these processes is part of ecological connectivity
Coastal zone flux	Natural flows in the coastal zone (e.g., migrating shorebirds, freshwater ecosystems) can be disrupted by human activities, representing a loss of ecological connectivity
Global climate change	Species may need to shift their ranges to adapt to climate change; such range shifts are part of ecological connectivity
Spatially dependent evolutionary processes	Long-term ecological connectivity is needed to maintain the evolutionary potential of biodiversity

*Based on Soulé et al. (2004).

that their index could be applied to quantify landscape connectivity by quantifying the connectedness of vegetation patches. However, it also may be used to quantify habitat connectivity or habitat isolation for individual species if it appropriately reflects the links between populations of a given species (Knaapen et al. 1992)—which depends on the habitat preferences, biology, and dispersal abilities of that taxon (see Chapter 6).

The preceding equations are just two of numerous indices that can be employed to calculate and quantify landscape connectivity (see Haines-Young and Chopping 1996; Tischendorf and Fahrig 2000a; McAlpine et al. 2002a), but most are based on similar principles of distances between vegetation patches, the numbers of physical connections that link them, and sometimes also the sizes and shapes of those patches (e.g., Schtickzelle and Baguette 2003). A meta-analysis by Moilanen and Nieminen (2002) recognized three broad kinds of connectivity measures: (1) simple nearest neighbor measures, (2) buffer measures where occupied patches within a limited neighborhood of a given patch are considered equally (with no effect of distance), and (3) complex measures that take account of distances to all possible population sources. Simple measures such as nearest neighbor ones performed poorly in predicting the response of what is important for particular species, and often failed to detect significant effects in comparison with more sophisticated metrics (Moilanen and Nieminen 2002). These findings highlight that landscape connectivity as defined by humans is sometimes only loosely related to particular species. On this basis, Moilanen and Nieminen (2002) recommended the use of complex measures that take into account the movement abilities of the species of interest, thus attempting to measure aspects of "habitat connectivity" rather than focusing only on "landscape connectivity" (see Figure 12.2). An example of a complex measure that uses a negative exponential dispersal kernel is:

$$S_i = \sum_{j=1} \exp(-\alpha d_{ij}) A^b_j$$

where S_i is the connectivity of patch i, d_{ij} is a measure of the distance between patches i and j, A is patch area of patch i, α is a parameter for scaling the effect of distance on migration, and $1/\alpha$ is the average migration distance.

Importantly, some caution is needed when applying landscape indices because there can often be considerable redundancy with other indices that essentially capture similar information about patterns of vegetation cover (see Chapter 17). In addition, some landscape connectivity measures can produce misleading perceptions. For example, for some indices, higher (and misleadingly "better") connectivity values are sometimes obtained in more modified landscapes than in intact ones because a larger number of small patches can result in greater patch interception rates (Tischendorf and Fahrig 2000b).

Negative Effects of Reduced Landscape Connectivity

Landscapes that retain more connections between patches of otherwise isolated areas of vegetation and therefore have higher levels of landscape connectivity are assumed to be more likely to maintain populations of various species that inhabited the original landscape (Brown and Kodric-Brown 1977; Haddad and Baum 1999). Conversely, a lack of landscape connectivity can have a range of negative impacts on assemblages. It may result in vegetation patches remaining unoccupied for suites of species (Villard and Taylor 1994; Robinson 1999), meaning that the spatial distribution of these taxa may not directly correspond to the spatial distribution of available habitat for them (Stenseth and Lidicker 1992a; Wiens et al. 1997). This is illustrated by some forest bird taxa that are unable to cross gaps and avoid open areas (Martin and Karr 1986; van Dorp and Opdam 1987; Bierregaard et al. 1992; Desrochers and Hannon 1997). Similarly, sets of species in patches of remnant vegetation where the surrounding matrix is unsuitable for foraging are more likely to suffer extinction than those assemblages where the matrix provides landscape connectivity (Laurance 1991; Viveiros de Castro and Fernandez 2004).

A particular case of reduced landscape connectivity is the "dissection" (sensu Forman 1995; see Chapter 2) of formerly continuous vegetation by roads. Roads can negatively affect a wide range of species, thus fundamentally altering landscape pattern and also reducing habitat connectivity for many individual species and changing ecological processes (reviewed by Spellerberg 1998; see Box 12.2).

Species whose primary habitat does not correspond to human-defined patches of vegetation can also be negatively affected by reduced landscape connectivity because of altered ecological processes. For example, Gray et al. (2004) examined the effect of landscape structure on two species of amphibians in wetland playas of the southern High Plains in central United States—the New Mexico spadefoot (*Spea multiplicata*) and the plains spadefoot (*S. bombifrons*). Reduced landscape connectivity in this landscape was associated with agricultural land use, which also increased sedimentation and decreased the hydroperiods of the playas. These changes to ecological processes, in turn, meant that habitat connectivity for both amphibian species was negatively associated with interplaya distance and positively associated with interplaya landscape heterogeneity (also see Chapter 14).

Finally, reduced landscape connectivity can alter ecological connectivity, thus leading to a range of cascading effects (see Figure 12.2). For example, the loss of landscape connectivity may alter the structure of food webs (Holyoak 2000) and disrupt ecological processes such as the decomposition of wastes (Klein 1989), seed dispersal (Cordeiro and Howe 2003), or polli-

Box 12.2. Roads and Disrupted Landscape Connectivity

There are an estimated 28 million kilometers of highways worldwide (Forman et al. 2002). Roads can have profound negative impacts on landscape connectivity, with potential negative consequences for the habitat connectivity of some species and, more generally, ecological connectivity (Figure 12.2). Roads have been found to impede the movements of a broad range of organisms, including invertebrates (Baur and Baur 1992; Haskell 2000), small mammals (Mader 1984; Burnett 1992), and large mammals (Clevenger and Waltho 2000; Forman and Deblinger 2000; Epps et al. 2005). Even rudimentary roads (tracks and log-skidding paths) can have significant negative impacts on the habitat connectivity of some species (Barnett et al. 1978; Bright 1998).

In addition to plant and animal mortality at the time of road construction, collisions with vehicles are a continuing and often substantial source of mortality for many animals living near roads (Bennett 1991; Trombulak and Frissell 2000). The problem is particularly acute where road systems intersect regular travel or migration routes for animal populations, which is often the case with amphibians moving to and from breeding habitats (Hels and Buchwald 2001), and many turtle species in the United States that regularly move through the landscape during parts of their annual life cycle (Gibbs and Shriver 2002). Reptiles often seek roads for thermal heating, and it has been estimated that approximately 5.5 million frogs and reptiles are killed annually on paved Australian roads (Ehmann and Cogger 1985). There are many examples of individual species for which significant levels of road mortality have been documented. Examples include the red-sided garter snake *(Thamnophis sirtalis parietalis)* in central Canada (Shine and Mason 2004), the eastern quoll *(Dasyurus viverrinus)* and koala *(Phascolarctos cinereus)* in Australia (Lee and Martin 1988; Jones 2000), and the hedgehog *(Erinaceus europaeus)* in the Netherlands (Huijser and Bergers 2000). Similarly, Huggard (1993) found that elk *(Cervus elaphus)* living close to the Trans-Canada Highway were 2.5 years younger than those more than 1 kilometer away. This effect was due to higher rates of animal mortality close to the highway (Huggard 1993).

nation (Paton 2000). Cascading effects of landscape change, including cascades triggered by a loss of landscape connectivity, are discussed in more detail in Chapter 15.

Features Contributing to Landscape Connectivity

There are three broad types of features that contribute to landscape connectivity: (1) wildlife corridors, (2) stepping stones, and (3) a "soft" matrix. Different features will result in increased habitat connectivity for different species and will maintain different aspects of ecological connectivity.

Wildlife Corridors

Wildlife corridors are physical linkages between patches of native vegetation (e.g., Bennett 1998). Wildlife corridors contribute to landscape connectivity and may facilitate increased habitat connectivity for some species (e.g., Bennett 1990b; Hewittson 1997; Beier and Noss 1998), but not others (e.g., Thomas et al. 1990). Many studies have attempted to examine the contribution that wildlife corridors can make to landscape connectivity. A detailed set of studies by Haddad (1999a, b) and colleagues (e.g., Tewksbury et al. 2002; Haddad and Tewksbury 2005; Levey et al. 2005) has explored the responses of a range of biota to the establishment of wildlife corridors in a plantation forest ecosystem in South Carolina, USA (Figure 12.3). Many interesting results have been generated from this pioneering work. For example, wildlife corridors directed the movement of various animal species, although some taxa also moved through the matrix (Haddad et al. 2003). Population densities of several groups of species were significantly higher in connected patches than in isolated ones (Haddad and Baum 1999). Perhaps most significantly, recent work demonstrated that

Figure 12.3. Wildlife corridor and landscape connectivity research site in South Carolina, USA (photo by United States Forest Service).

the landscape connectivity provided by corridors has the potential to enhance both the habitat connectivity of some species and the ecological connectivity of some key ecosystem processes (Box 12.3).

Wildlife corridors are believed to

- facilitate the movement of animals through suboptimal habitat (Palomares et al. 2000)
- provide habitat for resident populations (Lindenmayer et al. 1993; Bennett 1998; Mönkkönen and Mutanen 2003)
- enhance dispersal success (MacMahon and Holl 2001; Kirchner et al. 2003) such as by reducing mortality during dispersal (Beier 1993)
- prevent and reverse local extinctions by recolonization of empty patches (Brown and Kodric-Brown 1977; Fahrig and Merriam 1985; Burbrink et al. 1998)
- promote the exchange of genes between subpopulations (Aars and Ims 1999), increasing the effective population size and reducing genetic drift and inbreeding depression (Newman and Tallmon 2001)

Box 12.3. Wildlife Corridors: Landscape Connectivity, Habitat Connectivity, and Ecological Connectivity

Levey et al. (2005) studied seed dispersal by birds in relation to wildlife corridors in a forest ecosystem in South Carolina, USA. They set up eight experimental landscapes, each of which contained a mix of forest patches connected by wildlife corridors and unconnected patches. The study focused on the wax myrtle *(Myrica cerifera)* and one of its major seed dispersers, the eastern bluebird *(Sialia sialis)*. Observations of the behavior of the eastern bluebird suggested that the species was more likely to travel along the edge of wildlife corridors than cross the unforested matrix. This mode of corridor use inspired the "drift-fence hypothesis," which states that vegetation corridors intercept and direct the movement of species that may otherwise move through the matrix (Levey et al. 2005). In addition, Levey et al. (2005) were interested in finding where in the landscape seeds of the wax myrtle were dispersed. A particular question was whether seeds were more likely to be dispersed between patches that were connected by wildlife corridors than between unconnected patches. To answer this question, Levey et al. (2005) sprayed the fruits of the wax myrtle in some patches with a dilute solution of fluorescent powder. Using this method enabled defecated seeds of the wax myrtle to be identified in forest patches elsewhere in the landscape. The results demonstrated that on average, seeds were 37% more likely to be dispersed to connected patches than to unconnected patches. This study demonstrated that wildlife corridors can sometimes provide habitat connectivity—like in this example for the eastern bluebird and the wax myrtle. Finally, by maintaining an important ecological process—that is, seed dispersal—throughout the landscape, the study also demonstrated that wildlife corridors have the potential to enhance ecological connectivity.

- maintain the inherent species richness at the patch and landscape scale (Harris and Scheck 1991; Gilbert et al. 1998; Figure 12.4)

Some species may benefit from wildlife corridors that link suitable habitat (Gilbert et al. 1998; Haddad et al. 2003), including species that do not use the matrix or that avoid open areas (Martin and Karr 1986; Berggren et al. 2002) as well as those that disperse only through suitable habitat (e.g., Baur and Baur 1992; Nelson 1993).

Not all species use corridors (Lindenmayer et al. 1993; Hannon and Schmiegelow 2002), and their use may depend on the ecology of the species in question; for example, their scale of movement (Amarasekare 1994), patterns of behavior (Lidicker 1999), or social structure (Horskins 2004). Similarly, attributes of corridors such as their width and length, habitat suitability for a particular species, location in the landscape, and a range of other factors can affect corridor use by wildlife (Table 12.2).

Riparian Areas and Stream Buffers as Corridors

Riparian corridors or stream buffers are a particular type of corridor that can often be particularly effective at maintaining habitat connectivity for many species, as well as contributing to ecological connectivity (Kirchner et al. 2003; Hilty and Merenlender 2004). They provide habitat for large numbers of terrestrial and aquatic fauna and flora (Loyn et al. 1980; Naiman et al. 1993; Spackman and Hughes 1995). In addition, populations of several groups of species are more fecund in riparian areas (Soderquist and Mac Nally 2000), thereby providing more offspring to disperse to other less productive parts of the landscape. Riparian corridors may also act as dispersal routes for some terrestrial animals (Lindenmayer and Peakall 2000) and facilitate seed dispersal in plants (Kirchner et al. 2003). While riparian areas are useful for some terrestrial taxa, physical linkages outside the riparian zone are required to maintain landscape connectivity for other taxa (McGarigal and McComb 1992; Claridge and Lindenmayer 1994; see Box 12.3).

Caveats Associated with Wildlife Corridors

Wildlife corridors may have disadvantages (Simberloff et al. 1992). For example, corridors may facilitate the spread of genes that break up coadapted gene complexes in naturally isolated populations (Knopf 1992). They may also exac-

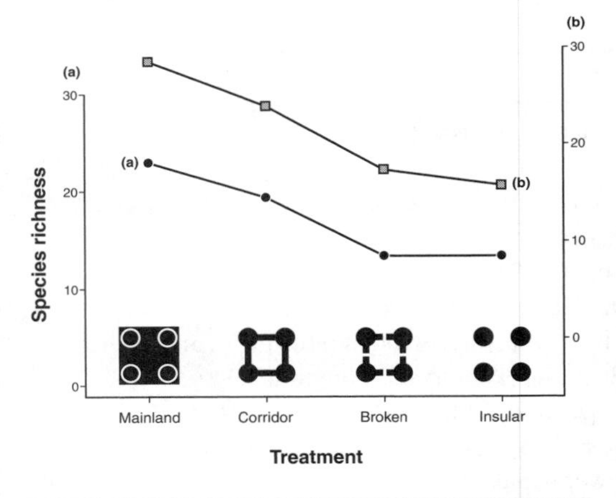

Figure 12.4. Species richness of microarthropods at (a) the patch scale and (b) the landscape scale in an experimental microecosystem consisting of four circular islands covered by moss with a diameter of 10 cm each. The islands were (1) located within continuous moss cover, (2) connected by corridors, (3) connected by interrupted corridors, or (4) not connected (redrawn from Gilbert, F., Gonzalez, A. and Evens-Freke, I. 1998: Fig. 1: The effect of connectivity treatments on diversity in experimental moss islands, p. 579, *Proceedings of the Royal Society of London Series B*, **265**, 577–582, Corridors maintain species richness in the fragmented landscapes of a microsystem, published by, and reproduced with permission from, The Royal Society of London).

erbate the spread of weeds, pest animals, diseases, and fires (Forney and Gilpin 1989). Corridors may be dominated by negative edge effects (Sisk and Margules 1993) or connect high quality habitat patches to areas of poor quality habitat that act as population "sinks" (Breininger and Carter 2003). Displaced individuals or nonbreeding animals may be attracted to and inhabit corridors (Soulé and Gilpin 1991). If individuals disperse when populations are below the carrying capacity of the local environment, sinks may be detrimental to metapopulation persistence (Breininger et al. 2006).

LIMITATIONS OF EXISTING CORRIDOR STUDIES

Although a number of important corridor studies have begun (e.g., La Polla and Barrett 1993; Andreassen et al. 1996; Gilbert et al. 1998; Haddad et al. 2003), many are at a relatively small scale (e.g., Aars and Ims 1999; Coffman et al. 2001; Berggren et al. 2002; Mabry and Barrett 2002), and it remains unclear whether the results can be readily extrapolated to larger spatial scales (Noss and Beier 2000; but see Haddad and Baum 1999). In addition, most corridor studies have focused on groups such as mammals and birds, and the relevance to other groups such as invertebrates is poorly known (Mönkkönen and Mutanen 2003; but see Berggren et al. 2002; Gonzalez and Chaneton 2002; Várkonyi et al. 2003). Moreover, many corridor studies have failed to sample the matrix and hence adequately to confirm animal presence and/or movement there (Cook et al. 2004). The usefulness of corridor studies could be further magnified by connecting estimates of effective interpatch dispersal with measures of matrix condition and landscape cover (including the presence of corridors or retained vegetation).

Stepping Stones

We define stepping stones as relatively small patches of native vegetation scattered throughout a landscape. They enhance landscape connectivity, and as their name suggests, they may facilitate habitat connectivity for a variety of species (Forman 1995; Dramstad et al. 1996). Table 12.3 summarizes a range of studies that have assessed the use of stepping stones by various groups of organisms. Notably stepping stones may also exist at scales larger than small patches within landscapes; for example, mountain meadows are used by the rufous hummingbird *(Selasphorous rufus)* during long-distance migration (Russell et al. 1994).

Table 12.2. Factors influencing wildlife corridor use*

Factor	*Examples*
Target species characteristics	Foraging strategy; colonial versus solitary social system (Recher et al. 1987; Andreassen and Ims 2001)
Gender	Gender differences in a given species (e.g., in dispersal or foraging) can significantly affect corridor use (Downes et al. 1997a, b; Aars et al. 1999; Várkonyi et al. 2003)
Biotic interactions	Aggressive interspecific behavior (Catterall et al. 1991), territorial defense and philopatry (Horskins 2004) and avoidance of humans (Beier 1995) can affect corridor use
Abiotic edge effects	Microclimatic conditions can reduce habitat suitability in corridors (Hill 1995), and interior conditions may be required to facilitate dispersal (Perault and Lomolino 2000)
Dispersal behavior	Random dispersal versus movement along vegetation gradients (Murphy and Noon 1992; Berggren et al. 2002), avoidance of open areas (Machtans et al. 1996; Várkonyi et al. 2003) or movement along corridors to areas with lower population density (Andreassen and Ims 2001)
Food availability	Food availability within a corridor can affect the rate at which animals move through a corridor (Bright 1998)
Vegetation attributes within corridor	Structural features can influence movement (Norton et al. 1995; Bowne et al. 1999; Merritt and Wallis 2004)
Corridor characteristics	Corridor width and length (Harris and Scheck 1991; Andreassen et al. 1996; Downes et al. 1997a, b; Bennett 1998; Sieving et al. 2000)
Topographic location and variation	Species richness may be greater in corridors spanning multiple topographic positions (Claridge and Lindenmayer 1994)
Vegetation gaps	Roads and tracks across corridors can pose barriers to movement (Bennett et al. 1994; Beier 1995)
Size of areas connected	Small connected patches may provide few dispersers to move through a corridor (Wilson and Lindenmayer 1996)
Number of corridors	The number of corridors can influence the chances of corridors being encountered during movement (Forman 1995)
Matrix condition	Clearcut versus selectively logged adjacent forest can influence corridor use (Rosenberg et al. 1997); some types of matrix facilitate movement of some species (Palomares 2001)

*Based on Lindenmayer (1994, 1998); Lindenmayer and Franklin (2002).

A "Soft" Matrix

A "soft" matrix is one characterized by a similar vegetation structure to patches of native vegetation (Franklin 1993; Lindenmayer and Franklin 2002). While corridors or stepping stones will be valuable for some taxa (Metzger 1997), maintaining habitat connectivity for other species requires retaining appropriate vegetation cover throughout the entire matrix (Murphy and Noon 1992; Franklin 1993; Rosenberg et al. 1997). The importance of matrix management for landscape connectivity is considered in more detail in Chapter 14. The maintainence of a soft matrix through additional tree retention and the retention of other vegetation elements was considered to be the most appropriate conservation management for the northern spotted owl *(Strix occidentalis caurina)* in the Pacific Northwest of the United States (Forest Ecosystem Management Assessment Team 1993; see Chapter 18).

Summary

We define "landscape connectivity" as the human-perceived connectedness of patterns of vegetation cover. We separately define "habitat

Table 12.3. Examples of studies that demonstrate the use of stepping stones by various organisms

Organism, location	*Description*	*References*
Butterflies, Europe	Stepping stones contributed to the connectivity in butterfly populations in Europe	Nève et al. 1996
Fruit pigeons and bats, Australia	Mobile species like fruit pigeons and bats are able to utilize resources in small habitat patches located many kilometers apart to help them move across the landscape	Date et al. 1991, 1996; Law et al. 1999
Plants, worldwide	Stepping stones may assist connectivity in plant populations as part of range shifts in response to climate change	Collingham and Huntley 2000
Fruit trees, Costa Rica	Scattered trees in the matrix function as stepping stones for genetic exchange between spatially isolated tree populations	Cascante et al. 2002
Birds, Costa Rica	Birds used scattered trees as stepping stones to traverse a modified farming landscape	Guevara and Laborde 1993
Rufous hummingbird (*Selasphorous rufus*), California	Observations of marked birds were used to investigate the quality of stopover habitats used by the rufous hummingbird while migrating	Russell et al. 1994
North Island brown kiwi (*Apteryx australis mantelli*), New Zealand	Radio tracking showed that the kiwi used small forest patches as stepping stones when traversing a modified landscape	Potter 1990
White-crowned pigeons (*Columba leucocephala*), Florida, United States	Radio-tracked pigeons used forest patches as stepping stones during dispersal	Strong and Bancroft 1994
Birds, eastern Australia	Birds used scattered trees as stepping stones to traverse a modified farming landscape	Fischer and Lindenmayer 2002b

connectivity" as the connectedness of habitat patches used by a given individual species; and "ecological connectivity" as the connectedness of ecological processes across multiple spatial scales. Although the three concepts are related, they are not synonymous (Box 12.4). The primary focus of this chapter is on landscape connectivity. Many metrics have been developed to quantify landscape connectivity; most of which consider the distances between vegetation patches, the number of links between them, or the amount of vegetation within a specified radius surrounding a site. The loss of landscape connectivity can affect the habitat connectivity of many species and even entire assemblages, especially for those dependent on native vegetation. It may also disrupt ecological processes, such as waste decomposition or seed dispersal, thereby reducing ecological connectivity. Three types of features contribute to landscape connectivity: (1) wildlife corridors, (2) stepping stones, and (3) a "soft" matrix.

Links to Other Chapters

Habitat connectivity was first defined in Chapter 6, which discusses the negative effects of habitat isolation on individual species. The loss of ecological connectivity can lead to cascades of ecosystem change, and potentially irreversible regime shifts. Such fundamental changes to ecosystem functioning are discussed in Chapter 15. Matrix management can be a key aspect of maintaining landscape connectivity and the habitat connectivity of some individual species. The importance of the matrix for this and other reasons is discussed in Chapter 14. Approaches to mitigate the loss of landscape connectivity are discussed further in Part V.

Box 12.4. Do Corridors Provide Connectivity?

Distinguishing between landscape connectivity and habitat connectivity can contribute to resolving the question of whether corridors provide connectivity posed by Beier and Noss (1998). Following our definitions, corridors always provide *landscape connectivity*, which is generally perceived to be an ecologically valuable property, given its likely contribution to ecological connectivity (Harrison and Bruna 1999; Noss and Beier 2000; Soulé et al. 2004). For many species, corridors also provide *habitat connectivity* (Beier and Noss 1998). However, for other species, such as the northern spotted owl (*Strix occidentalis caurina*) in the Pacific Northwest of North America (Murphy and Noon 1992) and Leadbeater's possum *(Gymonbelideus leadbeateri)* in southeastern Australia (Lindenmayer and Nix 1993), at least some types of corridors do not provide habitat connectivity because these species avoid them. The distinction between landscape connectivity and habitat connectivity is part of an ongoing theme in this book—that is, to distinguish between species-specific perspectives and human perspectives of landscapes (Fischer et al. 2004a; Manning et al. 2004a).

Further Reading

Noss (1991) gives a valuable outline of landscape connectivity and why it is important. Saunders and Hobbs (1991), Hobbs (1992), Smith and Hellmund (1993), Lindenmayer (1998), and Bennett (1998) provide detailed appraisals of issues related to maintaining landscape connectivity, particularly wildlife corridors. Soulé et al. (2004) discuss a wide range of ecological processes related to connectivity, which we summarized under the term "ecological connectivity," Franklin (1993), Franklin et al. (1997) and Lindenmayer and Franklin (2002) discuss various approaches to improve matrix management to enhance landscape connectivity. Reviews on the impacts of roads on landscape connectivity are provided by Spellerberg (1998) and Forman et al. (2002). An interesting set of analyses of landscape connectivity measures is presented by Moilanen and Nieminen (2002).

CHAPTER 13
Nestedness and Community Assembly

Species richness is often found to be higher in large patches of remnant vegetation (Herkert 1994; Smith et al. 1996; Nupp and Swihart 2000). Likely mechanisms for higher species richness in large patches include lower extinction rates, higher immigration rates, and greater habitat diversity (McGuinness 1984; Rosenzweig 1995; see Chapter 9). Also, in modified landscapes, increased landscape connectivity between vegetation patches can lead to higher movement rates of individuals between patches, thus reducing extinction risk and increasing the likelihood of recolonization following local extinction (Chapter 12).

In this chapter, we further investigate assemblages of species inhabiting patches of remnant vegetation. Rather than focusing only on species richness (Chapter 9), in this chapter, we also consider species composition of the communities inhabiting the patches within an assemblage. In particular, we discuss two lines of ecological theory that can contribute toward understanding species composition in patches of remnant vegetation: (1) nested subset theory and (2) community assembly rules.

Nested Subset Theory

Definition, Origin, and Examples

An assemblage distributed across a number of discrete sites is considered nested when the taxa present at relatively species-poor sites are subsets of those present at progressively more species-rich sites (Patterson and Atmar 1986). Whittaker (1998) likens the concept of nestedness to that of Russian dolls in which each doll has a smaller one inside it. Nested subset theory stems from observations of the composition of assemblages on islands (Darlington 1957) and mountain tops (May 1978). More recently, nested assemblages have also been observed in landscapes altered by human activities (e.g., Cornelius et al. 2000; Box 13.1), and

Box 13.1. The Application of Nested Subset Theory in Human-Modified Landscapes: A Case Study on Amphibians in Madagascar

Many recent studies have investigated the potential nestedness of faunal assemblages in human-modified landscapes where vegetation cover is patchy (Table 13.1). One illustrative study was conducted with a focus on amphibians in the central highlands of Madagascar (Vallan 2000). The region is approximately 1500 m above sea level and encompasses the Ambohitantely Nature Reserve. Although reserved, over a third of the area is severely modified, and the original high plateau rainforest that was once dominant has been converted to pseudosteppe (Figure 13.1). Vallan (2000) studied amphibians in seven rainforest remnants ranging in size from 0.16 to 1250 ha. Comprehensive amphibian searches ranging between 22 and 90 hours were conducted within each remnant, until each remnant's species accumulation curve had reached an asymptote. Twenty-eight amphibian species were recorded, and four additional species were observed only outside rainforest fragments (Figure 13.2). Using the nestedness calculator by Atmar and Patterson (1993), Vallan (2000) showed that large remnants had more species and that the species assemblage was significantly nested. In particular, some arboreal species appeared to be restricted to large remnants. Following a detailed discussion of amphibian ecology, Vallan (2000) concluded that both nested habitats and extinction pressure due to small population size had contributed to the nested distribution pattern of species; large rainforest remnants offered more types of different habitats, and supported higher amphibian populations than small remnants.

Table 13.1. Examples of nested assemblages in various ecosystem types, with particular emphasis on modified landscapes

Ecosystem type	*Example*	*Suggested mechanism*	*Reference*
Islands	Birds on New Zealand islands	Extinction	Simberloff and Levin 1985
	Lizards on islets near Guam	Extinction	Perry et al. 1998
	Small mammals on Thai islands	Extinction and immigration	Lynam and Billick 1999
Mountain tops	Mammals in the southwest United States	Extinction	Patterson and Atmar 1986
	Birds in the Andes	Immigration	Nores 1995
Modified landscapes	Plants in forest patches in Georgia, United States	Immigration	Kadmon 1995
	Small mammals in forest patches in Indiana, United States	Immigration and extinction	Nupp and Swihart 2000
	Small mammals in forest patches in Victoria, Australia	Extinction	Deacon and Mac Nally 1998
	Amphibians in forest patches in Madagascar	Extinction	Vallan 2000
	Birds in forest patches in the Afromontane mistbelt in South Africa	Extinction	Wethered and Lawes 2005
	Birds in forest patches in southeastern Australia	Extinction	Fischer and Lindenmayer 2005b
	Plants and fungi in forest patches in Sweden	Extinction, habitat nestedness	Berglund and Jonsson 2003
	Mammals in forest patches in Illinois, United States	Habitat nestedness, extinction	Rosenblatt et al. 1999
	Plants in forest patches in Belgium	Habitat nestedness	Honnay et al. 1999

in relatively unmodified patchy landscapes or regions (e.g., Patterson and Brown 1991). Some examples where nested assemblages have been observed are listed in Table 13.1.

A review of over 150 datasets from a range of climatic regions (including studies on vertebrates, mollusks, plants, and arthropods), concluded that nestedness was a common phenomenon for a wide range of assemblages (Wright et al. 1998). Worthen (1996) considered that the presence or absence of nestedness was a useful and fundamental descriptor for community composition. Although many assemblages are significantly nested (Atmar and Patterson 1995; Wright et al. 1998), it is important to note that very few assemblages are perfectly nested (sensu Figure 13.3). That is, although many assemblages can be arranged as roughly triangular species by sites matrices (Figure 13.3), most exhibit some unexpected presences ("outliers") or absences ("holes"; see Cutler 1991). The difference between perfect and statistically significant nestedness can have important implications for the use of nested subset theory in a conservation context, including in human-modified landscapes (Fischer and Lindenmayer 2005a; Box 13.2).

Mechanisms

In significantly nested assemblages, the occurrence sequences of species across sites are not random (Schoener and Schoener 1983; Simberloff and Levin 1985). Rather, some species occur in nearly all sites (e.g., species A and B in Figure 13.3), whereas others only occur in species-rich sites (e.g., species D and E in Figure 13.3). This nonrandom pattern suggests that one or several mechanisms structure the assemblage in a predictable way. Three ecological mechanisms are typically considered to be particularly important. These are selective extinction, selective immigration, and habitat nestedness.

If selective extinction is the dominant ecological mechanism, a nested assemblage corresponds to a set of predictable extinction sequences. Assuming extinction was the ecological mechanism causing the nested structure in Figure 13.3, once species E was extinct, species D would be the next most likely species to be lost from a given site. The presence of predictable extinction sequences is often deduced when an assemblage ordered by the area of individual sites is significantly nested (Lomolino 1996; Wright et al. 1998). When interpreted in relation to the size of individual sites, nested subset theory represents an extension of island biogeography, which predicts higher species richness in larger sites (Chapter 9). However, unlike island biogeography theory, nested subset theory contains information not only on species richness but also on the identity of the individual species inhabiting a given site (Doak and Mills 1994). Several authors believe that selective extinction from sites is a major driver of the nestedness of many assemblages distributed across semi-isolated sites (like islands, mountain tops, or forest fragments; e.g., Atmar and Patterson 1993; Wright et al. 1998; see Table 13.1). However, despite the vast literature on predicting extinction sequences from patterns of nestedness, in practice, the ability to predict accurately *which* members of a community are the ones likely to be missing from subsets of taxa is often limited (Saetersdal et al. 2005).

Box 13.2. Statistically Significant Nestedness versus Perfect Nestedness—an Important Difference for Conservation

Fischer and Lindenmayer (2005a) investigated the implications of imperfect nestedness in a conservation context in some detail. They compiled presence/absence data on birds in 43 native eucalypt forest patches scattered throughout an exotic softwood plantation in southeastern Australia. The bird assemblage was significantly nested when patches were sorted by their size. However, there were some "holes" and some "outliers"(i.e., unexpected absences and presences of birds at some sites). If nestedness had been perfect (i.e., if there had been no holes and no outliers) the single largest patch would have contained all bird species, including those of conservation concern. Arguably, perfect nestedness therefore would have indicated that the largest patch was disproportionately important for bird conservation. However, because the dataset was significantly, but not perfectly, nested, substantially more sites than the single largest patch alone were needed to capture most bird species. For example, less than half of the bird species classified as "sensitive to landscape change" co-occurred in any given patch. Using the largest patches only, more than a quarter of patches surveyed would have been required to capture 80% of sensitive species in at least one patch (Fischer and Lindenmayer 2005a). Thus, although small patches have many limitations, such as increased susceptibility to edge effects (Lovejoy et al. 1986) and limited food resources (Zanette et al. 2000), this case study illustrates that even when an assemblage is significantly nested in relation to patch size, small patches can make an important complementary contribution to large patches and should not be neglected in conservation management.

Another frequently cited mechanism potentially related to nestedness is selective immigration (see Table 13.1 for examples). Darlington (1957) believed this was the underlying mechanism for why many island archipelagos were nested (but see Wright et al. 1998). For example, if Figure 13.3 was structured by selective immigration, species A would be the species with the best dispersal ability, whereas species E would be the species with the worst dispersal ability. For assemblages distributed across semi-isolated sites, nestedness in relation to immigration is frequently assessed by sorting sites relative to their isolation (Lomolino 1996). A difficult problem in this context is that isolation is difficult to quantify in an ecologically meaningful way, especially for multiple species at the same time (Lomolino 1996; Chapter 6; Chapter 12).

Finally, habitat nestedness has been suggested to lead to predictable occurrence sequences of species (e.g., Honnay et al. 1999; see Table 13.1). Under the proposition of habitat nestedness, some types of environments

Figure 13.1. Rainforest fragments in the Ambohitantely area in Madagascar (photo by Denis Vallan).

contain suitable habitats for more species than other environments. For example, if Figure 13.1 was the result of nested habitats, site I would contain suitable habitats for all five species in the assemblage, whereas site V would contain suitable habitat only for species A. Habitat nestedness might occur where structurally simple environments contain a subset of those species found in structurally more complex environments.

Necessary Preconditions for Nestedness

Not all assemblages are nested (see Figure 13.3). For example, Graves and Gotelli (1993) studied mixed-species flocks of birds in Amazonia and found that ecologically similar congeneric species often replaced one another in different flocks. Thus species were not added or lost from the assemblage in a predictable sequence, but rather, different sets of species occurred in different flocks. Because of the resulting pattern in species by sites matrices, this type of assemblage is often referred to as a checkerboard pattern.

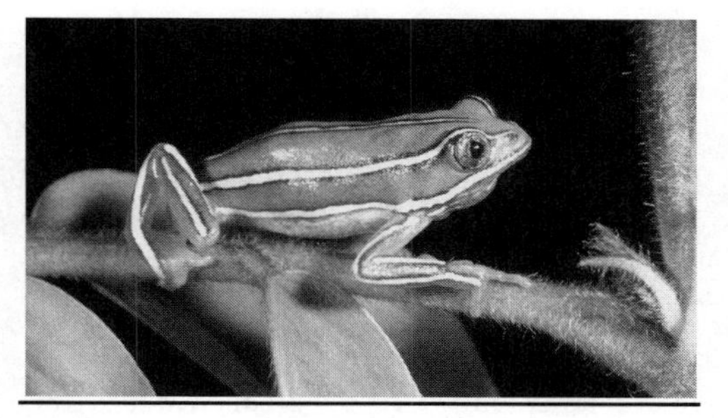

Figure 13.2. The amphibian *Heterixalus rutenbergi.* This is one of few amphibian species in the Ambohitantely area that occur in the open areas between the rainforest fragments (photo by Denis Vallan).

Patterson and Brown (1991) identified three necessary (but not automatically sufficient) preconditions for nestedness to occur. First, sites need to have a common biogeographic history. Second, contemporary environments at all sites need to be broadly similar. Third, niche relationships of species need to be organized hierarchically. The first two conditions are important to ensure that all sites have a common pool of potentially co-occurring species. The third condition is important to ensure that the sequence of presences of species at a series of sites varies in a predictable and monotonic way (Bruce Patterson, pers. comm.).

Departures from these conditions can mean that nestedness is weakened or absent. For example, Fischer and Lindenmayer (2005b) found that the assemblage of lizards in a fragmented landscape in southeastern Australia was not nested. The main reason for the lack of

Perfectly nested assemblage

Site I spp: A, B, C, D, E
Site II spp: A, B, C, D
Site III spp: A, B, C
Site IV spp: A, B
Site V spp: A

Species

Sites	A	B	C	D	E	*SUM*
I	p	p	p	p	p	5
II	p	p	p	p	-	4
III	p	p	p	-	-	3
IV	p	p	-	-	-	2
V	p	-	-	-	-	1

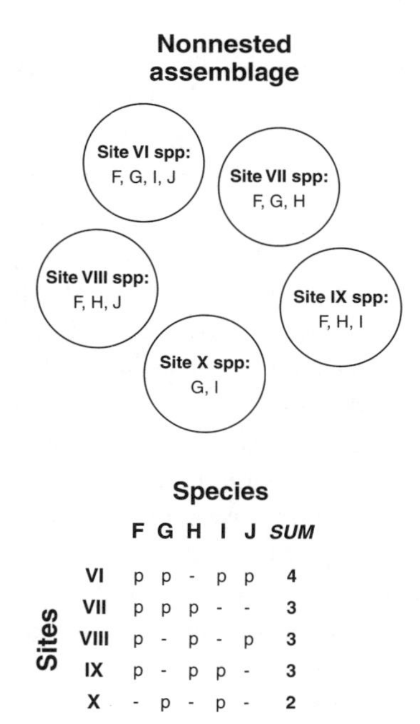

Sites	F	G	H	I	J	*SUM*
VI	p	p	-	p	p	4
VII	p	p	p	-	-	3
VIII	p	-	p	-	p	3
IX	p	-	p	p	-	3
X	-	p	-	p	-	2

Figure 13.3. Examples of two assemblages containing species A, B, C, D, E and species F, G, H, I, J. For a given species, "p" denotes its presence and "-" indicates its absence at a given site. The left assemblage is perfectly nested, i.e. the species present at species-poor sites are subsets of those present at progressively richer sites. The right assemblage is not nested.

nestedness in this case was that the pool of potentially occurring species was not the same at all sites—high elevation sites supported a different suite of lizard species than low elevation sites (see Fischer and Lindenmayer 2005c).

Nestedness in a Conservation Context

Like island biogeography theory (MacArthur and Wilson 1967; Chapter 9), nested subset theory has also been used to examine issues such as patterns of species distribution in human-modified patchy landscapes (Doak and Mills 1994; see Table 13.1). Because nestedness analyses are a way to study species composition as well as species richness, nested subset theory has also received attention as a potential tool to answer whether a single large reserve or several small ones are preferable for species conservation (Patterson 1987; Cutler 1991; Boecklen 1997). Most recently, nestedness has been suggested as a tool to pick surrogate species in an attempt to provide a scientifically defensible shortcut for species conservation (Smyth et al. 2000; but see Fleishman et al. 2002).

In human-modified landscapes, nestedness analyses can be useful for providing an overview of likely conservation issues. For example, they can show if local immigration or extinction is a likely driver of community composition, and if large or well-connected patches are particularly important for conservation (Fischer and Lindenmayer 2005b). However, given the community-level nature of nestedness tests, statistically significant nestedness should be interpreted with caution before it is used to develop conservation guidelines (e.g., regarding what size patches to protect; see Patterson 1987; Rosenblatt et al. 1999; Berglund and Jonsson 2003). Particular caution is warranted in using nested subset theory as a means of identifying ecological indicators (cf. Honnay et al. 1999; Smyth et al. 2000) because important species-specific differences and idiosyncrasies may be masked by global tests of nestedness (Simberloff and Martin 1991; Box 13.2). Given this, the identification of nestedness does not necessarily lead to clear insights for maintaining subsets of species over time (Whittaker 1998).

Probably the most important caveat of nestedness analyses in modified landscapes is that they are based on the island model (Chapter 3). Like island biogeography theory (Chapter 9), nestedness analyses in modified landscapes sometimes (implicitly) consider patches of native vegetation as spatially discrete "islands" surrounded by "unsuitable land" (Doak and Mills 1994). This view can be overly pessimistic and may fail to recognize the potential value of conservation strategies in the areas surrounding vegetation remnants for at least some species in an assemblage (Chapter 14).

Quantification of Nestedness

Many tools are available to test if datasets are significantly nested. Two considerations are particularly important in the context of assessing the statistical significance of a potentially nested assemblage.

First, there are a range of different nestedness metrics (e.g., Patterson and Atmar 1986; Wright and Reeves 1992; Atmar and Patterson 1993; Lomolino 1996; Brualdi and Sanderson 1999). These metrics are used to summarize the extent to which data matrices of species by sites are nested. Some metrics are more directly interpretable than others, and they respond slightly differently to different types of data matrices. Many metrics are based on counting the unexpected presences (outliers) or absences (holes) in a potentially nested matrix array (Cutler 1991; Brualdi and Sanderson 1999); another frequently applied metric is derived from thermodynamics and is based on the quantification of the degree of "unexpectedness" of species by sites matrices (Atmar and Patterson 1993, 1995).

Second, an appropriate null model needs to be chosen. A null model aims to provide a neutral background against which actual data can be compared (Gotelli and Graves 1996). In the context of nested subset theory, null models are usually used to generate a large number of "random" species by sites matrices. The statistical significance of the actually observed data array can then be obtained by comparing the value of its nestedness metric against the statistical distribution of the nestedness metric calculated for randomly generated matrices. Different null models make different assumptions about what constitutes "random" species distribution patterns—for example, some null models consider that all species are equally likely to occur throughout a given landscape, whereas others recognize that some species are inherently uncommon (Lomolino 1996; Jonsson 2001; Fischer and Lindenmayer 2002d). The choice of null model strongly affects the likelihood of detecting statistical significance (Patterson and Atmar 1986; Wright et al. 1998). Often, it is preferable to use a null model that accounts for natural variation in different species' ubiquities. Otherwise, significant nestedness can result from statistical artefacts and may be easily misinterpreted as ecologically valuable information on sequences of extinction, immigration, or habitat nestedness. An awareness of the assumptions and limitations of analytical methods to assess nestedness is fundamentally important to interpret ecological patterns correctly.

Assembly Rules

Assembly rules are general principles considered to determine or regulate which species form communities, including, for example, how new species might be added to a community as a consequence of dispersal and the order in which they colonize an area (Belyea and Lancaster 1999). As an illustration, based on a study of desert rodent communities in North America, Fox and Brown (1993, p. 358) articulated a community assembly rule stating that:

> each [new] species entering a community will tend to be drawn from a different [functional] group until each group is represented, and then the rule repeats itself.

In the case of this particular assembly rule, Fox and Brown (1993) stressed that their general principle governed the functional group to which an added species belonged (e.g., specialist granivore, omnivore, micropredator, etc.), rather than the identity of the species per se.

Other assembly rules have been proposed. For example, one states that resource utilization within guilds will increase with increased interspecific competition (Diamond 1975b; M'Closkey 1978; Hanski and Cambefort 1991; Pacala and Tilman 1994). A similar one is that resource utilization across guilds or functional groups will tend to increase with increased interspecific competition (Fox 1987; Wilson and Roxburgh 1994; Morris and Knight 1996). A prey–predator assembly rule proposed by Holt et al. (1994) and Grover (1994) is that per capita predation risk tends to decrease due to competition among prey for areas that provide refuges from predators. Several other kinds of assembly rules have been proposed by various workers, including (among others) Jeffries and Lawton (1984, 1985); Mithen and Lawton (1986), Frelich et al. (1993), Shorrocks and Sevenster (1995), Holt (1977), and Durant (1998).

The Controversial Literature on Assembly Rules

Although many studies have claimed to identify assembly rules for groups such as small mammals (Fox and Brown 1993), lemurs (Ganzhorn 1997), and plants (Wilson and Whittaker 1995; Wilson et al. 1995), the extensive literature on community assembly rules is a highly controversial one punctuated by major disagreements that have arisen since Diamond (1975b) originally proposed the concept for the birds of New Guinea. Although many workers acknowledge that groups of species clearly do not assemble randomly (reviewed by Whittaker 1998), some authors have argued that the null models used to test for the presence of assembly rules are unrealistic (Wilson 1995; Gotelli et al. 1997). Others claim that the evidence for assembly rules is weak (Wiens 1980; Stone et al. 1996), and other factors such as predation and disturbance may be more important in determining species coexistence (Gotelli et al. 1997). A further criticism has been that so-called rules are in fact trivial tautologies with limited ability to be generalized across assemblages and environments (Connor and Simberloff 1979). Finally, many authors are clearly frustrated by the loose and imprecise use of terminology associated with the study of assembly rules (Belyea and Lancaster 1999).

Disassembly Rules

Many issues and problems associated with assembly rules remain unresolved. In the context of this book on landscape modification and its impacts on species assemblages, a key issue is whether coherent disassembly rules might exist. That is, is there a predictable sequence (e.g., based on species-specific traits, membership of a given functional group, etc.) that might govern the sequence in which species are lost from a community in response to human-derived landscape change (Ostfeld and LoGuidice 2003)? Notably, as discussed by Belyea and Lancaster (1999), the concept of community disassembly involves not simply the decline in the number of species but the breaking down of interrelationships between species in a community. Such work is particularly important in the context of the impacts of landscape modification and the potential for cascading effects on key ecosystem functions (Chapter 15). As noted by Ostfeld and LoGuidice (2003), the pattern of community disassembly may not be important for ecosystem function if the activities of other species are able to compensate for those of the taxa that have been lost—paralleling the ecological redundancy notion of Walker (1992), which, in turn, contributes to ecosystem resilience (Walker 1995). Conversely, a lack of resilience and rapidly impaired ecosystem function may be associated with community disassembly where interspecies compensation and ecological redundancy are limited (Walker 1992, 1995; see Chapter 15)—as is believed to be the case in examples of trophic cascades where community disassembly follows the loss of top predators (Berger et al. 2001; Soulé et al., 2003, 2005; Ripple and Beschta 2005).

Unfortunately, far less work has been conducted on the topic of predictable community disassembly than the corollary of assembly rules (Mikkelson 1993). However, Belyea and Lancaster (1999) noted that communities are unlikely to disassemble in the reverse order to which they assembled (cf. Fox 1987), in part because species accumulation in a community is typically thought of as a slow evolutionary process, whereas disassembly could be a rapid and highly transitory process (Mikkelson 1993).

Summary

Two lines of ecological theory are potentially useful to analyze species composition in modified landscapes: nested subset theory and community assembly rules. Nested subset theory can be applicable to human-modified mosaic landscapes where patches resemble "islands." An assemblage is defined as nested when the species present in relatively species-poor sites are subsets of those present at progressively more species-rich sites. Three main ecological mechanisms are thought to explain why assemblages are nested: (1) species become extinct in a predictable order (e.g., with a reduction in patch size); (2) species colonize patches in a predictable order (e.g., because of their distances to a source of dispersers); and (3) the habitat types represented in the patches may be nested (e.g., because different types of habitat are added in a predictable sequence as structural complexity increases). In a conservation context, it is important to be aware that, even in assemblages that are significantly nested by patch size, small patches and the matrix can make important complementary contributions. This is, in part, because perfectly nested assemblages are very rare even though statistically significant nestedness is a common phenomenon. The main limitation of nested subset theory is that it is largely based on the island model of modified landscapes (see Chapter 3). Viewing patches of native vegetation as islands in a hostile sea can be inappropriate, especially in heterogeneous landscapes (Chapter 3, Chapter 14).

Community assembly rules are general principles hypothesized to regulate how new species might be added to a community and affect one another via species interactions. For example, one assembly rule states that species entering a community will be drawn from different functional groups until all functional groups are represented, at which point the rule will repeat itself. Community assembly rules may have less predictive power than nested subset theory. However, they can shed light on community composition and the implications of losing some types of species from modified landscapes. For instance, the loss of an entire functional group is likely to cause cascading ecosystem changes affecting many other species as well.

Links to Other Chapters

Nested subset theory is closely linked to the island model of modified landscapes (Chapter 3) and island biogeography theory (Chapter 9). Despite its value in some situations, the island model has important limitations. These are particularly apparent in heterogeneous landscapes where the matrix plays important ecological roles (Chapter 14). The loss of functional groups discussed in this chapter in the context of assembly rules can lead to cascading effects of ecosystem change. Such cascades are discussed in more detail in Chapter 15.

Further Reading

Key papers on nestedness include those by Patterson and Atmar (1986), Lomolino (1996), and Wright et al. (1998). The application of nested subset theory to modified landscapes is discussed by Patterson (1987). Recent illustrative case studies on nestedness in modified landscapes include Lynam and Billick (1999), Vallan (2000), and Berglund and Jonsson (2003). There is also a large literature on assembly rules, much of it highly theoretical. The seminal early work on assembly rules was by Diamond (1975b). Articles engaging in subsequent debates on assembly rules include those by Connor and Simberloff (1979), Wilson (1995), Stone et al. (1996), and Gotelli et al. (1997). An excellent overview of assembly rules is provided by Belyea and Lancaster (1999). Ostfeld and LoGuidice (2003) is a thought provoking simulation modeling paper on how community composition may change in response to landscape modification.

CHAPTER 14
The Matrix and Landscape Heterogeneity

In some situations, the spatial distribution of assemblages can be described well by the patch–matrix–corridor model where the matrix is considered to be the largest and often most highly modified patch type in a given landscape (sensu Forman 1995). In those situations, the processes affecting many single species (Part II) are mirrored closely by the relationship between landscape patterns and species assemblages (Part III). For example, loss of native vegetation may equate to habitat loss (Chapters 4, 9); loss of structural complexity and edge effects may translate into habitat degradation (Chapters 5, 10, 11); and large distances between vegetation patches may equate to increased habitat isolation for most species within an assemblage (Chapters 6, 12). In such situations, community composition within predefined vegetation remnants is often predictable to some extent (Chapter 13).

The analysis of modified landscapes as "island-like" patchy systems represents the historical origin and traditional stronghold of the "fragmentation" literature (Haila 2002). Sometimes such island-like behavior of modified terrestrial ecosystems is observed in modified landscapes; for example, in areas dominated by intensive agriculture (e.g., Rosenblatt et al. 1999; Brooker 2002). Although direct parallels between the response of a single species and the response of an assemblage simplify both science and management, there are instances where thinking about modified landscapes as mosaics of patches and corridors within a relatively hostile and uniform matrix is not appropriate (Wiens 1994). For example, different groups of organisms or species within a given assemblage may contrast strongly in their response to landscape change (Greenberg et al. 1994; Mac Nally et al. 2002). Although some species may be restricted to remnants of native vegetation, many species may not be (Daily 2001).

In this chapter, we examine situations where it would be overly simplistic to think of modified landscapes as aggregations of island-like patches within a hostile and homogeneous matrix. We highlight that the matrix can in fact play several important ecological roles. We also show that landscapes characterized by heterogeneous land cover patterns can provide habitat for many species. A key theme throughout this chapter is that there are situations where the traditional "fragmentation toolbox of concepts" breaks down. In such situations, the consideration of patches, corridors, and edge effects can be effectively complemented by an explicit recognition of the value of the matrix and landscape heterogeneity.

Ecological Roles of the Matrix

Forman (1995) defined the matrix as the largest background patch in a landscape, which exerts a dominant role on ecological processes (see Chapter 3). The matrix has many important roles. For example, it can affect the population dynamics of individual species, including population recovery after disturbance (Lamberson et al. 1994), the exchange of individuals between patches (Leung et al. 1993), and occupancy rates of vegetation patches (Villard and Taylor 1994). In the following section, we discuss three particularly important roles

Figure 14.1. Aerial view of the Tumut landscape in southern New South Wales, where patches of native eucalypt forest are embedded within a matrix of radiata pine (photo by David Lindenmayer).

of the matrix in some detail—the provision of habitat for individual species and sets of taxa, landscape connectivity (as defined in Chapter 12), and an ecological context for patches of native vegetation.

Habitat

In many modified landscapes, the notion of a matrix that is inhospitable to most species is not useful (Barrett et al. 1994). Examples where the matrix provides habitat for a range of species come from both forestry and agricultural landscapes. In southeastern Australia, Lindenmayer et al. (2002a, 2003a) and Fischer et al. (2005a) studied bird and lizard assemblages in native eucalypt forest patches located within an intensively managed matrix of the introduced radiata pine *(Pinus radiata)*. In this landscape, many species were not restricted to native forest patches, and some even used the matrix extensively (e.g., eastern yellow robin *[Eposaltria australis]*; garden skink *[Lampropholis guichenoti]*; Figure 14.1). Thus, although some native birds and lizards were genuinely confined to native forest fragments, many were able to use the matrix as well.

Similar findings have been made in many other parts of the world. For example, in the Una region of Brazil, Pardini (2004) studied small mammals in Atlantic Forest remnants. She showed that many small mammal populations extended from the native forest remnants into the matrix. Pardini (2004) partly attributed the relatively high regional mammal diversity to the fact that, despite extensive landscape change, the region was still dominated by forest, with nearly half of the matrix being covered by trees (e.g., regrowth forest, shade-grown cocoa plantations). Similarly, Gascon et al. (1999) found that over 70% of bird species inhabiting remnant patches of rainforest in the Manaus area of the Amazon also used the surrounding matrix to various extents. In many Australian grazing landscapes, scattered trees in the matrix are used by a majority of bat species (Law et al. 1999; Lumsden and Bennett 2005) and many bird species (Fischer and Lindenmayer

Box 14.1. An Experimental Study of Landscape Context Effects

A longitudinal study of the impacts of landscape context on species assemblages has been under way since 1997 at Nanangroe in southern New South Wales, Australia (Figure 14.2). The work is focusing on mammals, birds, and reptiles inhabiting 56 patches of remnant eucalypt woodland—where woodlands are composed of widely spaced trees where the crowns do not touch (Hobbs 2000; cf. Peterken 1996). These patches were formerly surrounded by grazing land and scattered trees where livestock grazing was the predominant land use. Since the ("pretreatment") beginning of the study, large areas formerly dominated by scattered trees have been cleared and replaced by dense stands of exotic radiata pine *(Pinus radiata)*. The study also includes a matched set of 50 eucalypt woodland remnants where the surrounding landscape context of scattered trees has remained unchanged (Lindenmayer et al. 2001).

Work over the past 9 years has indicated that bird communities inhabiting the woodland remnants surrounded by maturing plantation pine stands are indeed changing rapidly. A suite of species more characteristic of forest (rather than open woodland) environments has been attracted to the pine stands and many of these are "spilling over" into the more open adjacent patches of eucalypt woodland. In addition, vegetation attributes in the woodland patches (such as the amount of grass cover and density of the understory vegetation) have changed since the commencement of the study, and these physical changes have advantaged some taxa but disadvantaged others. The long-term trends in detection probabilities of birds taking place as a consequence of landscape transformation will continue to be quantified until at least 2020.

2002a; Manning et al. 2004b), and they provide useful resources for a range of reptiles (Fischer et al. 2005b). All of these examples demonstrate that the notion of species being restricted to unmodified patches of native vegetation is often incorrect. In many modified landscapes, parts of the matrix provide valuable habitat for a range of native species, especially if the structural contrast between the matrix and remnant patches of native vegetation is mimimized.

Despite the potential value of the matrix as habitat, it is important to recognize that not all species will survive in all types of matrix (Renjifo 2001). A given species' survival in the matrix depends on the interaction of its particular requirements with the environmental conditions in the matrix (Laurance 1991; Andrén 1997; Chapter 8).

Landscape Connectivity

In addition to providing permanent habitat, the matrix can be a movement conduit for some organisms, thus reducing the negative effects of habitat isolation. In this role, it is complementary to stepping stones and wildlife corridors (Chapter 12). A long-term study on rainforest fragments in the Amazon showed that vegetation structure in the matrix was a key factor affecting the ability of organisms to move between native forest patches (reviewed by Gascon et al. 1999; Laurance et al. 2002). Gascon and Lovejoy (1998) likened the matrix to a filter, whose pore size determined its permeability for different types of species. This notion has some empirical support in other modified ecosystems (e.g., Bender and Fahrig 2005). In a study on butterflies in Colorado (USA), Ricketts (2001) demonstrated that four out of six species examined moved more readily between meadows through a matrix composed of willow forest rather than conifer forest. Another interesting example highlighting the value of the matrix to reduce habitat isolation comes from Maine (USA), where Joyal et al. (2001) found that two threatened turtle species relied on favorable matrix conditions for their regular movements between geographically isolated wetlands. In this example, the maintenance of suitable matrix conditions was important to prevent the extinction of the turtles.

Ecological Context

A final and related key role of the matrix is to provide the ecological context for native vegetation patches located within it. Conditions in the matrix can have an important effect on which species are present in patches of native vegetation. For example, Diamond et al. (1987) demonstrated that the bird fauna in a single isolated Javan woodland could not be adequately described as a forest "island." Rather, the bird species found within it mirrored the species that were present in the surrounding landscape matrix. Context effects become particularly noticeable when species composition is compared between patches composed of similar vegetation types but located in different matrix types (Box 14.1; Figure 14.2). In southeastern Australia, Watson et al. (2005) found that patches were more island-like for woodland birds in landscapes where the matrix was less hospitable. Similarly, in Chile, native forest patches within a plantation of introduced pine species supported different suites of birds, depending on the types of tree species surrounding a particular native forest patch (Estades and Temple 1999).

Detectable context effects of the matrix do not occur for all species groups in all ecosystems (Edenius and Sjöberg 1997), and in other examples, the matrix can have a strongly negative context effect. For instance, Ås (1999) examined ants in Swedish forest patches and found that small patches were dominated by matrix-tolerant species, and only large patches contained forest-specialized ants. Similarly, Antongiovanni and Metzger (2005) studied birds in remnant patches of Amazonian rainforest surrounded by cattle ranches in Manaus, Brazil. They classified species according to their response to landscape modification as highly sensitive species, moderately sensitive, or positively affected by landscape change. The latter group of species was able to utilize the landscape matrix and often invaded small rainforest remnants (Antongiovanni and Metzger 2005). Such invasions of small patches by generalist species represent a type of biotic edge effect (see Chapter 11).

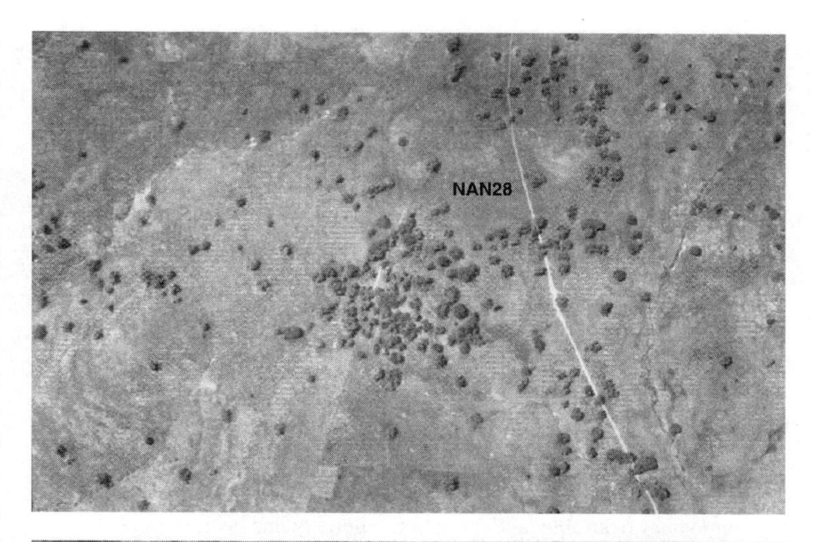

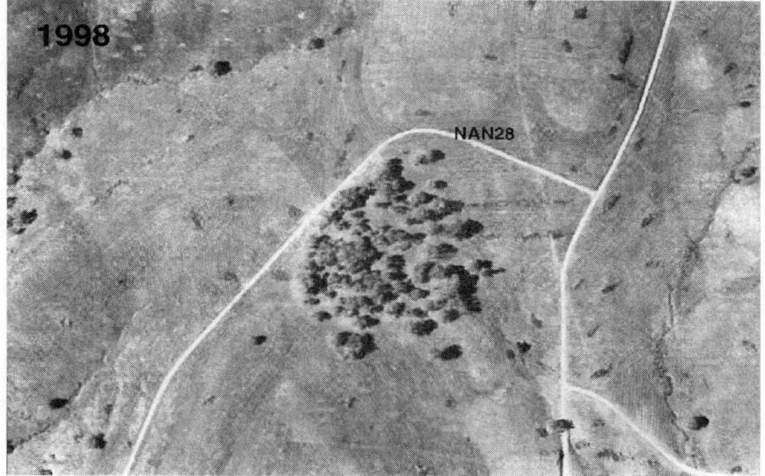

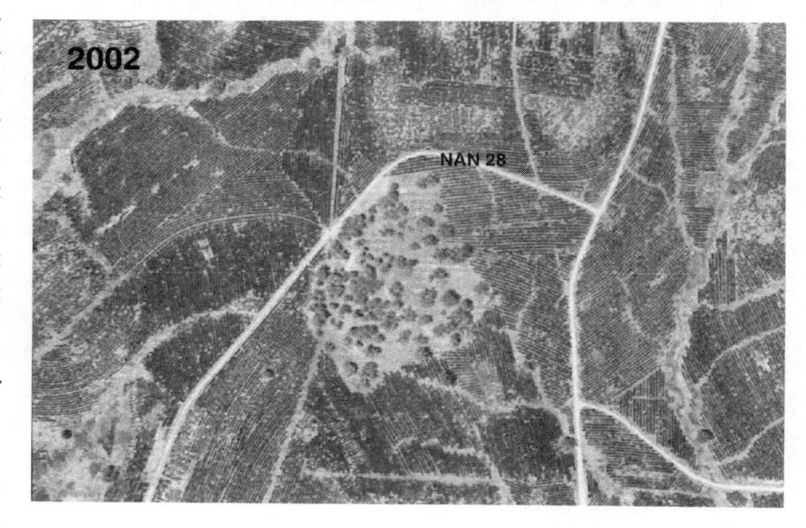

Figure 14.2. Sequence of images showing the changing landscape surrounding remnant woodland patches at Nanangroe, southeastern Australia (photos by State Forests NSW).

Box 14.2. Landscape Heterogeneity in a Forestry Context

Great care should be taken in assessing the extent and type of landscape heterogeneity required for matrix management. In a native forest context, this topic has frequently been misrepresented and abused by proponents of intensive wood production (Scientific Panel on Ecosystem Based Forest Management 2000). For over a century, production foresters have favored the "fully regulated forest" in which there is a perfectly balanced age distribution of forest stands (varying in age up to the rotation time). The aim of the fully regulated forest is to provide an even flow of wood products (Oliver et al. 1997). The notion of creating such "sustained-yield landscapes" that maximize timber production persists today, albeit in modified forms. For example, in southeastern United States, supposedly desirable landscape heterogeneity has been represented as 60 different age classes of managed pine forest (*Pinus* spp.) (e.g., Boyce 1995). The potential lack of stand level structural complexity within any given age class resulting from silvicultural management could result in the loss of key old-growth elements that take prolonged periods of time to develop, resulting in severe negative impacts on biota accumulating across entire landscapes. Similarly, Oliver and Larson (1996) and Oliver et al. (1997) proposed generic forest policies based on the creation of four or five structural types of managed forest over the entire forest estate. Such regulated or equilibrium forest landscapes composed of many structurally simplified stand age classes can have significant negative impacts on native species in the matrix, particularly when it is assumed that essentially all commercial forestland will eventually be harvested. These are not the types of "heterogeneous landscapes" needed to promote nature conservation. Many negative consequences of such oversimplified approaches have been documented, ranging from the complete elimination of some forest structures (Linder and Östlund 1998) to species loss (Lindenmayer et al. 1999a).

Landscape Heterogeneity

Some landscapes are so diversely patterned that the delineation of patches within a dominant background matrix (which itself may be highly nonuniform) becomes difficult or arbitrary. August (1983) differentiated between horizontal landscape heterogeneity and vertical structural complexity. Landscape heterogeneity is related to the extent to which a landscape viewed from the air is characterized by a diversity of environmental gradients or patch types. On the other hand, as outlined in Chapter 10, structural complexity is a local-scale phenomenon primarily concerned with the variation in vertical attributes encompassing the ground level and other layers of vegetation. Although this distinction is useful in theory, in practice many heterogeneous landscapes also have a relatively high degree of structural complexity (Box 14.2). Examples discussed following here include variegated grazing landscapes as opposed to fragmented cropping landscapes in southeastern Australia, organic farms as opposed to conventional farms (especially in Europe), and a case study of the Las Cruces farming landscape in Costa Rica.

Landscape Models in Heterogeneous Landscapes

In some heterogeneous landscapes, the patch–matrix–corridor model can become meaningless because the definition of patch boundaries can be somewhat arbitrary. In particular, patch boundaries defined from a human perspective may be irrelevant for many organisms, including some of conservation interest. In the terms of Manning et al. (2004a; see Chapter 3), the Umwelt relevant to most species is likely to differ from the Umwelt perceived by humans. Given this difference in perception between humans and other species, the terminology typically used for modified landscapes—including patches, corridors, and edges—can be of limited utility in highly heterogeneous landscapes. An alternative way to conceptualize highly heterogeneous landscapes is to apply the landscape contour model (Chapter 3), which recognizes that each species has its own individualistic patterns of distribution. In heterogeneous landscapes, different vegetation types occur side by side, structural complexity may be high, and many different types of niches are available that can be used by different organisms. Heterogeneous landscapes therefore often support more species than otherwise similar but less heterogeneous landscapes.

Benefits of Heterogeneity: Case Studies

Several case studies help to illustrate the advantages of landscape heterogeneity for conservation (Box 14.3). Fischer et al. (2004b) surveyed reptiles in the Nanangroe grazing landscape in southeastern Australia (see Box 14.1). The landscape was best described as variegated (sensu McIntyre and Hobbs 1999; see Chapter 2). Trees covered approximately 15% of the study area, and individual trees and small clumps of trees were scattered throughout the landscape. The majority of reptile species in the Nanangroe landscape were not restricted to woodland patches (Fischer et al. 2005b). Although species differed substantially in their habitat preferences (Fischer et al. 2004b), many benefited from tree-related habitat features present throughout the landscape, including tree hollows, fallen timber, and leaf litter on the ground.

In contrast, Driscoll (2004) surveyed reptiles several hundred kilometers to the west of Nanangroe. Tree cover in his study area was approximately 10%. However, unlike the Nanangroe landscape, this landscape was dominated by wheat fields. Scattered trees between clearly delineated woodland patches were extremely uncommon, and distances between woodland patches often exceeded several hundred meters (Don Driscoll, pers. comm.). Despite substantial survey effort, Driscoll (2004) recorded only two individuals of one reptile species in the matrix. Although some caution is needed in comparing the two studies because of some differences in climate, vegetation, and land use practices, the vastly different patterns of reptile distribution probably reflect the benefits of landscape heterogeneity for reptile diversity.

Benefits of heterogeneous farming systems have also been documented in many other cases. Hole et al. (2005) reviewed empirical studies that had compared various measures of biodiversity between organic and conventional farming systems. In contrast to large-scale intensive farming systems, organic farming is characterized by a reduced use of pesticides and inorganic fertilizers, the sympathetic management of noncropped areas, and mixed farming systems (Bengtsson et al. 2005). Hole et al. (2005) found 76 studies that compared various measures of biodiversity between the two types of farming systems, including studies of birds, mammals, butterflies, spiders, and several other groups. The review concluded that there were many instances where organic farming was associated with higher species abundances or richness, and only few examples where the opposite effect was detected (see also Bengtsson et al. 2005). Interestingly, Hole et al. (2005) acknowledged that it remained unclear if similar conservation benefits could also be achieved by

Box 14.3. The Value of Heterogeneity—Microexperimental Evidence

The conservation value of landscape heterogeneity becomes particularly evident when landscape heterogeneity is lost. An illustrative example comes from microexperimental work on plant diversity in North America (Tilman 1987), which examined relationships between species diversity and productivity in vegetation successional stages of new fields to forests. Productivity was artificially increased by adding nitrogen to the study plots. After some years, more than half of the plant species were lost from high nutrient treatments. In this system, natural heterogeneity was most probably underpinned by spatial variation in soil microenvironments. Such variation enabled the copersistence of plant taxa characteristic of low and high soil nutrient microenvironments. Adding nutrients erased spatial variation in soil nutrient patterns, and plant species loss accompanied the loss of heterogeneity (Tilman 1987). Although there are major problems in extrapolating the outcomes of this microexperimental work to larger spatial scales, the work nevertheless highlights how spatial heterogeneity and species richness can be correlated.

Box 14.4. Countryside Biogeography

Work in the Las Cruces farming landscape in Costa Rica (Figure 14.3) has been conducted under the label of "countryside biogeography." This term was coined by Daily (1999, 2001) to create an overarching framework under which various aspects of nature conservation in human-modified landscapes could be addressed. Countryside biogeography reocognizes that most of Earth's land is dominated by human settlements, agriculture, and forestry (Foley et al. 2005). Because human-modified "countryside" dominates terrestrial ecosystems, the future of many species will hinge to a large extent on how these modified ecosystems are managed. Conservation outside protected areas presents many new challenges to ecologists, including some important questions that go beyond their traditional comfort zone of scientific enquiry. Key questions include (partly adapted from Daily 1999, 2001):

- What is the capacity of human-dominated areas to support biota?
- Which traits of species enable them to survive in modified areas?
- What are some potential consequences of agricultural intensification?
- What are the links between biodiversity and ecosystem functioning?
- How can conservation and production (forestry, agriculture) be integrated?
- Which species should society try to conserve?
- What are the scientific and ethical arguments for conservation in modified landscapes?

targeted prescriptions in a conventional farming context (e.g., agri-environment schemes; see Kleijn and Sutherland 2003; Weibull et al. 2003; Vickery et al. 2004). In a separate review of farmland biota that did not differentiate between organic and conventional farms, Benton et al. (2003) argued that the loss of landscape heterogeneity due to agricultural intensification had been the key factor leading to species decline in European agricultural landscapes. Restoring heterogeneity within and between fields therefore was recommended as a key step toward ecological restoration by these authors (Benton et al. 2003; see also Part V of this book).

Some of the most compelling evidence for the benefits of landscape heterogeneity for native species conservation comes from agricultural landscapes in the tropics. One of the best studied examples is the Las Cruces landscape in Costa Rica (Daily et al. 2001; Mayfield and Daily 2005; Box 14.4; Figure 14.3). Extensive landscape modification in this area began in the 1960s, and to date, the continuous tropical moist forest once covering the area has largely been cleared (Mayfield and Daily 2005). An exception to this general pattern is the Las Cruces forest reserve; at 227 ha, this reserve is the largest remaining forest remnant within a circle of 15 km radius (Ricketts et al. 2001). Current land cover in the surrounding matrix includes coffee grown in full sun, shade coffee (planted among monospecific stands of shade trees), cattle pastures with scattered native trees, mixed farms, and remnant patches of native forest. The landscape is characterized by a high level of heterogeneity and variability in structural complexity, with a single land use type typically covering contiguous areas of no more than 30 ha (Ricketts et al. 2001). The maintenance of heterogeneity is likely to be of key importance to maintain the remaining species and prevent further extinctions in the Las Cruces area (Luck and Daily 2003; Mayfield and Daily 2005). Many taxonomic groups are still species-rich and occur throughout the Las Cruces landscape. For example, Daily et al. (2001) suggested that over 90% of bird species originally inhabiting the area still survived. Although many species were only found in forest fragments (Daily et al. 2001), nearly every second species used agricultural land in some way (Hughes et al. 2002). Scattered fruit trees in pastures, for example, attracted a range of bird species (Luck and Daily 2003). For moths, Ricketts et al. (2001) suggested that the majority of species appeared to move frequently between forested and agricultural areas, thus ac-

Figure 14.3. The Las Cruces farming landscape in southern Costa Rica. The photo shows a coffee plantation on a hillside with small patches of rainforest in the background and a "live fence" running up the left hand side of the hill (photo by Gary Luck).

tively using the heterogeneous landscape pattern. Similarly, species richness of butterflies was particularly high in areas where coffee and forest patches occurred close together (Horner-Devine et al. 2003). Despite these positive examples, not all taxonomic groups fared equally well. For instance, the mammal assemblage has shrunk from about 60 species originally to approximately 35 native species (Daily et al. 2003). A key issue in the Las Cruces landscape and others like it will be the conservation of species that do not fare well under altered ecological conditions and the need for focused action to ensure their persistence.

Summary

Preceding chapters in Part III have discussed assemblages in relation to patches, landscape connectivity, and edges. These topics have been central themes in the literature on species conservation in modified landscapes. However, in some situations, assemblages may not closely correspond to human-defined patch boundaries, corridors, or edges. The matrix (i.e., the largest and often most highly modified patch type in a given landscape), can be important because it can provide habitat, landscape connectivity, and a context for patches of native vegetation. These important roles of the matrix are illustrated in this chapter by examples on birds, turtles, and butterflies from a range of locations throughout the world. Some landscapes are characterized by spatial patterns that are so complex that a meaningful delineation of patch boundaries becomes difficult. In spatially heterogeneous landscapes, the distribution patterns of many species will not conform to human-defined patch boundaries. Heterogeneous landscapes often support more species than otherwise similar, but less heterogeneous landscapes. This is illustrated by examples of Australian reptiles, European agricultural systems, and a case study of the Las Cruces farming landscape in Costa Rica.

We conclude that well-understood and popular concepts about patch sizes, landscape connectivity, and edge effects must be complemented by considering the ecological roles of the matrix and landscape heterogeneity.

Links to Other Chapters

The matrix is discussed in the context of landscape models in Chapter 3. Approaches to enhanced matrix management and the maintenance of landscape heterogeneity are discussed in Chapter 18.

Further Reading

The role of the matrix is reviewed in a forestry context by Lindenmayer and Franklin (2002). Laurance et al. (2002) and Gascon and Lovejoy (1998) provide valuable overviews of a long-term investigation of landscape change in the Amazon basin, in which matrix conditions were found to be critically important. The role of landscape heterogeneity for European farming systems is reviewed by Benton et al. (2003). Reviews of the benefits of organic farming for native species were prepared by Hole et al. (2005) and Bengtsson et al. (2005). An extensive body of work by Daily and her coworkers discusses the highly heterogeneous Las Cruces landscape in Costa Rica (e.g., Daily et al. 2001, 2003; Ricketts et al. 2001; Hughes et al. 2002; Horner-Devine et al. 2003; Luck and Daily 2003; Mayfield and Daily 2005).

CHAPTER 15
Synthesis: Cascading Effects of Landscape Change

The preceding chapters in Part III have discussed a range of changes to landscape pattern typically associated with landscape modification, and how they can affect assemblages of species. Key patterns discussed included the loss of native vegetation (Chapter 9), the loss of structural complexity and vegetation deterioration (Chapter 10), edge effects (Chapter 11), the loss of landscape connectivity (Chapter 12), and the role of the matrix and landscape heterogeneity (Chapter 14). The separate discussion of these themes was useful to break down Bunnell's (1999a) "fragmentation panchreston" into its subcomponents—that is, to bring some conceptual clarity into an area of research where the term "fragmentation" is often applied too loosely (see Chapter 1). Although treating these themes separately was convenient to organize information in a logical sequence, typically, in real ecosystems, multiple aspects of landscape pattern change simultaneously (e.g., Gustafson 1998). This chapter introduces the notions of cascading effects of landscape change and regime shifts in ecosystems, particularly in response to simultaneous changes in vegetation cover, structural complexity, and species composition. Such cascades affecting entire ecosystems are analogous to our earlier treatment of interacting threatening processes affecting single species in Chapter 8. Following a range of examples of cascading effects of landscape change, we briefly discuss why the maintenance of functional and resilient ecosystems is important for the ongoing provision of ecosystem goods and services.

Cascading Effects of Landscape Change

Various effects of landscape modification typically interact. When these effects reinforce one another, they can trigger cascades of ecosystem change that can be difficult or impossible to reverse. Such cascades are sometimes referred to as regime shifts. Regime shifts occur when interrelationships between key variables in an ecosystem change fundamentally—they can be thought of as transitions where an ecosystem "flips" from one state to another (Scheffer et al. 2001; Folke et al. 2004; Walker and Meyers 2004). The following sections outline examples of cascading effects of landscape change in relation to three key aspects of landscape modification discussed in earlier chapters: changes to the extent and spatial configuration of vegetation cover, structural complexity, and species composition (Figure 15.1).

Vegetation Cover and Associated Disturbance Regimes

Landscape modification changes the extent and spatial configuration of vegetation in a given landscape (Chapter 2). For example, at the landscape scale, large proportions of native ecosystems may be lost (Chapter 9). Similarly, at a local scale, remnant patches of natural ecosystems may be too small (Chapter 9) or not well enough connected (Chapter 12) to support all species characteristic of their original state. In addition, landscape modification typically creates edges between different types of land use, which influence the remnant patches adja-

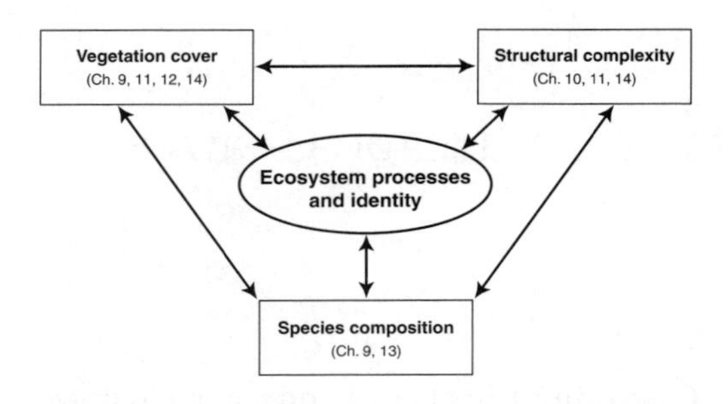

Figure 15.1. Conceptual summary of how three key ecosystem properties (vegetation cover, vertical structural complexity [see Chapter 10], and species composition) relate to ecosystem processes and ecosystem identity, with references to preceding chapters in Part III of the book.

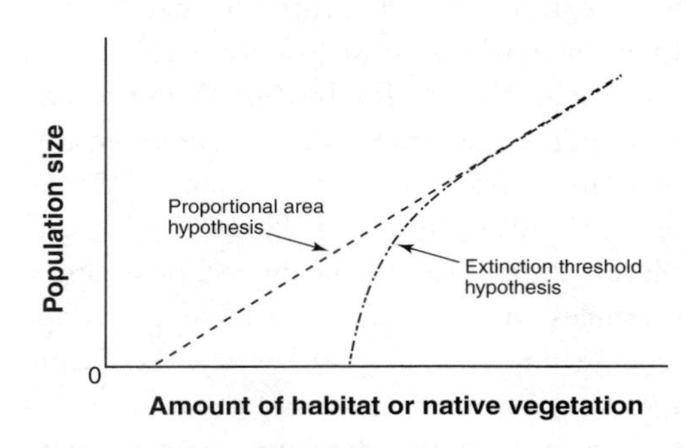

Figure 15.2. Comparison of rates of species loss proportional to the remaining amount of native vegetation and the threshold hypothesis where species are lost faster when vegetation cover is reduced be below approximately 30% (redrawn from Fahrig, L. 2003: reprinted, with permission, from the *Annual Review of Ecology, Evolution, and Systematics*, Volume 34© 2003 by Annual Reviews www.annualreviews.org).

cent to them via abiotic and biotic edge effects (Chapter 11).

Cascading effects on assemblages arising from the simultaneous losses of the amount of native vegetation and landscape connectivity have received considerable attention. Andrén (1994) examined empirical studies on birds and mammals and analyzed population sizes of individual species as well as species richness in relation to the proportion of relatively unmodified vegetation in a given landscape. Both for single species' populations and for species richness, Andrén (1994) found that these variables declined proportionally with vegetation cover until approximately 30% of native vegetation remained. Below this level of vegetation cover, both population sizes and species richness appeared to decline faster than predicted by a linear relationship (Figure 15.2). A likely explanation for the cascade of ecosystem change predicted for vegetation cover levels below 30% is that the loss of native vegetation per se is increasingly compounded by the loss of landscape connectivity, thus exerting particular pressure on individual species and species assemblages.

The possibility of a broad rule of thumb about how much native vegetation is needed to avoid regime shifts—for example, a 30% threshold—has received considerable attention by both scientists and workers in the policy arena (e.g., see review by Huggett 2005). The "30% rule" has been examined for individual species (see Chapter 8), assemblages (Parker and Mac Nally 2002; Drinnan 2005; Radford et al. 2005), and also broader measures of ecosystem integrity (e.g., tree health; see McIntyre et al. 2002). The concept has also been used to recommend benchmarks for land management (McAlpine et al. 2002b) (e.g., for native vegetation retention on farms) (Barrett 2000; McIntyre et al. 2000).

With respect to species richness of assemblages, it is important to note that the 30% rule is based on a binary classification of land into patches of native vegetation that are deemed suitable for all species in a given assemblage versus a matrix of universally unsuitable land. Although this classification may be a reasonable approximation of assemblage organization in some landscapes, often, the matrix and nonnative vegetation and structural features within it can play important ecological roles (Chapter 14). Thus, although the 30% rule is appealing and potentially a useful management guideline in some landscapes, it cannot be applied uncritically to all ecosystems and assemblages (see Box 15.1). Perhaps the most useful, albeit general, conclusion from the literature on vegetation thresholds to date is that cascades of negative effects on native species assemblages arising from the simultaneous losses of native vegetation and landscape connectivity are more likely to occur at low levels of native vegetation cover (Radford and Bennett 2004; see Figure 15.2).

Changes to the spatial and temporal patterns of disturbance regimes can also play an important role in cascades of ecosystem change. Fire is a major agent of disturbance in many ecosystems (Agee 1993; Cary et al. 2003). Landscape alteration can have significant effects on fire frequency, fire intensity, fire size, and other elements of fire regimes (sensu Gill 1975) (Box 15.2; Figure 15.3). For example, in South America, the spatial juxtaposition of cleared fire-prone pastures can significantly increase wildfire risk in adjacent rainforest fragments (Kauffman and Uhl 1991). Drying of more open and exposed vegetation following logging can also increase risks of a conflagration in rainforest environments such as those in southeast Asia (van Nieuwstadt et al. 2001). Similarly, relatively nonflammable patches in Yellowstone National Park burned when the surrounding altered landscape matrix supported high-intensity fires (Romme and Despain 1989). This change in fire regimes had major flow-on effects on the biota, with plant species composition and distribution patterns being dramatically altered (Spies and Turner 1999; Turner et al. 2003).

> **Box 15.1. Limitations of the "30% Rule"**
>
> The existence of a simple threshold amount of native vegetation below which ecosystems undergo regime shifts is appealing from both a scientific and a management perspective. However, several important caveats of the notion of such thresholds should be considered, particularly in the context of a simple threshold postulated to apply to many landscapes, like the "30% rule." First, a hypothesized threshold amount of native vegetation below which species richness declines more rapidly hides the fact that species can go extinct with native vegetation loss even above a postulated threshold level—albeit at a slower rate (see Figure 15.2). Second, a hypothesized threshold does not consider that land clearing tends to take place primarily in the areas with high primary productivity; thus the remaining native vegetation is rarely a representative sample of the original land cover (Chapter 2). Third, not all species respond to a given pattern of native vegetation in the same way (Mönkkönen and Reunanen 1999). Therefore, different levels of native vegetation may be suitable for different species (Homan et al. 2004; see Box 8.1 in Chapter 8), and some species may have continuous or linear rather than threshold relationships with native vegetation cover (Bunnell et al. 2003). For species richness—an attribute of ecosystems that emerges from many species' individualistic responses—this means its particular relationship with native vegetation will be highly situation-specific and will vary between taxonomic groups and landscapes (see Parker and Mac Nally 2002; Drinnan 2005; Radford et al. 2005). Thus, although there are cases where regime shifts have been associated with the 30% rule (e.g., Andrén 1994), the foregoing considerations highlight that the rule (like all ecological principles) cannot be applied uncritically in all landscapes (Lindenmayer et al. 2005c).

Structural Complexity

Landscape change not only leads to changes in horizontal patchiness but also causes changes in the vertical structural complexity of ecosystems (see Figure 15.1). Key mechanisms affecting structural complexity typically associated with landscape change include management practices leading to landscape-wide structural deterioration (Chapter 10) and edge effects (Chapter 11). In many cases, the nature of the landscape matrix will play a major role in the stability of ecosystems and their overall susceptibility to cascading effects of landscape change (Chapter 14). The extent to which resources supplied by native ecosystems are also available in the matrix will affect many species,

Box 15.2. The Disruption of Fire Regimes as a Result of Vegetation Loss and Subdivision

Vegetation loss and subdivision can have profound impacts on key ecosystem processes, with cascading negative effects on biota. An example is illustrated by forests dominated by stands of longleaf pine *(Pinus palustris)* in southeastern United States. Continuous stands of these forests once covered almost 30 million hectares. However, due to extensive clearing and logging, remaining forests are primarily regrowth, with only limited old growth still persisting. Several threatened species occur in old-growth longleaf pine forests, particularly stands characterized by large living cavity trees. For example, the viability of populations of the red-cockaded woodpecker *(Picoides borealis)* in longleaf pine stands is an important conservation issue (Sharitz et al. 1992; Rudolph and Connor 1996). This species excavates nests in large living trees that have significant wood decay caused by brown rot *(Fomes pini)* but which still provide significant sap flows that thwart the movement of snakes up the trunks of trees and deter predation by them (Jackson 1978, 1994).

Fire is the key agent of natural disturbance in longleaf pine stands—without it they are replaced by hardwood stands (Connor and Rudolph 1989). Historically, continuous stands of longleaf pine were burned approximately every 3 years by fires ignited by lightning strikes that spread rapidly among adjacent and highly connected stands. In particular, the highly flammable ground cover of old-growth stands facilitated the spread of fires. However, due to extensive landscape alteration, remaining stands of longleaf pine are now spatially separated such that the probability of lightning striking any particular area is greatly reduced, and the ground cover of regrowth stands does not carry fire as well as in old-growth stands. Thus a key ecological process in longleaf pine—regular natural disturbance by fire—has been disrupted. Extensive, regular, and costly prescribed burns are now required in longleaf pine stands to (1) maintain required stem densities, (2) eliminate competing hardwood species, (3) stimulate the development of understory plants, and (4) ensure the development of suitable habitat for a range of individual species.

and vegetation structure can be a key attribute in this context. For example, in the Tumut region of southern New South Wales, many native reptiles and birds are able to supplement their habitats associated with native forest patches by using the matrix of introduced pine forest for foraging or interpatch movement (Lindenmayer et al. 1999b; Fischer et al. 2005a). Importantly, in this forest-dominated area, several key structural elements, such as scattered native trees and shrubs and logs remaining from previously cleared native forest, are present throughout the landscape.

In contrast, the loss of key structural elements can lead to cascades of species extinctions. Tews et al. (2004) defined elements of structural complexity that are particularly important to a large number of species as "keystone structures." Tews et al. (2004) suggested that temporary wetlands in agricultural fields in northeastern Germany are good examples of keystone structures. These wetlands occur in conventionally ploughed depressions when they fill up with water after heavy precipitation in spring or winter. The temporary flooding of these depressions gives rise to a locally diverse and structurally complex vegetation layer including forbs, grasses, and dwarf-rush communities. Carabid beetles, in turn, respond strongly to this local diversity of habitats, and temporary wetlands contribute disproportionately to the species richness of beetles in agricultural fields (Tews et al. 2004). It follows that the loss of temporary wetlands would result in substantial losses of vegetation communities and beetle species.

Compelling evidence for the existence of keystone structures that are fundamental to ecosystem functioning also comes from African savannahs, where scattered trees are focal points for a wide range of plant and animal species (Box 15.3; Figure 15.4). Other kinds of keystone structure include large old trees with hollows that are used by a plethora of cavity-dependent vertebrates around the world (Fischer and McClelland 1983; Gibbons and Lindenmayer 2002) and mistletoe, which has critical food, shelter, and other roles for biota in many parts of the world (reviewed by Watson 2001, 2002).

Although some elements of structural

Figure 15.3. Longleaf pine stand in southeastern USA (photo by Jerry Franklin).

Figure 15.4. Dry savannah with scattered trees in the Serengeti, Tanzania (photo by Rob Heinsohn).

complexity are particularly important because they directly provide habitat for many species, keystone structures may also play crucial roles in regulating ecosystem processes. For example, trees in the agricultural areas of temperate Australia play a crucial role in the maintenance of hydrological processes. In temperate Australia, tree clearing for agriculture has been extensive over the last 200 years, with many billions of trees having been removed from areas such as the Murray-Darling Basin (Walker et al. 1993). However, where trees are removed and replaced by plants with lower rates of transpiration (e.g., crop and pasture plants), rates of groundwater recharge increase and groundwater levels rise. Water then moves upward through salt-laden sublayers of the soil. The resulting movement of salt to the surface can cause the severe salinization of ecosystems. The cascading effects of tree removal and the resulting salinization are often severe (Stirzaker et al. 2002). Salinization threatens agricultural productivity, water quality, the persistence of remnant vegetation, and many associated plant and animal species (State of the Environment Report 2001; Briggs and Taws 2003). McKenzie et al. (2003) demonstrated that, as areas of woodland vegetation contract due to increased salinity, communities like small ground-dwelling animals that depend on such habitats are also reduced. In the Western Australian wheatbelt, the Department of Conservation and Land Management (2000) estimated that over 450 endemic plants and 200 species of aquatic invertebrates were at risk of extinction from salinity. Thus the loss of woodland trees has triggered a dramatic cascade of ecosystem change, where the loss of structural complexity is further exacerbated by the disruption of key ecosystem processes. The extent of the salinity problem in Australia is severe. More than 5.7 million ha of land are currently threatened by dryland salinity, and this figure may further increase to 17 million ha by 2050 (Commonwealth of Australia 2001, 2002).

Species Composition

In addition to vegetation cover and vertical ecosystem structure, the loss of species per se can lead to cascading effects of landscape change. No species exists in isolation—rather, species interact with one another through processes like competition, predation, and mutualisms (Chapter 7). In Chapter 7, we briefly highlighted some examples of how landscape change could disrupt interactions between species and thus may lead to the decline or loss of some species. Here, we revisit the interconnected roles of individual species and ecosystem processes in the context of cascading effects of landscape change. Important questions in this context include: (1) When does the decline or loss of individual species have negative cascading effects on other species? (2) What is the relationship between ecosystem function and species richness?

An early but influential analogy for the relationship between species loss and cascading changes of ecosystems is the "rivet popping" metaphor by Ehrlich and Ehrlich (1981). These authors likened global ecosystems to "Spaceship Earth," where Earth's species represent the rivets that support the wings of Spaceship Earth. Without enough rivets, the wings of Spaceship Earth would fall off, causing a fatal crash—similarly, without enough species, Earth's ecosystems will collapse. This intuitive analogy suggests that ecosystems with many species may be more stable (Peterson et al. 1998) or resilient (Elmqvist et al. 2003). "Resilience" refers to a system's capacity to "absorb disturbance . . . so as to retain essentially the same function, structure, identity, and feedbacks" (Folke et al. 2004, p. 558; also see Holling 1973); that is, the ability of a system to rebound from disturbance to its prior state. However, just like with rivets connecting the wings to a spaceship, the loss of any single species will not necessarily lead to collapse. Rather, the likelihood of species loss causing irreversible ecosystem change is linked to the functional redundancy of species. That is, some species are functionally more replaceable (or irreplaceable) than others (Walker 1992).

Walker's (1992) initial paper on "redundancy" caused some debate because it was incorrectly interpreted by some as an excuse to give up on the conservation of so-called redundant species. As Walker (1995) later clarified, this was not what he had meant to imply. Indeed, recent work has indicated that maintaining as many "redundant" species as possible that fulfill a particular ecosystem function is

Box 15.3. The Keystone Role of Scattered Trees in African Savannahs

Savannahs may be broadly defined as ecosystems where an open tree layer and a grass layer coexist, often due to the interplay of fire, grazing, and rainfall (Mistry 2000). Scattered trees in African savannahs are good examples of keystone structures because ecosystems would be fundamentally different without them (Tews et al. 2004). Scattered trees are important focal points for both plant and animal life. Under their canopy, there is more shade, and nutrient concentrations are higher than in the surrounding grassy areas because of the accumulation of leaf litter and animal dung (Belsky 1994). In addition, scattered trees facilitate improved water infiltration, thus reducing water limitation in an otherwise dry environment (Vetaas 1992). In combination, these changes to the physical environment create favorable conditions for the recruitment of many plant species, including understory shrubs. Plant diversity therefore is often higher under scattered trees than in surrounding grassy areas (Vetaas 1992; Belsky 1994). Scattered savannah trees are also key habitat features for many animals. Dean et al. (1999) showed that a wide range of animal species in the southern Kalahari desert used scattered trees for food or shelter. Animals include raptors that build nests in the canopy, several herbivores like the springbok *(Antidorcas marsupialis)* that shelter in the shade of scattered trees, and arboreal mammals like the tree rat *(Thallomys paedulcus)*, which nests in cavities of large trees. Given the range of ecological roles fulfilled by scattered trees, their continued existence is widely regarded as a fundamental part of nature conservation in African savannahs (e.g., Jeltsch et al. 1996).

Figure 15.5. Gopher tortoise (photo by Tracey Tuberville).

highly desirable to maintain functional and resilient ecosystems (Walker et al. 1999; see also Beattie and Ehrlich 2001). Elmqvist et al. (2003) highlighted that multiple species fulfilling similar ecological functions provide an important insurance against management mistakes or unforeseen environmental change (e.g., cyclones; Elmqvist et al. 2002). This is because the likelihood of at least some species within a given functional group persisting after a severe disturbance increases as the number of species within the functional group increases. Elmqvist et al. (2003) suggested that managing ecosystems to maintain a high level of "response diversity" within a given functional group was therefore important to maintain resilience.

Although species typically overlap in their contribution to ecosystem function, some species provide unique functions and are disproportionately important. For example, in the wheatbelt of Western Australia, the plant *Banksia prionotes* is the only one that provides suitable nectar resources for local species of honeyeaters at certain times of the year (Lambeck 1997). Honeyeaters, in turn, are important bird pollinators for many local plant species. Hence, the continued existence of *Banksia prionotes* is fundamentally important for maintaining ecosystem function in this area (Lambeck 1997; see also Walker 1995).

Similar situations occur in other ecosystems throughout the world. Another prominent example is the gopher tortoise *(Gopherus polyphemus)* in the southeastern United States (Figure 15.5). The gopher tortoise lives in extensive subterranean burrows in dry upland areas of longleaf pine *(Pinus palustris)* sandhills, xeric oak (*Quercus* spp.) hummocks, scrubs, dry prairies, and coastal dunes. More than 360 species of other vertebrates and invertebrates are known to use the burrows made by the gopher tortoise (Jackson and Milstrey 1989), including several such as the Florida mouse *(Podomys floridanus)* and many endemic invertebrates that are obligate burrow users and cannot survive without access to burrows. Landscape modification through urban development, forestry practices, and altered fire regimes, coupled with disease and collisions with motor vehicles, is reducing populations of the gopher tortoise. Because of the importance of the species' burrows, its decline is having major cascading effects on many other taxa (Gopher Tortoise Council 2005; Reed Noss, pers. comm.).

Species like *Banksia prionotes* in the Western Australian wheatbelt and the gopher tortoise in the southeastern United States are often referred to as keystone species (e.g., Knapp et al. 1999). Power et al. (1996, p. 609) defined a keystone species as "one whose impact on its community or ecosystem is large, and disproportionately large relative to its abundance." The term was originally coined by Paine (1969), who noted the disproportionate effect of removing the carnivorous starfish *Pisaster ochraceus* from rocky intertidal areas between Alaska and Baja California. Because of ongoing confusion over the precise definition of what constituted a keystone species, Soulé et al. (2003, 2005) coined the alternative and often interchangeable term "strongly interacting species." The loss of strongly interacting species is likely to lead to cascading effects for many other species (Soulé et al. 2005). Soulé et al. (2005) identified four ways in which strong ecological interactions can be most evident: (1) habitat enrichment, (2) predation, (3) competition, and (4) mutualisms. Numerous examples of strongly interacting species and the processes threatening them are listed by Soulé et al. (2005). North American examples discussed by these authors include the beaver *(Castor canadensis)*, which exerts a strong influence on the habitat of many other species; the gray wolf *(Canis lupus)*, which is an important predator affecting deer abundance and therefore indirectly affecting vegetation condition; the coyote *(Canis latrans)*, which outcompetes the domestic cat *(Felis catus)* in southern California with positive flow-on effects for the local avifauna (Box

Box 15.4. The Loss of a Keystone Species Can Trigger Cascading Effects of Landscape Change: Coyotes, Cats, and Birds in Suburban Scrub Patches in Southern California (also see Figure 15.6)

Because of their position at the top of the food chain, large carnivores often play a disproportionately important role in controlling ecosystem function (Soulé and Terborgh 1999; Coppolillo et al. 2004). The decline or loss of large carnivores can lead to cascading effects at lower levels of the food chain. An elegant example of such a cascade was presented by Crooks and Soulé (1999), who studied the relationships between the coyote *(Canis latrans)*, domestic cat *(Felis catus)*, and scrub-breeding birds in coastal southern California. Here, landscape modification and urbanization have led to the widespread destruction of native sage scrub. Crooks and Soulé (1999) studied 28 patches of sage scrub, ranging in size from 2 to 102 ha. The coyote was found to be sensitive to landscape modification and was less abundant in smaller patches than in large patches. The decline of the coyote in smaller patches, in turn, created more favorable conditions for medium-sized "mesopredators," like the domestic cat, partly because coyotes occasionally attack cats. In patches where the coyote was rare or absent, domestic cats were abundant. Questionnaires to hundreds of residents whose properties were adjacent to scrub patches found that nearly one-third of residents owned domestic cats. These cats, in turn, frequently preyed on scrub-breeding birds and other small animals. Crooks and Soulé (1999, p. 565) estimated that "cats surrounding a moderately sized fragment (~100 residences) return about 840 rodents, 525 birds, and 595 lizards to residences per year." Predation pressure on birds was considered unlikely to be sustainable, and bird populations were predicted to decline further in small patches. The case study illustrates how landscape change caused the decline of one species, thus leading to a range of cascading ecosystem changes (see also Figure 15.6).

15.4; Figure 15.6); and the mutualistic relationship between Clark's nutcracker *(Nucifraga columbiana)* and whitebark pine *(Pinus albicaulus)*, which are dependent on one another for food and seed dispersal, respectively (see Soulé et al. 2005 and references therein). The role of keystone or strongly interacting species is a rapidly developing research area, and it is likely that both concepts and terminology in this area will continue to be refined for some time into the future.

In summary, species-rich ecosystems are more likely to be resilient and are less likely to experience cascades of ecosystem change or regime shifts as a result of landscape modification. However, the extent to which the loss of any given species will affect other species and overall ecosystem integrity depends on its functional role and on the response diversity of species that fulfill similar functions. The loss of highly interactive species that fulfill unique functional roles is likely to lead to particularly severe cascades of ecosystem change for many other species.

Why Avoiding Regime Shifts Is Important

There are many reasons for avoiding further species decline and maintaining functional and resilient ecosystems. Following Gaston and Spicer (2004), two broad categories of reasons for maintaining biodiversity are summarized in Table 15.1: (1) present-day use values and (2) nonuse values (also see Daily 1997). Present-day use values include marketable commodities as well as ecosystem services (Costanza et al. 1997). BirdLife International (2000) estimated that the biophysical environment currently returns an estimated $US 33 trillion in goods and services to human society each year. Forest products are among the most important commodities. Wood products have been estimated to contribute approximately $US 400 billion annually to the world market economy (or about 2% of total gross domestic product; World Commission on Forests and Sustainable Development 1999). Important ecosystem services are also derived from natural ecosystems, including the regulation of stream flow, soil protection (Zedler 2003), and the provision of species that are important to maintain agriculture (Kearns et al. 1998).

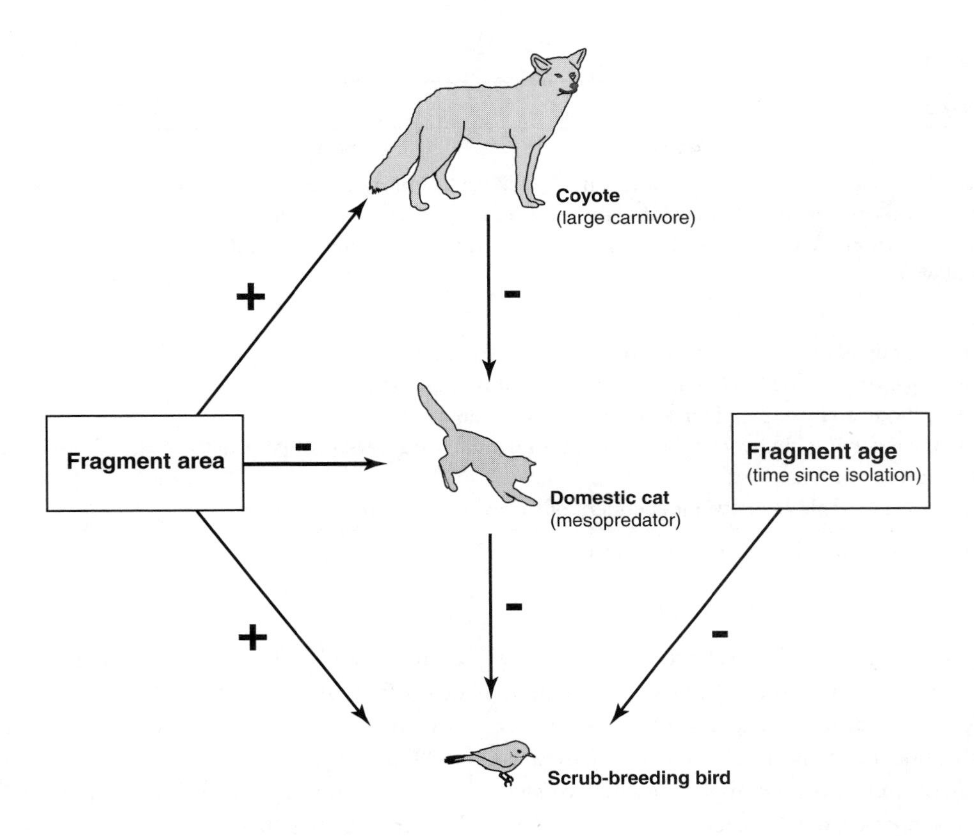

Figure 15.6. Cascading effects from the loss of the coyote in sage-scrub patches in southern California (redrawn from Crooks, K.R. and Soulé, M.E. 1999: reprinted by permission from Macmillan Publishers Ltd [Nature, 400, 563–566], 1999). The figure illustrates that patch size was negatively related to coyote abundance; the decline of the coyote, in turn, led to an increase of smaller predators like the domestic cat; increased predation pressure by cats led to a decline in scrub-breeding birds.

For example, Ricketts et al. (2004) suggested that Costa Rican coffee farmers profited from maintaining native forest habitat because this was used by pollinators, which were important to sustain the coffee industry. Ricketts et al. (2004) estimated that pollination services by native wildlife were worth approximately $US 60,000 per farm per year.

The deterioration of functional ecosystems can lead to major losses in ecosystem services (Daily 1997). For example, Sekercioglu et al. (2004) analyzed the ecological and societal consequences of anticipated global losses of bird species. Their analysis suggested that by 2100, 7 to 25% of the world's bird species would either be extinct or (due to their small population sizes) would contribute insignificantly to ecosystem functioning. This decline of the global avifauna, in turn, was predicted to have far-reaching negative consequences for the provision of essential ecosystem processes such as the decomposition of organic matter, pollination, and seed dispersal (Sekercioglu et al. 2004).

Nonuse conservation values are less tangible than present-day use values. However, some authors believe that nonuse values are at least equally important. Soulé (1985) recognized that the discipline of conservation biology was fundamentally based on the normative postulate that diversity of life is good in its own right. Thus species that are not pivotal to ecosystem functioning may still be highly valued for intrinsic reasons. Most threatened species, for example, contribute negligibly to ecosystem services (Sekercioglu et al. 2004), but "interspecific altruism" toward such species may nevertheless be seen as an important moral duty (Ehrenfeld 1976; Rolston 1985).

Table 15.1. Reasons for maintaining biodiversity*

Value	*Explanation*
Present-day use values	
Direct-use values	Values for the consumption or production of marketable commodities, including food, medicine, biological control agents, raw materials, recreational harvesting, tourism
Indirect-use values	Values of various ecosystem services such as nutrient cycling, climate control, provision of clean air and water
Nonuse values	
Option value	Value of biodiversity for the future, both for use or for nonuse
Bequest value	Value of passing on the biodiversity resource to future generations
Existence value	Value of biodiversity to people irrespective of use or nonuse
Moral/ethical value	Value of conserving biodiversity because of moral and ethical reasons that give the right of species in addition to humans to exist
Intrinsic value	Inherent value of biodiversity, independent of the value placed on it by people

*Modified from Gaston and Spicer (2004); Lindenmayer and Burgman (2005).

Avoiding cascading effects of landscape change, or regime shifts in ecosystem functioning, is critical for maintaining biodiversity. By definition, once regime shifts do occur, they are difficult to reverse, or impossible to reverse if genes, species, or ecosystems are permanently lost. Given this consideration, it is generally more efficient to avoid undesirable regime shifts in the first place, rather than let systems degrade to a state where major restoration efforts are required (Milton et al. 1994).

Other Regime Shifts

Many ecosystem processes other than the ones already described here can be severely impaired as a result of landscape alteration (McIntyre et al. 2000, 2002). Examples include the decomposition of organic matter (Klein 1989; McGrady-Steed et al. 1997), pollination (Prance 1991; Robertson et al. 1999; Cunningham 2000), seed dispersal (Cordeiro and Howe 2003; Bodin et al. 2006), and the formation of mycorrhizal associations (Maser et al. 1978; Perry 1994). In many of these cases, particular elements of the biota need to be maintained at functionally effective levels to maintain key ecosystem processes and, in turn, to limit negative flow-on effects to other species (Conner 1988; Soulé et al. 2003).

In addition, regime shifts may occur due to human activities that are not primarily related to landscape modification, such as pollution or hunting. Some examples of regime shifts associated with human activities other than landscape modification are summarized in Table 15.2.

Summary

Human landscape modification can result in extinction cascades. Such cascades can arise for a multitude of different reasons. Three common reasons include: (1) the excessive loss of native vegetation and associated altered disturbance regimes, (2) the loss of key structural elements (e.g., tree hollows in many forests), and (3) the loss of species that fulfill ecological functions that are not fulfilled by other remaining species (e.g., predation, pollination, and many other functions). In general, ecosystems are more likely to recover from an external shock if functional groups are species-rich.

Table 15.2. Examples of anthropogenic regime shifts that are not directly related to landscape change*

Original state	*Cause of shift*	*Trigger of shift*	*New state*
Clear-water lake	Phosphorus accumulation in agricultural soil and lake mud	Flooding, warming, overexploitation of predators	Turbid-water lake
Coral-dominated reef	Overfishing, coastal eutrophication	Disease, bleaching, hurricanes	Algae-dominated reef
Grassland	Fire prevention	Heavy rains, continuous heavy grazing	Shrub-bushland
Grassland	Hunting of herbivores	Increased prevalence of disease	Woodland
Kelp forests	Functional elimination of apex predators	Thermal event, storm, disease	Sea urchin dominance
Pine forest	Microclimate and soil changes, loss of pine regeneration	Decreased fire frequency, increased fire intensity	Oak forest
Tropical lake with submerged vegetation	Nutrient accumulation during dry spells	Nutrient release with water table rise	Floating-plant dominance

*The table shows the transition of ecosystems from their original state (column 1) to their modified state (column 4) as a result of underlying causes (column 2) and more immediate triggers (column 3). Modified from Folke et al. (2004), which has more details and additional references.

This is because high species richness within a given functional group increases the likelihood of at least some species surviving an external shock to the ecosystem and being able to compensate for the loss of other, functionally similar species. Keystone species fulfill unique functional roles and exert a disproportionate effect on ecosystem functioning relative to their abundance. Their conservation is particularly important to maintain functioning ecosystems.

There are no simple recipes for how to avoid extinction cascades in modified landscapes, but a key first step is to identify that extinction cascades are indeed taking place; then it is possible to determine why they are happening. Broadly speaking, patterns that result in low levels of native vegetation cover and low landscape connectivity are more likely to trigger extinction cascades than patterns with extensive native vegetation; reduced structural complexity is more likely to be linked to cascading effects than high structural complexity; and species-rich systems with many species fulfilling a given ecosystem function are less likely to suffer from cascading effects than ecosystems where functional groups are represented by only a few species. Avoiding extinction cascades is important to maintain essential ecosystem services (such as the supply of freshwater and nutrient cycling) that humans and their production industries rely upon. Other equally important reasons to maintain biodiversity include present-day use values, future use values, and nonuse values.

Links to Other Chapters

Cascading effects of landscape change often arise as a result of the various changes discussed in Part III of this book to landscape pattern (Chapters 9, 11, 12), structural complexity (Chapters 10, 11), and species composition (Chapter 13). Interrupted species interactions can trigger regime shifts, and some examples of such interruptions are discussed in Chapter 7. The loss of particular species may also trigger regime shifts. The threats leading to the extinction of individual species are summarized in Chapter 8.

Further Reading

Key papers on regime shifts are those by Scheffer et al. (2001), Folke et al. (2004), and Walker and Meyers (2004). Andrén (1994) and Fahrig (2003) reviewed the relationships between the amount of native vegetation and species richness. A special section in volume 124, issue 3 of *Biological Conservation* (e.g., see Lindenmayer and Luck 2005) considers a range of issues associated with thresholds and species diversity. The notion of keystone structures was introduced by Tews et al. (2004). Scattered trees in African savannahs are an excellent example of keystone structures, and they are discussed briefly by Tews et al. (2004) and in more detail by Dean et al. (1999). Ehrlich and Ehrlich (1981) introduced the now famous rivet popping metaphor, and Walker (1992, 1995) discussed the notion of ecological redundancy within a given functional group. A detailed review of stability/species richness issues is provided by Peterson et al. (1998). Key papers on resilience were written by Holling (1973) and Elmqvist et al. (2003). Useful papers on keystone species are Paine (1969), Power et al. (1996), Soulé et al. (2003, 2005), and Ripple and Beschta (2005). Daily (1997) is a seminal book on ecosystem services. Many textbooks summarize the reasons for maintaining biodiversity; a short and accessible overview is given by Gaston and Spicer (2004).

PART IV

Studying Species and Assemblages in Modified Landscapes

The preceding chapters and parts in this book have focused on how individual species and assemblages respond to the altered ecological processes and spatial patterns of vegetation cover that accompany landscape modification. Part IV moves away from the examination of ecological responses and outlines some of the ways in which the impacts of landscape change can be studied. However, Part IV is not completely divorced from the "fragmentation panchreston" theme that has permeated the preceding parts and chapters. Rather, careful consideration of the subcomponents of the panchreston can help focus studies on key problems and questions and promote progress toward better mitigation of the impacts of landscape modification.

This section of the book contains two chapters (Figure IV.1). One explores field-based approaches to studying species- and assemblage-level responses in modified landscapes; these are methods where new data are gathered. The second concerns desktop methods that involve reanalyzing, synthesizing, or value-adding to existing field data. Both broad categories of approaches can be employed to examine single-species problems as well as the response of multiple species and assemblages to landscape modification.

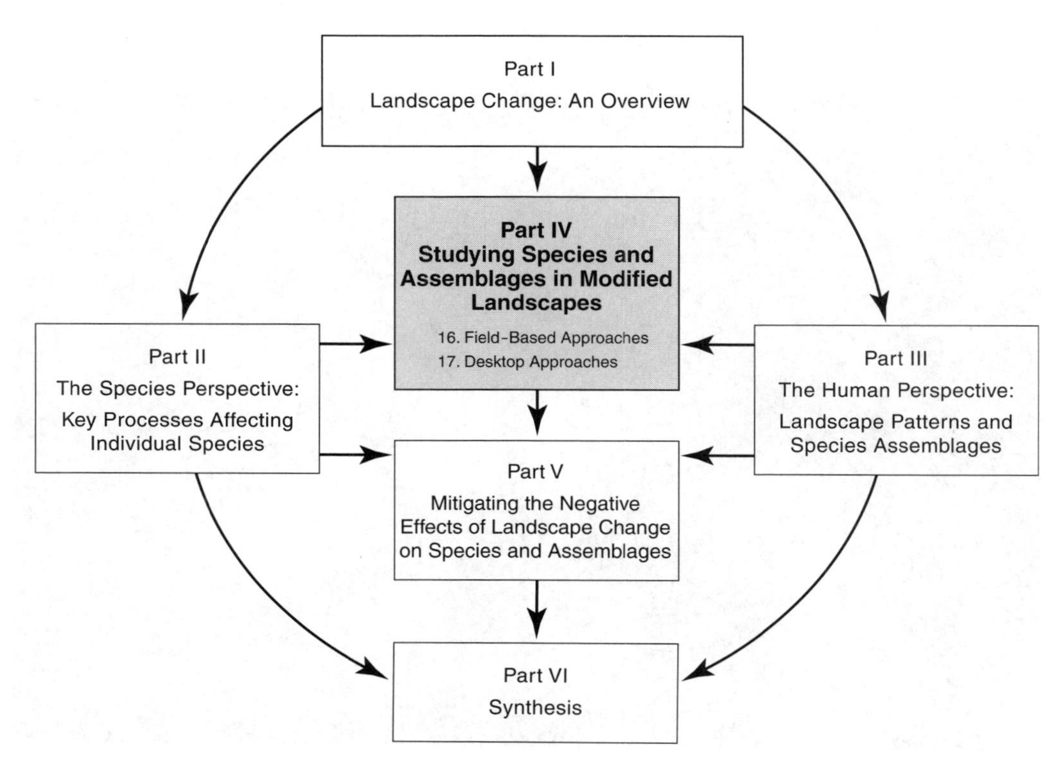

Figure IV.1. Topic interaction diagram showing links between the different parts of the book, and the topics covered in the chapters of Part IV.

We emphasize that our treatment of both kinds of approaches is, at best, very cursory and is designed to provide readers with a very brief taste of the ways in which they might tackle particular problems in studying landscape modification. Indeed, entire books have examined topics that we only briefly touch on in Part IV, such as radio tracking (e.g., White and Garrot 1990), habitat analysis (e.g., Morrison et al. 1992), the design of experiments (Underwood 1995; Quinn and Keough 2002), and mathematical and simulation modeling (Starfield and Bleloch 1992; Burgman et al. 1993). We also readily acknowledge that we have not presented many other possible approaches that could be employed to study species and assemblages, such as hierarchy analysis (Allen and Starr 1988; Allen and Hoekstra 1992) and path analysis (Shipley 2000).

Figure IV.2. Natural landscape in central Iceland (photo by David Lindenmayer).

CHAPTER 16

Field-Based Approaches to Studying Species and Assemblages in Modified Landscapes

The knowledge gained about landscape change, habitat loss, and habitat subdivision depends on the methods used to study these phenomena. One possible way to categorize empirical studies of the effects of landscape change on biota is to distinguish between process-oriented and pattern-oriented approaches. Process-oriented approaches tend to investigate ecological processes in detail and often deal specifically with the biology and ecology of individual species and their responses to landscape change. Examples of such approaches include radio tracking and habitat analysis.

Pattern-oriented approaches relate ecological data to landscape pattern. Pattern-oriented approaches are often used to examine aggregate measures of species occurrence (like species richness), particularly in response to the patterns of vegetation cover that arise when landscapes are modified. Pattern-oriented investigations can be true experiments, natural experiments (sensu Diamond 1986), and observational studies. Usually, these kinds of studies attempt to quantify patterns in the first instance, and then infer which ecological processes may have given rise to the observed patterns. We readily acknowledge that our strict demarcation between these two broad kinds of approaches is somewhat arbitrary. However, it allows for a straightforward way of thinking about, and presenting, the wide range of empirical investigations that characterize the "fragmentation" literature.

In this chapter, we briefly examine some examples of process-oriented and pattern-oriented approaches to study the impacts of landscape alteration on biota. In addition, we discuss some of the challenges that make field studies on landscape change difficult and that need to be considered when designing or interpreting field studies.

Process-Oriented Approaches for Single Species

Process-oriented approaches for studying the impacts of landscape alteration often focus closely on the biology and ecology of a single species. In most cases, they are applied in the study of processes affecting a single species directly, and they help improve understanding about the patterns of distribution and abundance of that species. Many factors influence the distribution and abundance of species, and determining which factors are particularly important for a given taxon is a fundamental task in ecology (Elton 1927; Woodward 1987). Field survey work is a fundamental and obviously critical part of establishing where a species occurs and why it occurs where it does. It is well beyond the scope of this book to discuss the wide methods that can be deployed in field sampling. Many books and scientific articles have been written on field methods for particular taxonomic groups or sets of species within those groups (e.g., Robbins 1978; Recher 1988; Sutherland 1996; Wilson et al. 1996; Ralph et al. 1997). It is also well beyond the scope of this book to discuss the many complex issues associated with the design and implementation of field surveys, and, again, many books and articles focus extensively on these issues (see Usher 1986; Margules and Austin

1991; Thompson et al. 1998). Rather than explore field sampling and survey methods and survey design per se, in the following sections, we touch on a small subset of approaches that can be highly relevant to studies of landscape change, including habitat analyses, radio tracking, and studies of animal movements.

Habitat Analyses

One of the most critical factors for all species is the suitability and availability of habitat (Morrison et al. 1992). Moreover, understanding what constitutes habitat for a species is an essential part of understanding the impacts of landscape change on that species (Clark and Shutler 1999). This is especially important because habitat loss is the most prevalent cause of decline and loss of species around the world (Gibbons et al. 2000; Sala et al. 2000; Primack 2001; Chapter 4).

All methods for identifying habitat are based on detecting associations between environmental attributes and the presence or abundance of species. Typically, habitat measures relate to the biology of the species. Attributes such as large cavities in trees are critical for animals like forest owls (Fischer and McClelland 1983); skinks and invertebrates may depend on coarse woody debris and rocky outcrops (Harmon et al. 1986; Goldsborough et al. 2003); and amphibians may require suitable breeding ponds within a specific landscape context (Hazell et al. 2001; Guerry and Hunter 2002).

Habitat analyses need to take account of the fact that the distribution and abundance of a given individual species are influenced by factors at multiple scales (Gutzwiller and Anderson 1987; Schneider 1994; Jaquet 1996; see Box 1.3 in Chapter 1). As an example, Diamond (1973) showed how the distribution of birds in New Guinea was influenced by multiscaled processes ranging from broad geographic factors to branch sizes of individual trees. With (1994) demonstrated how nymphal stages of a species of grasshopper moved and interacted with the patch structure of landscapes differently than larger, faster-moving adults.

A variety of methods may be used to combine variables into an estimate of habitat suitability. Subjective methods such as habitat suitability indices (HSIs) link species and habitat variables using mathematical functions created by people with ecological knowledge of the species. Statistical methods use explicit models of empirical data to link the occurrence of a species with habitat parameters.

Habitat Suitability Indices

HSIs are based on the habitat evaluation procedure (HEP) developed by the United States Fish and Wildlife Service in the early 1980s (United States Fish and Wildlife Service 1980, 1981), and aim to indicate habitat quality for a given species in a particular area. The approach has been applied extensively in North America (see van Horne and Wiens 1991 for a review) and indices have been constructed as part of the assessment of the habitat requirements of many vertebrate groups (Burgman et al. 2001). HSIs have been used as predictive tools to evaluate the quality of habitat for wildlife (O'Neill et al. 1988; Rand and Newman 1998) and to predict the effects of various kinds of landscape alteration such as logging, vegetation clearance, and dam construction on a given species or set of taxa (Gray et al. 1996).

HSI models are simple to construct, and they make use of whatever knowledge is available. Experience and expert judgment are difficult to incorporate in more explicit statistical procedures (but see Martin et al. 2005). Often, the tacit knowledge of experienced scientists and resource managers is all that is available for decision making, and this knowledge can be immensely important (Fazey et al., 2006). HSI models can be useful in this context because they assist scientists and resource managers to draw together available information on a species. This can be done relatively quickly, helping to identify gaps in knowledge. Thus the method can help the development of hypotheses associated with species–habitat relationships and the impacts of environmental changes that can subsequently be tested more rigorously (Minta and Clark 1989; van Horne and Wiens 1991). In other cases, HSI-like approaches can be used to generate variables for input into statistical habitat models (Catling et al. 2001).

Despite the potential value of HSIs, they are not without limitations. For example, HSIs do not account explicitly for correlations among the independent variables. The process of combining variables and assigning weights may account for nonindependence, but the procedures for deciding on the model structure and the weights are not explicit (Minta and Clark 1989; Reading et al. 1996). Similarly, the procedure for specifying the function relating habitat suitability to the value of the environmental variable is usually not explicit. Burgman et al. (2001) noted that, because the HSI method is based on expert opinion, there is rarely empirical evidence linking habitat suitability to the demographic success of the species. Therefore, indices may be highly uncertain.

Statistical Approaches for Habitat Analyses

There are numerous statistical methods for examining habitat relationships (reviewed by Guisan and Zimmermann 2000; see Box 16.1). Here, we explore the use of two kinds of regression analysis that are particularly well suited to habitat studies and have been widely applied (e.g., Austin 2002): logistic regression and Poisson regression. They are part of a set of procedures called generalized linear models (McCullagh and Nelder 1988; Crawley 1993; Burgman et al. 2005).

Often a species is recorded as either present or absent, with no indication of abundance. Vegetation quadrats, mammal hair tubes, scat samples, spotlight surveys, frog and bird call recordings, and bird sighting transects often result in such binary records of presence/absence data (Sutherland 1996). These data may be used together with habitat attributes to build a model of the relationship between a species and its environment. Because the response (or dependent) variable in the relationship is binary (the species is either present or not), logistic regression is used in place of ordinary regression (Collett

Box 16.1. General Guidelines for Model Building and Some Limitations

It is important to ensure that the assumptions associated with a statistical method of habitat analysis are satisfied. Assumptions are rarely tested, even though in many cases tests are available (e.g., Nicholls 1989). In logistic regression, it is important to test for dependence between explanatory variables, interactions between variables, and the form of the relationship between the explanatory and response variables (e.g., whether it is linear or some other form; see Austin et al. 1994a, b). Confidence limits on predictions should be reported routinely, and extrapolations of predictions beyond the extremes of the independent variables should be avoided. Attempts should be made not to overfit models by testing large numbers of potential explanatory variables (Manning 2004). Statistical models can generate spurious relationships between dependent and independent variables. For example, a relationship may arise simply because a species occurs in places characterized by particular features (e.g., rocky outcrops), although these features may not necessarily be needed as a part of the habitat of that species.

A model with good explanatory power will not necessarily predict reliably. For example, in a study of the rufous treecreeper (*Climacteris rufa*) in Western Australia, Luck (2002b) found that a model of nest site use could not make robust predictions because of the large number of unused (but apparently suitable) nest sites. The best way to evaluate a model's predictive power is to test it with new data (e.g., Elith 2000; Pearce and Ferrier 2000). A statistical model is more likely to have good predictive power if both statistical and biological understanding are part of model formulation (Austin 2002).

There can be other important caveats and critical thinking associated with habitat modeling. For example, species–habitat relationships, like many other aspects of ecology, can be scale sensitive (Schneider 1994), and caution is required in extrapolating between different spatial scales.

1991). Explanatory variables may be continuous (such as daily temperature) or discrete categories (sometimes called factors; for example, discrete types of vegetation associations).

Some species are ubiquitous, so that it is not possible to discriminate between sites where the taxon is present and where it is absent. In these circumstances, the abundance of the species might be a more useful indicator of habitat suitability. Poisson regression is a form of generalized linear model in which the response variable is a count of the number of incidents, such as the abundance of animals (Vincent and Haworth 1983). A special form of Poisson regression, zero-inflated Poisson regression, deals with rare species that are absent from many sites (see Welsh et al. 1996, 2000).

Quantitative habitat analyses result in hypotheses that can subsequently be tested in the field (Block and Brennan 1993). However, relationships between a dependent (response) variable and independent (explanatory) variables do not imply causality. There is no guarantee that the results of statistical analyses will correctly identify the most important underlying ecological processes (van Horne 1983). This may be particularly important for species subject to a range of threatening processes where populations occur in particular locations not because those places are preferred habitat but because those locations are where threatening processes do not manifest (Caughley and Gunn 1996). Similarly, Haila et al. (1996) demonstrated that habitat models for some bird species in Finland varied between years. Thus a given year's habitat model may not correctly identify the preferred habitat for a given species per se. Haila et al. (1996) argued that some bird species appeared to "avoid the worst" habitat, rather than be restricted to "optimal habitat" in any given year.

Finally, species–habitat relationships do not provide information about how much habitat is needed to support a population in the long term. Other tools, such as population viability analysis (PVA), help with this type of question, although habitat information is essential for spatially explicit forms of such models (e.g., McCarthy et al. 2000; Carroll et al. 2003, 2004).

Other Methods

As previously outlined, we have touched on just two kinds of statistical models that can be applied in studies of habitat. Many others exist; a detailed exploration of them is well beyond the scope of this book, but they include approaches such as generalized linear mixed models (Schall 1991; Searle et al. 1992) and generalized additive models (Hastie and Tibshirani 1990) (e.g., Elith 2000; Guisan and Zimmermann 2000; Manning 2004; Rushton et al. 2004). Bayesian methods are also increasingly being recommended for use in habitat analyses, and there is a rapidly expanding literature on such approaches together with an increasing range of computer packages for implementing Bayesian statistics (e.g., Spiegelhalter et al. 2003). Other useful methods are developing, such as receiver operating characteristic (ROC) curves that can provide a convenient way of summarizing the predictive performance of habitat models (Swets et al. 2000; Elith, 2000; Mackey et al. 2002). Finally, it is increasingly being recognized that different combinations of explanatory variables may explain similar levels of variability in a dataset, and a good understanding of a species' biology together with methods such as Akaike's information criterion (AIC) (Akaike 1973) can be important in comparing plausible statistical models of a species' habitat.

Habitat Selection

Some kinds of habitat analysis focus on the broad patterns of distribution and abundance of a species and document, for example, which areas of vegetation are occupied or left vacant. In other cases, habitat analyses may be focused at a smaller spatial scale in an effort to identify the attributes of particular structural features of vegetation, which make them suitable nesting, sheltering, or foraging sites (Johnson 1980; Pope et al. 2004). In these cases, habitat suitability may be inferred from the frequency of use of particular structures by a given individual or group of animals. Habitat selection work usually quantifies cases where a species uses resources disproportionately, relative to their availability at a given location (e.g., Luck 2002a, b). Habitat selection work has focused on a wide range of features, such as the suitability of cavity trees as nesting sites (Lindenmayer et al. 1996; Webb and Shine 1998), the use of parts of the vegetation as feeding areas (Cunningham et al. 2004; Goldingay and Quin 2004), and the characteristics of microhabitats used by reptiles (Blazquez 1996; Díaz 1997; Martín and Salvador 1997). The various kinds of qualitative and quantitative methods already outlined for habitat analyses can apply equally to finer-scaled habitat selection work.

Studies of Movement Patterns

The movement patterns of a species in a landscape can provide important insights into how and why it responds to human-derived landscape modification (e.g., Barbour and Litvaitis 1993). For example, home range studies can provide valuable information on habitat selection preferences, and the use of space (White and Garrot 1990; Harris et al. 1990) and how they may be altered in landscapes subject to human modification (Pope 2003). Similarly, studies of dispersal can be critical for predicting species persistence patterns in modified landscapes (e.g., Sarre et al. 1995). In the following section, we discuss the roles of radio tracking and dispersal studies as two of many field methods for exploring animal movements in modified landscapes.

Radio Tracking

Sometimes the aim of a study is to determine the area over which an animal moves or to identify which parts of its home range it uses most frequently—such as in a habitat selection investigation. One of the most effective ways to do this is to use radio tracking. Radio tracking has often been used in modified landscapes to determine how species respond to altered patterns of vegetation cover (e.g., Redpath 1995; Jackson 2000; Sweanor et al. 2000; Haughland and Larsen 2004). Radio tracking involves the use of small radio transmitters attached to an animal, usually via a collar fixed around its neck or some type of harness, although in the case of some animals (such as snakes), transmitters need to be fitted internally (e.g., Brown and Weatherhead 2000). A receiving system can then be used to determine the precise location of the collared animal. Repeated "fixes" of an animal's location over a set period can provide information on the location of its den sites, the size and shape of its home range, or how much its territory overlaps with those of other collared individuals (e.g., White and Garrot 1990; Mizutani and Jewell 1998). Radio tracking may also be used to study the dispersal of animals from their natal territories (see next section), with the resulting data assisting management strategies such as the design and location of wildlife corridors, overpasses, and underpasses (Mansergh and Scotts 1989; Sutherland 1996).

Although radio tracking is one of the most widely used field methods in wildlife biology (Sutherland 1996), it is not without limitations. For example, probabilistic models of home range estimation, such as the kernel method, assume that location observations of radio-tracked animals are independent of each other (Swihart and Slade 1985). However, independence of observations is often difficult to achieve in radio-tracking studies (De Solla et al. 1999), even using observation intervals of 1 to 6 weeks (e.g., McNay et al. 1994; Rooney et al. 1998). The preferred minimum time between location fixes to achieve noncorrelated data can be estimated by the time it would take for an animal to traverse its home range or to arrive at any point within it (Newdick 1983). However, given the logistical constraints of most field studies, data are usually gathered in clusters over shorter time periods, and therefore autocorrelated data are unavoidable (Harris et al. 1990). Indeed, autocorrelation of data is often inherent in radio-tracking studies (De Solla et al. 1999) and may reflect an inherent pattern of home range usage where animals choose to move in a nonindependent fashion (Lair 1987). However, highly autocorrelated data may underestimate home range size (Swihart and Slade 1985), thereby influencing descriptions of home range use and habitat selection (Cresswell and Smith 1992). Increasing the sample size can increase the accuracy of home range estimation, but there is generally a tradeoff with autocorrelation of data (Swihart and Slade 1985). Methods for achieving independence by eliminating observations or subsampling the data (as recommended by Swihart and Slade 1985) may still result in underestimation of home range (Rooney et al. 1998) as well as reducing sample size, with the subsequent loss of potentially important biological information (De Solla et al. 1999).

Finally, special care needs to be taken in radio-tracking studies to prevent animals being injured. For example, radio transmitters need to be fitted appropriately so they are not too tight and lead to skin lesions or wounds. Conversely, if a radio collar is too loose, an animal can jam a leg underneath it and subsequently die. The weight of radio collars also needs careful consideration because it is possible that transmitters heavier than 3 to 5% of an animal's body weight might lead to changes in behavior and even death. Some studies in the Northern Hemisphere have shown that radio collars can have significant negative impacts on animals. For example, as a result of being fitted with radio collars, dispersal distances in the black-footed ferret *(Mustela nigripes)* were altered (Oakleaf et al. 1993).

In some cases, movement and other kinds of valuable data can be gathered by using marking techniques to individually identify animals rather than using radio tracking. For example, color banding of birds is a widely applied method to study movement patterns and habitat use (Saunders and de Rebeira 1991; Russell et al. 1994; Breininger et al. 2006). Mark–recapture methods can be similarly useful (e.g., Bennett 1987; reviewed by Lebreton et al. 1992). However, like all field methods, color banding and mark–recapture techniques are not without problems because they can alter patterns of animal behavior (Baptista and Gaunt 1997) and in some cases significantly increase risks of mortality (Morton 1984; McCarthy and Parris 2004).

Dispersal

As outlined in Chapter 6, for some species, landscape alteration can result in the subdivision of habitat patches and thus the isolation of populations within remaining habitat patches. Understanding and mitigating the effects of habitat isolation on a species often require data on the species' dispersal ability and patch recolonization (Johnson and Gaines 1990). However, a paucity of data on the effects of patch isolation often arises because the key processes of dispersal and patch recolonization are extremely difficult to study using standard field techniques (Chepko-Sade and Halpin 1987) such as individual color banding of birds or the radio tracking of individuals (Oakleaf et al. 1993; Baptista and Gaunt 1997). Often the most effective way of quantifying dispersal and patch recolonization is to explore effective dispersal; that is, to quantify the number of individuals that successfully breed following interpatch movement (Horskins 2004). Genetic markers combined with knowledge of movement patterns can be valuable in this regard (e.g., Peakall et al. 2003; Horskins 2004). The most useful studies in this context are often multidisciplinary and bridge demographic and genetic spheres of research (Clarke and Young 2000; Berry et al. 2005).

Removal experiments are another method that can be used to generate an improved understanding of patterns of dispersal and interpatch movements of individuals in modified landscapes. In these studies, populations of a given species are removed from their habitat patches, and the process of population recovery is monitored to provide information on the rate of recovery (Stickel 1946; Bondrup-Nielsen 1983; How et al. 1984; Verts and Carraway 1986). Immigration into patches where animals were previously removed has been observed in several studies of small mammals (e.g., Middleton and Merriam 1981; Bender and Fahrig 2005) and birds (e.g. Grey et al. 1997). When demographic data on population recovery are coupled with genetic data, it is possible to determine the source of individuals that have helped reverse the "localized extinction" (Lindenmayer et al. 2005a).

Although removal experiments can be useful, one of the potential problems with them is that some residents remain in the "cleared" area and when trapped later are incorrectly classified as recolonists (Stenseth and Lidicker 1992b). The logistics of conducting removal experiments at landscape scales that are meaningful for developing useful insights into the impacts of landscape modification also present a major challenge (but see, for example, Grey et al. 1997), as do ethical concerns about what to do with individuals removed from patches.

Other Useful Approaches for Single Species

There are numerous other kinds of process-oriented studies that can be useful for examining the responses of individual species to landscape alteration. For example, careful observations of the flocking behavior, foraging patterns and food availability (Zanette et al. 2000), or levels of breeding success of a given species can be very informative about the reasons for decline or other kinds of responses to landscape alteration (e.g., Gardner 2004). Similarly, observations of dispersal behavior can be critical for understanding why a given species may or may not cross a particular matrix (e.g., Desrochers and Hannon 1997; Graham 2001; Cooper and Walters 2002) or why it is confined to certain vegetation types while moving (Holekamp 1984; Garrett and Franklin 1988; Kindvall and Ahlen 1992).

Other kinds of studies that highlight interspecific interactions such as competition (Piper et al. 2002), predation (Kareiva 1987; Lahti 2001), or mutualism (Elmes and Thomas 1992; Kearns et al. 1998; Cordeiro and Howe 2003) can also be valuable. For example, Chapter 11 contains a range of examples of biotic edge effects that have been uncovered through studies of interspecific interactions such as altered levels of nest parasitism and nest predation among bird populations within human-modified landscapes. Direct observations of predation and parasitism have been made in some studies, whereas others have surveyed rates of nesting success or quantified losses of quail eggs from surveys of artificial nests (see Box 11.1 in Chapter 11).

Pattern-Oriented Approaches

Pattern-oriented approaches to studying biota in modified landscapes often investigate aggregate measures of species occurrence like species richness or assemblage composition in relation to landscape pattern. A common aim of such studies is to infer which ecological processes were the likely causal mechanisms of the observed patterns. In the following section, we describe several classes of pattern-oriented approaches that can be used to explore the impacts of landscape alteration on species or assemblages, including experiments, natural experiments, observational studies, and microcosm studies. Each of these investigations can be longitudinal, cross-sectional, or both. Longitudinal studies investigate changes in landscape pattern and associated ecological patterns through time. However, the effects of landscape change can take a long time to

Table 16.1. Experimental design issues associated with "fragmentation" studies*

	Level of question asked		
	Indication of change in biota	*Separation of modification effects from other temporal change*	*Separation of modification effects from site effects*
Count before and count after modification	Yes—but choice of time for post-modification sample may mean a change is missed	No	No
Count before and after landscape change in a modified and a control site	Yes	Yes—but confounding interaction of season and modification effects if timing of samples not carefully planned	No
Count before and after modification in randomly allocated replicate modified and control sites	Yes	Yes	Yes

*Adapted from work on fire studies by Whelan (1995, p. 139). The table outlines various approaches that could be used to make inferences about impacts of landscape modification on biota. It also indicates the appropriateness of different approaches for each question asked, and points to possible shortcomings.

manifest, and many research projects cover relatively short time scales of only a few years (Fazey et al. 2005a). Cross-sectional studies attempt to overcome this problem by substituting space for time. Typically, cross-sectional studies in a given landscape or region compare locations that were severely modified a long time ago with locations that were modified more recently, or that have not been modified extensively.

Experiments

True experiments in field ecology typically include (1) replicates of the particular treatment in question (e.g., replicates of specified patch sizes), (2) controls in which a treatment is not applied for purposes of comparison against sites where treatment is applied, and (3) adequate description of pretreatment conditions (see Oksanen 2001). It is well beyond the scope of this book to discuss experimental design fully. In most cases, a useful first step for conservation biologists is to consult with a statistician when designing an experiment. Experiments with adequate replication of treatments, adequate sample sizes (and adequate statistical power), quantification of before-treatment conditions in each replicate, and replicated "control" sites matched to the treatments, can generate robust data and outcomes. Table 16.1 outlines various experimental design approaches that could be used to make inferences about the impacts of landscape modification on biota. The table indicates the appropriateness of different approaches for a range of potential questions.

True experiments examining the effects of the loss or subdivision of patches of native vegetation (or habitat) are comparatively rare, perhaps because it is difficult, time consuming, and expensive to manipulate large areas and difficult to find sufficient replicates, particularly for large patches. Debinski and Holt (2000) identified 20 such experiments in terrestrial ecosystems, 6 in forests and 14 in grasslands or old fields. Most of these were located in North America or Europe (Figure 16.1). These experiments have examined two hypotheses (after Debinski and Holt 2000): (1) that species richness and abundance increase with increasing

Figure 16.1. Calling Lake Logging Experiment (photo by Fiona Schmiegelow).

patch area (Chapter 9), and (2) that movement and species richness increase with increased landscape connectivity between patches (Chapter 12). Debinski and Holt (2000) found many species-specific responses with many inconsistent results across different studies. Arthropods showed the best fit with theoretical expectations. For example, larger fragments supported greater numbers of species (e.g., Margules et al. 1995). However, many other groups did not respond in ways predicted from theory, including birds, mammals, early successional plants, mobile generalists, and long-lived species (Debinski and Holt 2000). Indeed, about half of the studies examined by Debinski and Holt (2000) found no effect of patch size.

Notably, all of the experiments reviewed by Debinski and Holt (2000) were underpinned by a view of landscapes from the perspective of the patch–matrix–corridor model (sensu Forman 1995; see Chapter 3), which has major limitations for practical interpretation, generality, and extrapolation to other landscapes (see the section Why Field Studies in Modified Landscapes Are Difficult).

Although experiments can provide a powerful method to investigate the impacts of landscape change on biota, most experiments occur at spatial scales that are too small to be realistic. Although small-scale experiments can reveal interesting findings in some situations (e.g., Golden and Crist 1999; Lenoir et al. 2003), they can be of limited value for other situations, such as studies of wide ranging bird and mammal species. Conducting true landscape-level experiments at spatial scales relevant to mobile groups such as birds and bats is a difficult challenge (Wiens 1994). Some taxa will respond to the details of the landscape mosaic (Chapter 14) but the effects of complexity created by interacting landscape components will be extremely difficult to test (Wiens 1999). Moreover, extrapolating results of small-scale experiments to large-scale ecosystems can be highly error prone (Blackburn and Gaston 2003) because most spatial processes are scale dependent (Carpenter et al. 1995; Thrush et al. 2000).

A rare example of a large-scale experiment comes from the Canadian province of Alberta. The study was designed to quantify the impacts

Box 16.2. The Tumut Fragmentation Study

The experimental design for the natural experiment at Tumut was developed after prolonged consultation with a team of professional statisticians, and the study encompasses 166 sites in three broad classes: (1) 86 eucalypt fragments stratified across four patch size classes, five forest types, and two ages of surrounding pine forest; (2) 40 sites located in large areas of continuous native forest; and (3) 40 sites dominated by stands of radiata pine *(Pinus radiata)*, which form the matrix surrounding the eucalypt fragments. The design features include:

- Stratification of the eucalypt remnants to ensure the full environmental space of the study region was represented
- Random selection of the 86 eucalypt remnants within strata to minimize the chance of bias, enabling averaging over random factors
- Locating sites in large areas of continuous eucalypt forest to provide "control" areas, large enough to support viable populations of the many species that once occurred in the study region
- Using data on climate, forest type, and geology to ensure the range of environmental conditions were matched across the three broad classes of sites
- Selecting sites dominated by radiata pine stands so that potential habitat value of the landscape matrix surrounding the fragments was not ignored (see Chapter 14).

The work at Tumut found that small to intermediate-sized patches (0.5–3 ha) supported considerably higher numbers of vertebrate species than anticipated, highlighting the importance of remnants as small as 0.5 ha. Many species of native animals occurred in stands of radiata pine, but their presence was often related, in part, to nearby remnant eucalypt forest. Thus a mosaic of remnant native vegetation and softwood stands will have significantly higher conservation value than a radiata pine monoculture. At Tumut, as in most natural experiments, there is no information on species occurrence patterns prior to the beginning of landscape alteration in the 1930s.

of landscape change on boreal forest birds (Schmiegelow and Hannon 1993). The design encompassed three broad kinds of sites: isolated patches of remnant boreal forest (where patches were isolated by clearcutting of the surrounding forest), patches of remnant boreal forest connected on one side to riparian areas, and control sites. The isolated patches were 1, 10, 40, and 100 ha, and the connected patches were 1, 10, and 40 ha. Each size class was replicated three times, and sites were carefully selected to ensure that the same forest type was sampled at each site (Schmiegelow and Hannon 1993). Importantly, sampling of birds commenced before and then continued after the experiment. The study identified many important findings (Schmiegelow et al. 1997). For example, no changes in species richness were identified except in the 1 ha patches, where it increased. However, changes in bird community composition were prominent and turnover rates were higher in isolated forest remnants than in connected remnants or control sites. Declines of particular species appeared to be related to their migration strategy. In isolated forest remnants, both neotropical migrants and residents declined. Neotropical migrants also declined in the connected forest remnants (Schmiegelow et al. 1997). An important interpretation of the experimental findings was that the impacts of short-term landscape change on the bird assemblage may have been less substantial than observed in other systems because boreal forests (and their associated biota) are subject to small- and large-scale natural disturbances (Schmiegelow et al. 1997).

Complex, interacting factors are difficult to examine in field experiments, despite the fact that they can have important influences on species persistence; examples include cumulative effects (sensu McComb et al. 1991; Burris and Canter 1997) and incremental long-term changes in landscape cover (Bennett and Ford 1997; also see Chapter 15). In addition, many factors are controlled in true experiments, making extrapolation of the results to other species, ecosystems, or ecological phenomena unreliable (Davies et al. 2001b; Oksanen 2001). Despite these problems, some ambitious landscape-scale experiments have commenced (van Jaarsveld et al. 1998).

"Natural" Experiments

Natural experiments overlay an experimental design on an ecosystem where change or active manipulation has occurred or is planned (Carpenter et al. 1995). They can be broadly similar to "true" experiments (Diamond 1986), but the changes are not controlled by the researcher. Usually, they occur at larger scales than true experiments. For example, the Tumut fragmentation study (Box 16.2) examines patterns of species richness and abundance in remnants of eucalypt forest embedded in an extensive exotic radiata pine *(Pinus radiata)* plantation. The work is taking place over a large area (100,000 ha), and the primary focus is on birds and mammals (Box 16.2).

Observational Studies

The vast majority of studies of landscape change are observational investigations that do not use active interventions (e.g., manipulation of sites) or replicated experimental designs to sample biotic response (or other response variables) to landscape change (reviewed by McGarigal and Cushman 2002; Fahrig 2003). Typically, observational studies sample a range of existing kinds of sites in a given landscape at one point in time. They usually lack the replication of site types and/or the "controls" that characterize well-designed experiments or natural experiments. Despite the inherent problems of observational studies, they can nevertheless produce important insights into the impacts of landscape modification on biota (McGarigal and Cushman 2002).

An example of an observational study comes from the Naringal area in southwestern Victoria, Australia. Here, Bennett (1990a) compared the historical and current status of mammals using historical and anecdotal records, collections of road kills, museum archives, and fauna surveys in 39 patches of 0.3 to 92 ha in size. At the time of European settlement there were 33 species of native mammals in the Naringal region. Of these, 6 are now regionally extinct, including the dingo *(Canis lupus dingo)*, the tiger quoll *(Dasyurus maculatus)*, the eastern quoll *(Dasyurus viverrinus)*, the common wombat *(Vombatus ursinus)*, the koala *(Phascolarctos cinereus)*, and the eastern pygmy possum *(Cercartetus nanus)*. The first four were hunted extensively because they were considered pests. Six introduced taxa have invaded: the house mouse *(Mus musculus)*, black rat *(Rattus rattus)*, red fox *(Vulpes vulpes)*, feral cat *(Felis catus)*, rabbit *(Oryctolagus cuniculus)*, and brown hare *(Lepus capensis)* (Table 16.2). Although the identity of taxa has changed dramatically, the total number of mammal species is unchanged. Bennett (1990a) showed that larger taxa were more vulnerable to landscape change and habitat fragmentation. For example, the southern brown bandicoot *(Isoodon obusulus)*, sugar glider *(Petaurus breviceps)*, and red-necked wallaby *(Macropus rufogriseus)* were absent from smaller patches.

Bennett (1990a) concluded that it was important to avoid further clearing and to maintain landscape connectivity and the suitability of habitat within existing remnant patches of native vegetation (Bennett 1990a). Even if landscape managers achieve these objectives, species might still be lost over time (the extinction debt; see Chapter 9). Andrew Bennett (pers. comm.) has begun revisiting his sites in the Naringal region to monitor the long-term impacts of landscape change on vertebrates.

The example of Bennett's (1990a) study outlined above refers to his observational investigation of mammal species inhabiting different vegetation remnants. However, in parallel with that study, Bennett (1987) completed a process-based investigation of how one species, the long-nosed potoroo *(Potorous tridactylus)* responded to landscape change. The study revealed that factors such as small body size, limited home range, and dispersal by both sexes contributed to the persistence of the long-nosed potoroo in small and scattered vegeta-

Table 16.2. Occurrence of mammals in remnant vegetation in the Naringal district, southwestern Victoria*

Patch size (ha)	›2	3–7	8–15	16–40	41–100
Number of patches	8	8	8	8	7
Species					
Rabbit (*Oryctolagus cuniculus*)	**63**	**100**	**100**	**100**	**100**
Bush rat (*Rattus fuscipes*)	38	**75**	**100**	**100**	**100**
Common ringtail possum (*Pseudocheirus peregrinus*)	25	**88**	**100**	**88**	**100**
Fox (*Vulpes vulpes*)	0	**63**	**100**	**100**	**100**
Echidna (*Tachyglossus aculeatus*)	13	50	**100**	**100**	**100**
Brown Antechinus (*Antechinus stuartii*)	13	50	**100**	**100**	**86**
Swamp wallaby (*Wallabia bicolor*)	13	13	**75**	**63**	**86**
Long-nosed potoroo (*Potorous tridactylus*)	0	13	50	**63**	**100**
Eastern gray kangaroo (*Macropus giganteus*)	13	50	13	**63**	**57**
House mouse (*Mus musculus*)	25	25	38	38	**57**
Cat (*Felis catus*)	25	38	50	38	43
Swamp rat (*Rattus lutreolus*)	0	13	13	25	29
Long-nosed bandicoot (*Perameles nasuta*)	0	13	13	0	43
Red-necked wallaby (*Macropus rufogriseus*)	0	0	0	38	29
Sugar glider (*Petaurus breviceps*)	0	0	13	25	29
Southern brown bandicoot (*Isoodon obesulus*)	0	0	13	13	14
Common brushtail possum (*Trichosurus vulpecula*)	0	0	13	0	14
Black rat (*Rattus rattus*)	13	25	0	0	0
Brown hare (*Lepus capensis*)	13	0	0	13	0

*After Bennett (1990b). Exotic species are underlined, and frequencies greater than 50% are in boldface.

tion remnants. Importantly, there was valuable complementarity between the kinds of studies. The pattern-based study highlighted some general changes in the faunal assemblages across the study area, whereas the work on the long-nosed potoroo illustrated the processes contributing to the occurrence of one of the species in the mammal assemblage. The studies by Bennett (1987, 1990a) underscore one of the major themes of this book; namely, the complementary nature of pattern- and process-based work in deriving an understanding of the response of biota to landscape change.

Experimental Model Systems and Microcosm Experiments

Some workers have sought to examine biotic responses to landscape change through highly controlled and replicated small-scale investigations. These are sometimes called experimental model systems (Wiens et al. 1993), and they have been used to test predictions from a range of theories associated with landscape alteration (e.g., Kareiva 1987; Quinn and Hastings 1987; Robinson et al. 1992). Collinge and Forman (1998) used a grassland mowing experiment to assess how landscape transformation affected abundance and species richness of invertebrates. The study revealed some unexpected responses, several of which did not corroborate predictions from theory (such as island biogeography theory; see Chapters 2, 9). In another small-scale mowing experiment, Parker and Mac Nally (2002) also tested for effects of different spatial patterns of micro-landscape change on invertebrates, and they uncovered some interesting findings. For example, species abundances and species richness differed significantly between the cores and the edges of patches. Grassland habitats have been used in many small-scale studies of landscape modification (e.g., Barrett et al. 1995; Holt et al. 1995; Wolff et al. 1997; Braschler and Baur 2005). More than half of the "fragmentation experiments" reviewed by Debinski and Holt (2000) were mowed grass experiments.

Other kinds of microexperimental approaches have generated valuable insights into the impacts of landscape transformation on biota. The persistence of populations of microarthropods such as Acari, Collembola, and Tardigrades living on moss-covered limestone blocks was found to be promoted by a rescue-effect (sensu Brown and Kodric-Brown 1977; Chapter 6) facilitated by the presence of corridors (Gonzales et al. 1998; Hoyle 2005). Similarly, a laboratory-based study of *Drosophila* spp. found that the movement of animals along a "corridor" reduced demographic stochasticity and contributed to the persistence of populations (Forney and Gilpin 1989). Other microecosystem studies also support the positive effects of corridors in reversing negative habitat isolation effects for some species (Gilbert et al. 1998).

Although these kinds of small-scale experiments and studies are undoubtedly valuable and provide useful tests of theories and the predictions made using those theories, the feasibility of accurately extrapolating the results of such highly controlled small-scale studies to large-scale landscape mosaics remains questionable (Wiens 1999; Blackburn and Gaston 2003), particularly for wide-ranging species such as bats and birds (but see Srivastava et al. 2004).

The Underlying Landscape Models Used in Pattern-Oriented Approaches

As outlined in detail in Chapter 3, the conceptual landscape model applied to characterize a landscape can have a major impact on how a system or set of species is studied and what the outcomes of a particular study are. Although each species varies in how it perceives a landscape (Manning et al. 2004a; Chapter 3), the vast majority of studies of landscape modification are designed on the basis of human perspectives of landscapes. Indeed, all of the experiments, natural experiments, observational studies, and microcosm studies outlined in the preceding sections are underpinned by the patch–matrix–corridor model (sensu Forman 1995; see Chapter 3; Box 16.3).

Although the island model and the patch–matrix–corridor have been used to design most experiments, natural experiments, observational studies, and microcosm studies that have been described in this book, it is possible to design studies on the basis of landscape attributes that are less closely related to a human perception of landscapes and that are sometimes more directly relevant to particular species and assemblages. For example, Fischer et al. (2004b) used attributes such as aspect and topographic position in addition to vegetation cover to design a study of the impacts of landscape modification on reptiles in southeastern Australia. In this study, reptiles responded to a wide range of habitat features at multiple spatial scales. Human-defined patch boundaries were no more important than a wide range of other ecological variables contributing to overall landscape heterogeneity (Fischer et al. 2004b; see Chapter 14).

The problem of a simple focus on a human perspective of landscapes is highlighted by insights from a range of authors who argue there has been an excessive focus on vegetation remnants while ignoring key issues associated with the surrounding landscape context, particularly the role of the surrounding matrix (see critiques by Simberloff et al. 1992; Laurance and Bierregaard 1997; Haila 2002) and the substantial contributions to population persistence it can make (Gascon et al. 1999; Chapter 14). Wiens (1989) noted that many studies miss "the point that it is often the structure of an entire landscape mosaic rather than the size or shape of individual patches [that matters]."

Box 16.3. The Changed Trajectory of a Large-Scale Study of the Impacts of Human Landscape Modification on Biota

The Biological Dynamics of Forest Fragments Project (BDFFP) was established more than 2 decades ago in tropical moist forests of central-north Brazil. The project illustrates neatly how a change in the perception of how a landscape functions (and hence what is the underlying landscape model) can significantly alter the way in which studies are conducted and results interpreted. The BDFFP began under a different name—the Minimum Critical Size of Ecosystems Project (Lovejoy 1980; Lovejoy et al. 1986). This was because the initial goal of the experiment was to provide empirical data on the minimum area needed to conserve rainforest ecosystem processes and biodiversity (Lovejoy and Oren 1981). This was to be done by tracking the loss of species from rainforest remnants of different sizes after they were isolated by clearing the surrounding vegetation for cattle grazing (Bierregaard and Stouffer 1997). The proposed study design was a large-scale experiment with replicated forest remnants in logarithmically increasing size classes (1, 10, 100, 1000 ha). One of the project's initial goals was to address what were then very controversial issues flowing from the theory of island biogeography (MacArthur and Wilson 1967) such as the SLOSS debate—Will more species be conserved by a *s*ingle *l*arge or *s*everal *s*mall reserves? (Wilson and Willis 1978; Simberloff and Abele 1982).

Due to a variety of circumstances, not all size classes were established, and replication was limited for those size classes that were established. The extent of clearing (and subsequent cattle grazing) in the surrounding matrix was variable, resulting in matrix conditions adjacent to the rainforest remnants ranging from pasture to advanced regrowth moist forest. While the initial aim of the project was to examine species–area relationships, the enforced changes to the experimental design meant the project subsequently focused more intensively on other aspects of landscape modification than initially intended, and as reflected in its name change. Perhaps fortuitously, this modified design forced the BDFFP to focus much more on the interrelationships between the landscape matrix and dynamics within rainforest remnants (Bierregaard et al. 1992) than on the SLOSS question (Gascon et al. 1999). These biological dynamics included edge effects (Kapos 1989) and interrelationships between the biota of rainforest remnants and matrix conditions (Bierregaard and Stouffer 1997; Gascon et al. 1999). In addition, while the loss of primary forest led to the loss of some primary-forest-dependent bird species, the modified landscape mosaic composed of primary rainforest remnants and regrowth forest had significant conservation value. Information about the interrelationships between remnants of native vegetation and the matrix from the BDFFP has had considerable generic value for highlighting some of the key impacts of landscape modification on biota that characterize many other regions around the world (Forman 1995). This case study illustrates the influence that underlying landscape models can have on empirical work. What started out as a project based on the island model (see Chapter 3) through its own findings shifted to an investigation based on the patch–matrix–corridor model (Forman 1995; see Chapter 2). This case study also highlights how the establishment of major research infrastructure led to many different kinds of studies and a large and growing list of high-quality publications that are yielding valuable new insights.

We believe that a better understanding of the effects of landscape change requires (1) greater application of the different kinds of landscape models that can be used to characterize modified landscapes and design investigations of landscape alteration (Ingham and Samways 1996); (2) greater acknowledgment of the fact that landscape context is a critical factor influencing species responses in modified landscapes (Enoksson et al. 1995; Lindenmayer et al. 2002a); (3) an appreciation that many taxa use a range of landscape components, including both vegetation remnants and the matrix in which they are embedded (Åberg et al. 1995; Flather and Sauer 1996; see Chapter 14); and (4) the use of "whole" landscapes as the unit of study and analysis (rather than patches within landscapes) to better illustrate the key properties of landscapes themselves that influence biotic responses (Radford et al. 2005; Andrew Bennett pers. comm.).

Why Field Studies in Modified Landscapes Are Difficult

Many nonecologists remain perplexed as to why progress on understanding the impacts of landscape change has not been faster and why methods to mitigate such effects are not better developed. Part of the problem is that studies of the impacts of landscape change are extremely difficult for a wide range of reasons, often making it problematic to assign causality unequivocally to patterns that are observed and, in turn, readily translate the results of one idiosyncratic case study to another landscape or species assemblage. Following are some of the inherent difficulties associated with studies of the impacts of landscape modification on biota:

- Suitable habitat and responses to landscape change are species-specific, making it difficult to extrapolate results from one species to another or to the same species in different landscapes (Davies et al. 2000; Freudenberger 2001; Bakker et al. 2002; see Chapters 3, 4).
- Biological attributes of species (e.g., home range sizes, patterns of fecundity, and sex ratios) in continuous habitat may be markedly different from those in vegetation remnants (Barbour and Litvaitis 1993; Matthysen et al. 1995; Downes et al. 1997a; see Chapter 7). This means species can respond to altered landscapes in sometimes quite unexpected ways (Redpath 1995).
- Rare species can respond to landscape alteration in different ways from more common species, adding to the difficulties in accurately predicting responses.
- Species richness patterns (which are the basis of many investigations) are most strongly influenced by common species, not rare ones (Lennon et al. 2004).
- Landscape history can affect species distributions. Species respond to landscape processes that took place many years previously (Suckling 1982; Loyn 1987). For example, the present occurrence of beetles in hedgerows in France is closely associated with historical patterns of distribution of the species (Petit and Burel 1998).
- No two parts of any landscape are the same, making true replicates impossible.
- Many ecosystems are naturally patchy (Figure 16.2) and for almost all species in all such landscapes, the occurrence of suitable habitat will also be patchy (Hastings and Harrison 1994; Harrison and Taylor 1997; Hayward et al. 2003). A subsequent key problem is to untangle the impacts of human activities on biota that add anthropogenic spatial change to underlying spatial patterns (Cooper and Walters 2002).
- The trajectory of decline in landscape attributes or particular elements of the biota may be different from their recovery (Miller and Mullette 1985), giving rise

to complex patterns of asymmetry in landscape response.

- Productive areas often support more or different kinds of species (Braithwaite 1984, 2004) than more extensively cleared and subdivided, less productive ones (Armesto et al. 1998). This often confounds comparisons between unaltered and modified areas (see Chapter 2). Yet relatively unaltered "control" areas can be critical for many studies as appropriate baselines for comparison and valid interpretation—not only for aggregate measures like species richness but also for populations of individual species (Downes et al. 1997a, b) and their genetic structure (e.g. Peakall et al. 2003). Unfortunately, the nature of landscape modification often means that "control areas" in unmodified landscapes are very uncommon or simply do not exist—often, these areas are already cleared or have been extensively modified.
- Many processes within patches are linked to processes in the surrounding landscape (Box 16.4). Thus, landscape context can lead to otherwise similar habitat patches supporting markedly different communities (Harris 1984; Lindenmayer et al. 1999b; see Chapter 14). However, such relationships can be complex. For example, in a study of birds in South Africa, Wethered and Lawes (2003) found that small patches of native forest surrounded by exotic plantations supported more species than those surrounded by grasslands. However, large patches surrounded by grasslands were richer in species numbers than exotic plantations (Wethered and Lawes 2003).
- Many species respond to several threatening processes that have impacts at several spatial scales (Cale 1999; Lindenmayer 2000; Chapter 8). These can be difficult to tease apart because they are confounded, cumulative, or interactive. Indeed, as outlined in Chapters 4 and 6, many studies have not treated habitat loss and habitat fragmentation as separate processes and subsequently have not been able to distinguish their effects (Fahrig 2003). Moreover, the way the term "habitat fragmentation" is often used refers to all processes that affect the biota in human-modified landscapes (Haila 2002; Villard 2002)—the panchreston problem (Bunnell 1999a; see Chapter 1).
- Some attributes typically examined in landscape alteration studies are highly collinear, making it difficult to determine key causal factors. For example, many studies identify significant relationships between remnant area and species richness (Rosenzweig 1995; Gaston and Spicer 2004). However, many vegetation attributes like structural complexity are often correlated with remnant area; hence, species richness patterns may be a response to vegetation attributes rather than to area per se (Dobkin and Wilcox 1986; Haila 1986; Donnelly and Marzluff 2004; Chapter 9).
- The scale of many studies is not well defined. Scale issues matter because the scale of landscape change that may be appropriate for a study of, for example, invertebrates (Collinge and Forman 1998; Parker and Mac Nally 2002) will not be meaningful for wide-ranging vertebrates such as birds (Wiens 1997). Moreover, the majority of "landscape studies" examine patches within a single landscape (Fahrig 2003). Studies of whole landscapes have examined relationships of species richness with landscape patterns (e.g., Bennett and Ford 1997; Villard et al. 1999; Radford et al. 2005), and the occurrence of individual species across entire landscapes or regions; for example, the superb parrot *(Polytelis swainsonii)* in southeastern Australia (Manning 2004).

Figure 16.2. An aerial view of the the Brazilian Pantanal—naturally patchy landscape (photo by David Lindenmayer).

The difficulties associated with investigations of landscape modification make it impossible to design ideal or "perfect" studies (Hurlbert 1984; Huston 1997). Nevertheless, although studies of landscape alteration are difficult, this does not mean that they should not be attempted. Rather, it is critical to be aware of the problems and pitfalls that can arise, and the potential limitations and caveats that should be allied with any inferences made or conclusions drawn.

Summary

Conservation science, and particularly work on the impacts of landscape alteration on biota, must be underpinned by field data. We recognize two broad (but not necessarily mutually exclusive) categories of field-based approaches for gathering such data. Process-oriented approaches often focus directly on the ecological processes affecting individual species. Examples of such studies include investigations of habitat requirements and habitat selection analyses, and studies of movement patterns and dispersal. The other category of field studies consists of pattern-oriented approaches. These approaches often focus on biological variables in relation to landscape pattern. They often examine aggregate measures of species occurrence (e.g., species richness) or groups of taxa, but they can also be targeted at single species. Pattern-oriented approaches can include experiments, natural experiments, observational studies, and microcosm studies. Process-oriented and pattern-oriented approaches are equally useful and can be highly complementary. Both process-oriented and pattern-oriented studies can be difficult to conduct and interpret, especially because, in many areas, multiple interacting factors can be related to the responses exhibited by individual species and species assemblages. Despite these difficulties, data gathered in the field are fundamentally important, not only in their own right but also because they are essential for many kinds of desktop analyses like simulation modeling, which are examined in Chapter 17.

Box 16.4. Why Landscape Context Matters—People, Reserves and Species Persistence in Africa and the United States

Harcourt et al. (2001), Parks and Harcourt (2002), and Harcourt and Parks (2003) examined the effects of landscape context on species persistence within reserves in Africa and the United States. The work showed that small reserves in both Africa and the United States were in areas where the surrounding human population density was significantly higher than in areas surrounding large reserves. In Africa, local human population density correlated significantly with human-caused mortality of carnivores (Harcourt et al. 2001). In the United States, extinction rates of large mammals correlated significantly with local human population density, but not with reserve size; a result in contrast with other studies showing strong relationships between reserve size and mammal extinctions in North America (e.g., Newmark 1987). Parks and Harcourt (2002) concluded that landscape context is extremely important for small reserves because wildlife populations within them are susceptible to small population size (as a function of small reserve size) and high human population density in the surrounding area.

Links to Other Chapters

The critical role of habitat analyses as a key form of field-based study relates strongly to the underlying importance of habitat availability for species persistence and the loss of habitat for the decline of an individual species—topics that are the central theme of Chapter 4. Radio tracking and other kinds of work on movements become critical in landscapes subject to subdivision and isolation, which are explored in Chapter 6 for the habitat of individual species, and in Chapter 12 on landscape connectivity. The various ways of designing and conducting experiments, natural experiments, and observational studies and then interpreting the results of such investigations can be strongly influenced by the landscape model that is used to characterize a landscape (e.g., the patch–matrix–corridor model or variegation model). Chapter 3 contains an outline of several kinds of landscape models. More generally, Part II of this book tends to focus on ecological processes and how they affect individual species, whereas Part III investigates the relationships between landscape patterns and assemblages of species.

Further Reading

Morrison et al. (1992) and Guisan and Zimmermann (2000) are useful introductions to habitat analyses. Nicholls (1991) contains a valuable demonstration of the use of generalized linear models in modeling plant and animal distributions. Burnham and Anderson (1998) is an excellent discussion of statistical modeling, model selection, and inference and, in turn, applications for habitat analysis. A seminal paper on habitat selection is Johnson (1980). White and Garrot (1990) is a key text on radio tracking. Useful references on dispersal are the edited books by Chepko-Sade and Halpin (1987), Stenseth and Lidicker (1992a), and Bullock et al. (2002). Debinski and Holt (2000) provide a valuable review of the formal experiments on "habitat fragmentation" that have been conducted to date. An extensive review of a range of observational studies is provided by McGarigal and Cushman (2002). Crome (1994) gives a thought-provoking analysis of the limitations of studies of landscape change and ways to improve work in the future. The essay by Manning et al. (2004a) discusses the species-specific perception of landscapes, and Ingham and Samways (1996) and Fischer et al. (2004a) highlight why the kind of landscape model used can strongly influence the interpretation of the effects of landscape change. The studies by Bennett (1987, 1990a) are excellent ones for highlighting the complementary nature of pattern and process-based investigations and the gains that can be made by bringing them together.

CHAPTER 17

Desktop Approaches to Studying Species and Assemblages in Modified Landscapes

The previous chapter examined some of the many field-based approaches that can be employed to study single-species responses and species-assemblage responses to landscape modification. The empirical data gathered from these various kinds of studies can be valuable in many ways. Various kinds of desktop approaches can be employed to add further value to these data through, for example, simulation modeling, linking data with landscape indices, and gathering the results of many different field studies as part of reviews or meta-analyses. This chapter briefly explores a small subset of the many types of desktop approaches that can be used to promote understanding of the impacts of landscape modification on species or species assemblages.

Models and Modeling

Many different kinds of models can be used in studies of landscape modification and the responses of biota to such change. A crude taxonomy of models is presented in Box 17.1; all of the models in this list can make valuable contributions to understanding biotic responses to altered landscapes. The various kinds of models of landscape cover discussed in Chapter 3 such as the patch–matrix–corridor model (Forman 1995) and the landscape variegation model (McIntyre and Hobbs 1999) can be defined as conceptual models (model type 4 in Box 17.1). Statistical models (model type 1 in Box 17.1) were examined in some detail in the context of applications to quantifying the habitat requirements or patterns of finer-scaled habitat selection by single species (see Chapter 16).

In the remainder of this section, we explore two broad kinds of models in some detail—metapopulation models and models used in population viability analysis (PVA). Sometimes these kinds of models overlap, as in the case of spatially explicit models used to assess the viability of metapopulations (Beissinger and McCullough 2002). Depending on how these models are constructed and the problems they are designed to address, metapopulation models and PVA models can be classified as theoretical models, data-rich models, or conceptual models. We have chosen to focus on metapopulation and PVA models in particular because they have been used widely in studies of biotic responses to modified landscapes, especially as part of single-species investigations (Beissinger and McCullough 2002).

Population Viability Analysis and Metapopulation Modeling

Extinction risk assessment tools such as PVA explore issues associated with the viability of populations (Fieberg and Ellner 2001). PVAs are critical in conservation because the assessment of species extinction risk lies at the heart of conservation biology (Burgman et al. 1993; Fagan et al. 2001). The objective of PVA is to provide insights into how management can influence the probability of extinction (Boyce 1992; Burgman 2000; Possingham et al. 2001). It provides a basis to evaluate data and to anticipate the likelihood that a population will

Box 17.1. A "Taxonomy" of Uses of Models

The term "model" is employed widely in conservation biology and can have a range of different and sometimes confusing meanings (Burgman et al. 1993). Cockburn (1991) described a useful "taxonomy" of different types of models. These are characterized by markedly different limitations, assumptions, and required knowledge.

Type 1: Statistical models. These models are equations or relationships derived from analyses of data that typically include a response variable and explanatory variables (see Chapter 16).

Type 2: Data-free or theoretical models. In these models biological and ecological phenomena are explored using mathematics. These models may entirely lack data or be loosely underpinned by biological data. They may provide new insights into the possible behavior of ecological systems and open new avenues for other studies (Cockburn 1991). A good example is the early and highly simplified theoretical model of metapopulation dynamics developed by Levins (1970) in which all patches were equidistant and of identical size. This theoretical model was a progenitor of more detailed and realistic models used in studies of metapopulations (e.g., Hanski and Gilpin 1991; Hanski 1999a; see Chapter 6). Another example is the use of "neutral models" in landscape ecology, where simulated landscapes are generated to explore potential impacts on ecological processes (e.g., Franklin and Forman 1987) or particular species (Gustafson and Parker 1996; Turner et al. 2001).

Type 3: Data-rich models. In some circumstances, empirical data are available to describe some aspects of particular ecological phenomena. Models can then be developed on the basis of perceived understanding of these phenomena, and they can be useful for a range of purposes (Starfield and Bleloch 1992; Burgman et al. 1993; Tyre et al. 2001), including (1) identifying the extent of congruence between existing knowledge and theory, (2) determining if there are additional insights that can be gained from existing knowledge (for example, such models may add value to field data and statistical relationships that have been gathered; McCarthy et al. 2000), and (3) examining the outcomes of interactions between many factors that may influence particular ecological phenomena but which would otherwise be very difficult to track (e.g., see Mackey et al. 2002). A model of the dynamics of the endangered remaining population of the helmeted honeyeater *(Lichenostomus melanops cassidix)* is an elegant example of a data-rich model (McCarthy et al. 1994). Here, the model was based on a detailed understanding of the biology and ecology of the species, including its population size, breeding system, and habitat requirements. The model demonstrated the potential vulnerability of the population to demographic stochasticity. It provided new insights into factors that place the species at risk and highlighted ones that had previously been overlooked (McCarthy et al. 1994).

Type 4: Conceptual models. We add conceptual models to Cockburn's (1991) list. They are representations of ideas about how a system works. Conceptual models may be presented in natural language or as pictures, flow diagrams, or logic trees (Lindenmayer and Burgman 2005). They form the starting point for developing statistical and theoretical models. They should capture what a person believes to be true about the system in question, at a scale that suits the context and level of understanding (Burgman et al. 2005). Conceptual models may also be highly empirical, such as the fruit fly *(Drosophila melanogaster)*, which Cockburn (1991) noted has served as a useful model from which to explore key issues in genetics.

persist (Boyce 1992; Possingham et al. 2001). More generally, PVA may be seen as a systematic attempt to understand the processes that make a population vulnerable to decline or extinction (Gilpin and Soulé 1986; Shaffer 1990). In practice, "PVA" usually refers to building computer-based quantitative models of the likely fate of a population. Probabilities of extinction are estimated by Monte Carlo simulation (Lindenmayer et al. 1995b). The most appropriate model structure depends on the availability of data, the essential features of the ecology of the species, and the kinds of questions that managers need to answer (Starfield and Bleloch 1992; Burgman et al. 1993). PVA has been used widely around the world, and there are literally hundreds of examples of its use in studies of a wide range of species (Beissinger and McCullough 2002).

Metapopulations in a Population Viability Analysis Framework

Spatial structure and metapopulation dynamics can have important impacts on the persistence of species in modified landscapes (Chapter 6). If individuals of a given species move between patches, one cannot estimate the risk of extinction of an entire species based on the risk of extinction of its separate subpopulations because of the interdependency of, and the interactions among, the subpopulations. Metapopulation dynamics can be especially important for threatened or endangered species, many of which exist in small, semi-isolated populations that have resulted from habitat loss and habitat subdivision. Multiple populations also introduce new susceptibilities to threatening processes. In addition to impacts that decrease the mean survivorship or fecundity in a single population, metapopulations are sensitive to impacts that affect movement of organisms and increase the isolation of populations. Spatially and temporally explicit models like those developed by Possingham and Davies (1995) and Akçakaya and Ferson (1992) can be useful in modeling the viability of metapopulations in response to a range of phenomena, including various management options. There are numerous other applications of PVA in a metapopulation context (Possingham et al. 2001; Beissinger and McCullough 2002). However, such models are not problem-free. Models simplify the systems they portray, and metapopulation models are no different in this regard (Hanski 1999a, b).

Most metapopulation models such as Hanski's (1994b) incidence function model consistently favor simplicity over complexity (Ludwig 1999). Many complexities are frequently ignored—like spatial and temporal changes in the suitability of the matrix surrounding patches of habitat (see Pope et al. 2000; Carroll et al. 2003, 2004), or the influence of the matrix on the success of interpatch dispersal (Cale 1999). Wiens (1997) noted that most metapopulation models assume that "the matrix separating subpopulations is homogeneous and featureless." Metapopulation models can sometimes predict poorly when these complexities are ignored (Bender and Fahrig 2005). For example, in a plantation landscape in southeastern Australia, Hanski's (1994b) incidence metapopulation model predicted poorly the distributions of two species of arboreal marsupials that persisted in the matrix (Lindenmayer et al. 1999c). In the same landscape, McCarthy et al. (2000) found the fit between model-predicted and actual values for patch occupancy was poor for birds that were able to forage both in patches of native vegetation and in the surrounding matrix. In addition, metapopulation models generally do not accommodate changes in home range sizes that can follow habitat loss and habitat subdivision (e.g., Barbour and Litvaitis 1993), changed landscape mosaics (Milne et al. 1992), or frequent movements between many patches (e.g., to gather spatially separated food resources; Redpath 1995; Boone and Hunter 1996).

The complexities of species responses as outlined here indicate why it is vital to check

Box 17.2. The Predictive Ability of PVA Models in a Landscape Subject to Native Vegetation Loss and Fragmentation

Predictions of population size and patch occupancy were compared with field observations for several vertebrates and a range of PVA models for a highly modified forest landscape in southeastern New South Wales (Lindenmayer et al. 2003c). The predictive abilities of the models differed markedly between species. Predictions for the greater glider *(Petauroides volans)* were accurate, whereas for two species of small terrestrial mammals (the bush rat *[Rattus fuscipes]* and agile antechinus *[Antechinus agilis]*), the predictions from all the models were poor. These are some of the best-studied animals in Australia, and abundant, high-quality life history data are available for them. The results of model testing were attributed to factors such as species-specific relationships between patch size, quality, and animal abundance; complex spatial responses to the patch system that could not be readily captured by simple measures of patch size and patch isolation; and the fact that the life histories of the species in altered landscapes were different from those in continuous forests. On the basis of extensive model testing, it was possible to identify the kinds of models and sorts of populations for which predictions are likely to be accurate (see Table 17.1).

model assumptions before models are applied to real landscapes and real conservation problems (Doak and Mills 1994; Fahrig and Merriam 1994; Wiens 1994; Pope et al. 2000). Decisions that rely on models should be made cautiously, and considerable effort needs to be directed at both testing model assumptions and monitoring the outcomes of management actions (Possingham et al. 2001).

The Predictive Ability of Population Viability Analysis

Spatially explicit PVA models are sometimes criticized because they lack data and therefore often do not make reliable predictions (Doak and Mills 1994; Ludwig 1999; Ellner et al. 2002; Box 17.2). Although there are several examples where predictions from models and actual dynamics of populations have compared favorably (e.g., Puerto Rican parrot *[Amazona vittata]*, whooping crane *[Grus americana]*, and Lord Howe Island woodhen *[Tricholimnas sylvestris]* Lacy et al. 1989; Mirande et al. 1991; Brook et al. 2000), in general, PVAs have produced highly variable predictions (Coulson et al. 2001; Ellner et al. 2002).

Since Brook et al. (2000) assessed the predictive accuracy of PVA for 21 populations (8 bird, 11 mammal, 1 reptile, and 1 fish species; see Figure 17.1), discussion of the predictive accuracy of PVAs has increased (e.g., Brook 2000; Burgman 2000; Fagan et al. 2001; Fieberg and Ellner 2001; Beissinger and McCullough 2002; Brook et al. 2002; Reed et al. 2003; O'Grady et al. 2004). McCarthy et al. (2001, 2003) outlined methods for testing the accuracy of forecasts from PVAs and the reliability of relative predictions (Table 17.1). Perhaps most importantly, McCarthy et al. (2001) emphasized that:

> [The] role of model testing is not to prove the truth of a model, which is impossible because models are never a perfect description of reality. Rather, testing should help identify the weakest aspects of models so they can be improved.

Data availability often limits the predictive accuracy of most population viability analyses (Boyce 1992; Burgman et al. 1993; Caughley 1994; Taylor 1995; Ellner et al. 2002). Even the simplest models require more parameters than are usually available. Even so, PVAs can be valuable in several ways. For example, the models organize information for subsequent empirical tests (Walters 1986), especially if they summarize available data consistently and transparently (Burgman 2000; Brook et al. 2002). Their use can also help identify knowledge gaps (Burgman et al. 1993), highlight problems for which preemptive action may be beneficial (Tilman et al. 1994), and promote the understanding of complex ecological processes that might otherwise be overlooked in fieldwork (Gilpin and Soulé 1986; Temple and Cary 1988; Bender et al. 2003). In addition, through sensitivity analyses, PVA models allow the identification of which model structures and parameters are most important (Possingham et al. 2001).

Landscape Indices

Often, spatial patterns of vegetation cover created by landscape alteration are linked with the responses of particular elements of the biota. Metrics to characterize such patterns are termed landscape indices (Wegner 1994; Forman 1995; Smith 2000; Lindenmayer et al. 2002b), and they are used widely in studies of modified landscapes, particularly observational studies (Haines-Young and Chopping 1996; Gustafson 1998; McAlpine et al. 2002a).

Landscape indices attempt to provide descriptions of spatial landscape patterns, particularly "patchiness" and the size, shape, composition, juxtaposition, and arrangement of landscape units (e.g., vegetation patches). Such indices typically attempt to capture three broad groups of phenomena:

- Composition, or the identity and characteristics of the different types of patches

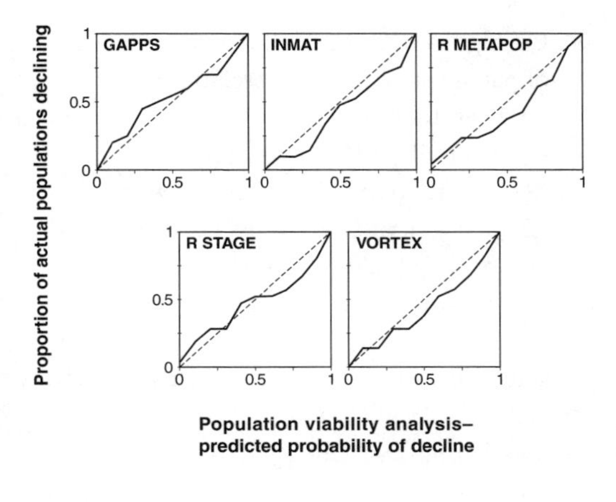

Figure 17.1. Population viability analysis (PVA)-predicted probability of decline versus the actual proportion of 21 real populations that were modeled which declined below the corresponding threshold size. The headings on the diagrams represent different computer packages that were employed in PVA (redrawn from Brook et al. 2000: reprinted by permission from Macmillan Publishers Ltd [*Nature*, **404**, 385–387], 2000).

Table 17.1. Contrasts between characteristics of populations whose fates are likely to be more accurately predicted by population viability analysis and those likely to be less accurately predicted*

More accurate population prediction	*Less accurate population prediction*
Single population	Metapopulation
Closed population	Open population
Discrete habitat boundaries	Diffuse habitat boundaries
Uniform habitat conditions	Heterogeneous habitat conditions
Constancy of life history attributes across habitat types and landscape conditions	Variation in life history attributes between habitat types and landscape conditions (e.g., fragmented versus unfragmented landscapes)
Constancy of species interactions across habitat types and landscape conditions	Variation in species interactions between habitat types and landscape conditions
Distance-related dispersal patterns	Habitat-related dispersal patterns
Simple social systems	Complex social systems

*Modified from Lindenmayer et al. (2003c).

- Configuration, the spatial arrangement of patches and temporal relationships between these units
- Connectivity, either from a human perspective (landscape connectivity; see Chapter 12), or from a species-specific perspective (habitat connectivity; see Chapter 6)

There are vast numbers of landscape indices (Turner 1989; Wegner 1994; reviewed by Haines-Young and Chopping 1996) and their use in studies around the world is increasing (Nicholls 1994; Short and Turner 1994; Incoll et al. 2000; Morgan 2001). Landscape indices may be valuable to help characterize a landscape and establish whether a pattern has changed (Wegner 1994; Smith 2000; McAlpine et al. 2002a). They may also be valuable for linking spatial patterns of landscape cover to species responses. However, when using landscape indices to describe landscapes and guide management actions, it is important to be clear about the reason for using them and the inherent limitations of using such an approach (Tischendorf and Fahrig 2000a, b). Cale and Hobbs (1994), Gustafson (1998), and Lindenmayer et al. (2002b) critically appraised the use of landscape indices. They noted:

- The methods used to develop indices are often not provided.
- It is not always clear what to do with landscape indices once they are generated because they do not necessarily link closely to land management.
- Few measures are used consistently in different investigations.
- Landscape indices generally fail to account for factors such as the vertical complexity of vegetation, which is known to be important for bird species richness and abundance (e.g., MacArthur and MacArthur 1961; Gilmore 1985; Brokaw and Lent 1999; see Chapter 14).
- Many indices provide sophisticated ways of highlighting intuitively obvious landscape patterns and have led to few new insights.
- Many landscape indices are highly correlated (and therefore provide similar or redundant information about the landscape).
- Many studies generate large numbers of metrics and hence many more than the degrees of freedom available to facilitate robust statistical analyses.
- Each species responds differently to the same spatial scale of landscape change and human disturbance (e.g., Davies and Margules 1998; Villard et al. 1999). Hence no single measure adequately reflects change for all biota. Landscape indices may be meaningless when species' perceptions of a landscape are unrelated to the way humans characterize and map that landscape (see Chapter 3).
- Most indices provide an instantaneous, static measure of landscape pattern, although temporal dynamics may be important, such as vegetation succession or the length of isolation of vegetation remnants (Loyn 1987; Bennett 1990a; Fahrig 1992; Gascon et al. 1999).
- Values for indices are often scale dependent, making it difficult to compare results from different landscapes and spatial scales. In addition, the scale of a species' movement is often not well linked to the scale at which landscape indices are generated.

In relation to the final point listed above, Cale and Hobbs (1994) noted that:

> Indices of [landscape] diversity can indicate that differences in heterogeneity exist when no actual change in habitat has occurred from the organism's point of view, or [they] can fail to detect important changes in habitat. The principal reason for this is the problem of matching the scale of measurement with the scale at which organisms perceive the environment.

Different indices have different strengths and limitations depending on context. The units and the context (e.g., the size of a "landscape") should be clearly defined. Finally, any system for the application of landscape indices should establish a framework for evaluating predictions and decisions that flow from the application of the indices (Lindenmayer et al. 2002b).

Reviews

Although there have been many valuable insights from the literally thousands of studies undertaken worldwide on landscape change, fragmentation, and vegetation loss (Debinski and Holt 2000; McGarigal and Cushman 2002), the vast majority of studies have been "one-off" projects. In addition to being one-off investigations, most ecological studies to date have (1) examined problems at a small scale or single scale, (2) focused on a single species or group of species, or (3) taken place within a single landscape (e.g., Fazey et al. 2005a). A key problem is that many studies yield results specific to a particular situation, and their outcomes are not readily transportable to others. Krebs (1999) noted that many researchers are disappointed when their "general findings" do not apply to systems beyond the ones they have studied. Marked complexity in ecological findings should be expected for at least three reasons. First, landscape change alters many ecological processes and can have cumulative impacts on different species (Kirkpatrick 1994; Paine et al. 1998). Second, each species responds differently to landscape change (Robinson et al. 1992; Davies et al. 2000; Chapter 3) and responds to a range of factors at multiple scales (Forman 1964; Diamond 1973; Chapter 3). Third, the same species can respond differently in different landscapes (Åberg et al. 1995), with factors such as habitat requirements varying between different regions; for example, as occurs in the Australian greater glider *(Petauroides volans);* see Lindenmayer 2002).

Box 17.3. The Use of a Systematic Review Process to Assess Landscape Alteration Impacts

Much has been written about the importance of evidence-based approaches to ensure best-practice human medicine. That is, information from a range of sources (such as randomized trials) is accumulated to determine, for example, the veracity of relationships between the use of a kind of drug and the response of the medical condition of a group of patients. A key part of an evidence-based approach in medicine has been the development of a systematic review process for reviewing studies on a given topic, particularly clinical trials (Sackett et al. 2000). In 1993, an organization known as the Cochrane Collaboration was established to guide the completion of objective systematic reviews and facilitate the communication of the results. Many improvements in human medicine occurred as a result.

Several authors have suggested that an evidence-based approach should be employed to better guide conservation efforts (e.g., Pullin and Knight 2001) in which the evidence relating, for example, the effectiveness of a particular conservation action (e.g., soft-release reintroduction) for a given target group is rigorously assessed using a systematic review process. Fazey et al. (2005b) examined the possible role that an approach similar to the Cochrane Collaboration might play in the completion of systematic reviews of environmental conservation. Although there are many major differences between the disciplines of medicine and conservation science (e.g., the number and types of studies conducted), Fazey et al. (2005b) concluded that an organization like Cochrane Collaboration could play a valuable role in systematically reviewing evidence on topics such as the magnitude of impacts of particular kinds of threatening processes or the effectiveness of a given impact mitigation strategy in human-modified landscapes.

It is perhaps not surprising then that many investigations have yielded outcomes specific to a particular landscape or taxonomic group in that landscape, which are not readily generalized to other locations or assemblages (Simberloff 1992; McGarigal and Cushman 2002). It is possible that more generic outcomes may come from syntheses of results of studies of different sets of taxa in the same landscape (e.g., Robinson et al. 1992; Gascon et al. 1999) as well as studies of a range of landscapes subject to different disturbance regimes.

Several kinds of reviews may assist attempts to find general principles about the impacts of landscape change on biota, which species are most and less extinction prone, and which mitigation strategies might be the most effective. One technique is the systematic review process that has been developed for medicine (Box 17.3). Another is to formally synthesize existing insights via meta-analysis (e.g., Murtaugh 2002; Lajeunesse and Forbes 2003), in which the results of a large number of similar studies are subject to formal (quantitative) analysis to search for evidence of general patterns (Osenberg et al. 1999). Meta-analysis has been used widely in medicine and agriculture, and there is considerable potential for its application to studies of landscape change (Rosenberg et al. 2000; Peek et al. 2003). There is a need to be aware of some potential problems with meta-analyses in ecology. Usually, original data are not used; instead, data are compendia of results from published work by other authors. The outcomes are heavily influenced by biases toward papers where strong results are reported, and particular journals from which papers are reviewed (Murtaugh 2002; Peek et al. 2003). Unfortunately, studies with null outcomes rarely find their way into the literature. In addition, past synthesis work has often been hamstrung by large between-study differences in field sampling protocols. Nevertheless, there have been valuable meta-analyses of several phenomena in human-modified landscapes. For example, Bender et al. (1998) conducted a meta-analysis of the effects of habitat patch sizes on species' populations.

Not all conservation information requirements can be met through a formal review procedure like meta-analysis or other types of systematic and quantitative review (see Box 17.3). Less formal or qualitative thematic reviews can also make a major contribution to an improved understanding of the impacts of landscape change on biota. Examples are the reviews of "fragmentation" effects by Saunders et al. (1991) and syntheses of studies of nest predation and brood parasitism among birds at patch boundaries (see Rudnicky and Hunter 1993; Kremsater and Bunnell 1999). Other examples of qualitative thematic reviews have addressed corridors and connectivity (Bennett 1990b, 1998; Hobbs 1992; Beier and Noss 1998) and fragmentation experiments (Debinski and Holt 2000). Qualitative thematic reviews can be useful because they can encompass a wide range of investigations and are not constrained by an inherent need to formally compare broadly similar studies (such as controlled experiments). Indeed, the very wide range of kinds of ecological studies of altered landscapes (see earlier discussion) means that a significant cross-section of investigations would not be suited for formal quantitative comparison as demanded by meta-analysis.

Thus both quantitative and qualitative thematic reviews will be important for determining if there are generic outcomes that can be identified to improve conservation management in modified landscapes. Reviews of existing knowledge are a key part of the scientific method. They help to identify general patterns and thus inform the types of questions asked in future studies (Figure 17.2; see also Pickett et al. 1994).

Summary

Extensive empirical data can be gathered by field-based methods (see Chapter 16). In many cases, it is possible to add further value to these empirical data through the application of desktop approaches and, in turn, to generate an improved understanding of the responses of biota to landscape modification. The insights generated from approaches in which no new field data are gathered can also inform decisions about the management strategies that are likely to be most effective for limiting losses of native species as an outcome of landscape

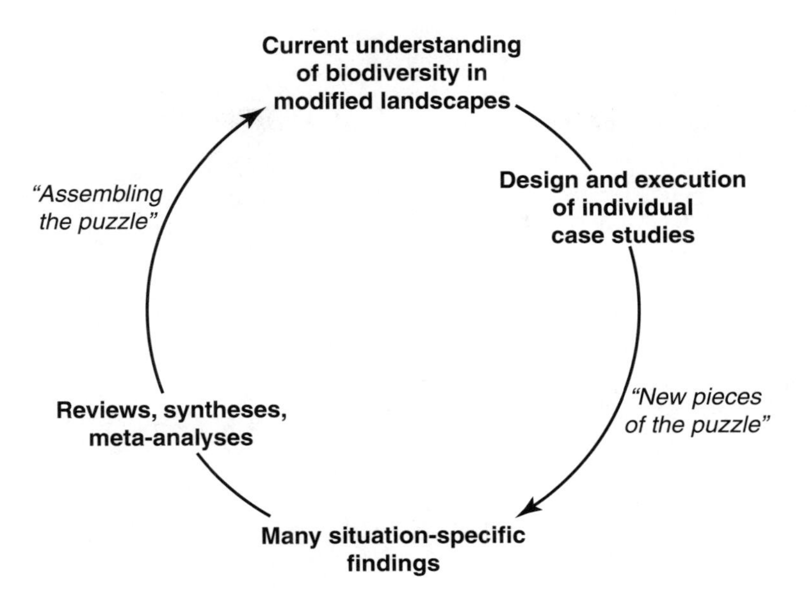

Figure 17.2. The complementary role of empirical case studies and reviews for understanding the processes shaping biotic responses modified landscapes.

alteration. Simulation models such as those used for PVAs, and landscape indices to characterize patterns of vegetation cover are frequently applied examples. Qualitative thematic and quantitative reviews are two additional and fundamentally important desktop approaches critical for improved understanding of biotic responses to landscape modification. They have a particularly valuable role in studies of landscape modification because they attempt to distill general patterns from what otherwise can often appear to be species-specific or landscape-specific findings that do not readily generalize to other taxa or landscapes.

Links to Other Chapters

Chapter 8 summarizes the processes that can threaten the persistence of particular species, which are often modeled as part of the PVA process. Spatially explicit metapopulation models are often used because of the negative impacts of habitat subdivision on individual species; this topic is discussed in detail in Chapter 6. Landscape indices are used to quantify landscape patterns as part of assessing landscape connectivity for assemblages (Chapter 12) and habitat connectivity for individual species (Chapter 6). Typical patterns arising from human landscape modification are discussed in Chapter 2 and Part III of the book.

Further Reading

Excellent reviews of models and modeling are provided by Starfield and Bleloch (1992), Burgman et al. (1993), McCallum (2000), and Williams et al. (2002). There are many texts and papers that have focused on PVAs, including Beissinger and Westphal (1998), Possingham et al. (2001), Beissinger and McCullough (2002), and Reed et al. (2002). Haines-Young and Chopping (1996) and Gustafson (1998) present detailed overviews of landscape indices. Cale and Hobbs (1994) and Lindenmayer et al. (2002b) discuss the limitations of landscape indices in studies of landscape change on biota. Osenberg et al. (1999), Murtaugh (2002), and Lajeunesse and Forbes (2003) discuss some important aspects of meta-analysis.

PART V

Mitigating the Negative Effects of Landscape Change on Species and Assemblages

Human landscape modification (e.g., for agriculture or forestry) can have a range of negative effects on both individual species and species assemblages (Parts II and III). A detailed investigation of individual species suggests that species respond individualistically to their environment (Gleason 1939; Chapter 3). However, species-specific responses to human landscape modification pose a difficult challenge to conservation management because it is impossible to study and manage every individual species separately (Simberloff 1998). In addition, no species exists in isolation. Rather, multiple species interact with one another through processes like competition, predation, or mutualist relationships (Chapter 7).

The "fragmentation panchreston" proposed by Bunnell (1999a) explicitly recognizes that landscape change encompasses a range of processes, alters vegetation patterns, has impacts on particular species, and has impacts on suites of taxa and assemblages. Throughout this book, we have argued that a constructive approach to understanding this complexity is to acknowledge the complementary roles of focusing on (1) individual species and ecological processes (Part II), and (2) human-defined landscape patterns in relation to species assemblages (Part III). We reiterate that this distinction between processes and patterns is a matter of research strategy rather than dictated by nature. We also acknowledge that there are cases where individ-

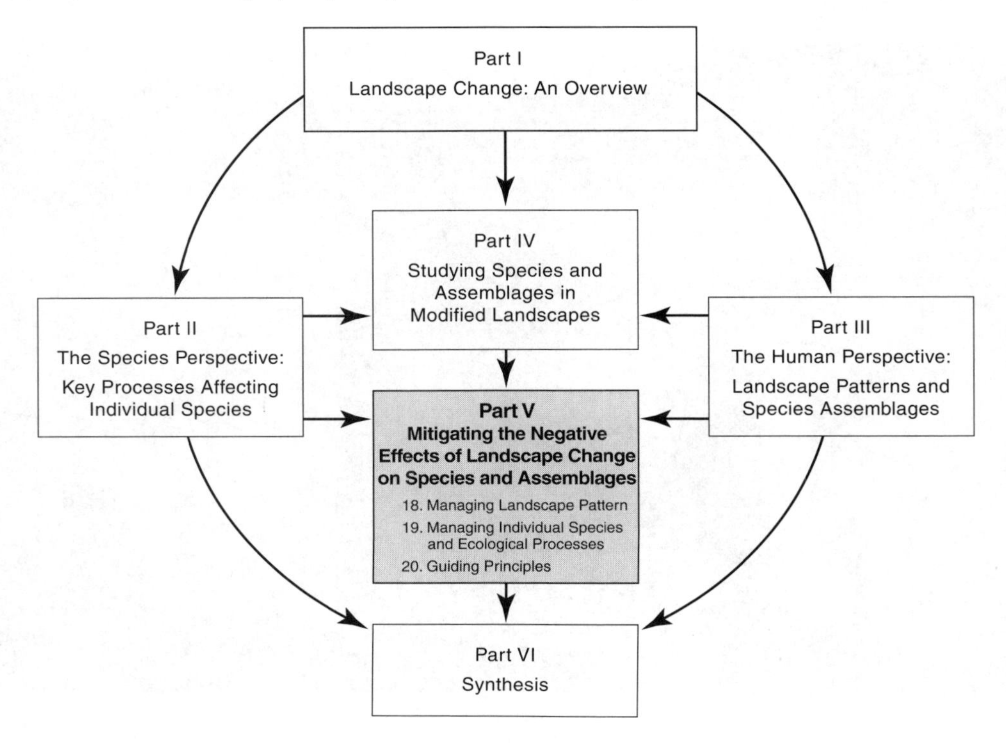

Figure V.1. Topic interaction diagram showing links between the different parts of the book, and the topics covered in the chapters of Part V.

ual species and species assemblages respond in similar ways to landscape modification; and that in some cases, human-defined landscape pattern may be a reasonable proxy for ecological processes.

Part V of this book discusses strategies to mitigate the negative effects of human landscape modification on individual species and assemblages of species. It contains three chapters. Chapter 18 outlines five mitigation strategies that are based on managing landscape pattern. These pattern-based strategies circumvent the need to analyze in depth every individual species and every single ecological process. Implementation of the pattern-based guidelines outlined in Chapter 18 will result in a landscape pattern that is likely to benefit a wide range of species simultaneously—despite potential differences between individual species, and despite an incomplete understanding of the ecological processes driving their individualistic distribution patterns. Although the pattern-based principles in Chapter 18 are likely to be a useful starting point for conservation in human-modified landscapes, they should be complemented by an assessment of particularly important ecological processes and individual species. Chapter 19 outlines five management strategies targeting ecological processes and individual species.

Together, Chapters 18 and 19 provide the basis for Chapter 20, which concludes this part of the book. Chapter 20 summarizes a list of 10 guiding principles for mitigating the negative effects of human landscape modification on individual species and species assemblages. These guiding principles are a succinct management synthesis not only of Chapters 18 and 19, but in many ways, of the entire book.

Notably, there is some overlap in the content presented in earlier parts of the book and Part V. A key difference between earlier parts and Part V is that the former parts of the book are a detailed scientific discussion, whereas Part V provides a shorter, more applied, and more tangible summary of key management recommendations. The relationship of Part V to other parts of the book is summarized in Figure V.1.

Figure V.2. Boreal forest landscape near Smithers, Central British Columbia, Canada (photo by David Lindenmayer).

CHAPTER 18

Managing Landscape Pattern to Mitigate the Decline of Species and Assemblages

Managing every individual species and every single ecological process presents an insurmountable challenge to conservation practitioners. A useful approach to mitigating the negative effects of human landscape modification on species and species assemblages therefore is to break up this problem into two steps. The first step is to manage landscape pattern in a way that will benefit many species simultaneously. This step is an efficient starting point for conservation management because even in the absence of detailed ecological knowledge, the implementation of a handful of pattern-based management strategies is likely to provide conservation benefits in many situations. This chapter outlines five management strategies that are likely to be useful for conservation in a wide range of modified landscapes. For each strategy, we briefly summarize its scientific basis with reference to previous chapters, and provide some examples. Whereas this chapter focuses on pattern-oriented management approaches, the next chapter focuses on complementary species-oriented and process-oriented management strategies.

Strategy 1: Maintain and/or Restore Large and Structurally Complex Patches of Native Vegetation

Rationale

The species–area curve is one of the few general principles in ecology (Rosenzweig 1995). That is, other things being equal, larger patches tend to support more species than smaller patches (Preston 1962; Chapter 9). In addition to the size of a given area of native vegetation, the vegetation structure of that area is also fundamentally important. Again, other things being equal, vegetation that is structurally complex tends to support more species than structurally degraded vegetation (MacArthur 1964; Chapter 10). Large and structurally complex patches of native vegetation therefore are an important part of mitigating the negative effects of landscape change on species and assemblages of species.

Protection

A critical first step in conserving terrestrial ecosystems is to protect formally as much native vegetation as possible, given other societal objectives (Margules and Pressey 2000). This has many major benefits because it will make it more likely that (1) the overall amount of native vegetation cover is maximized, and losses of species reliant on native vegetation will be minimized (Fahrig 2003); (2) more kinds of habitats are conserved for individual species; (3) conditions are retained for specialist species; (4) populations of many individual species will be larger and better buffered against external negative disturbance regimes; and (5) key ecosystem processes such as hydrological regimes will be retained (Noss and Cooperrider 1994; Lindenmayer and Franklin 2002).

A key issue associated with the maintenance of existing vegetation cover is that of nonreplaceability (Margules and Pressey 2000). That is, in the vast majority of circumstances, even highly sophisticated restoration programs may not be able to restore fully the complement of biota and ecological processes that originally occupied an area prior to human landscape modification. Thus, although restoration programs are important in many modified landscapes, a fundamental starting point is to ensure that remaining areas of native vegetation are formally protected (Salt et al. 2004). Notably, nature reserves alone cannot guarantee the protection of all native species because they are too few, too isolated, too static, and not necessarily safe from overexploitation (Liu et al. 2001; Bengtsson et al. 2003; Rodrigues et al. 2004). Given these limitations, off-reserve conservation in human-modified landscapes (e.g., agricultural or forestry-dominated landscapes) is an important conservation strategy in addition to setting aside reserves.

Developing a network of nature reserves is a regional-scale strategy to provide core habitat for many species (Noss and Harris 1986; Soulé and Terborgh 1999). At a smaller scale, patches of native vegetation scattered throughout a given human-modified landscape can fulfill similar ecological roles. Structurally characteristic patches of relatively unmodified native vegetation are a key part of conservation in both forestry and farming landscapes. For example, exotic softwood plantations in southeastern Australia often contain remnants of native forest or woodland, and these remnants provide important habitat for numerous species of native wildlife (Recher et al. 1987; Lindenmayer and Hobbs 2004). Similarly, McIntyre et al. (2000) synthesized ecological research in subtropical grazing landscapes in eastern Australia and recommended that 30% of a given farm should support native woodland, with individual woodland patches ideally being at least 5 to 10 ha. Similar examples can be found throughout the world. For example, patches of native vegetation have been shown to provide conservation benefits in farming landscapes in Europe (Farina 1995; Hinsley et al. 1999), Africa (Cordeiro and Howe 2003; Bodin et al. 2006), Central America (Graham 2001; Luck and Daily 2003), South America (Handford 1988), and North America (Rosenblatt et al. 1999; Nupp and Swihart 2000).

Restoration

In addition to the protection of patches of native vegetation, in some modified landscapes, structurally characteristic patches need to be restored. Ecological restoration means different things to different people (Dobson et al. 1997; Robertson et al. 2000; Higgs 2005). For the purposes of this book, we consider restoration to mean guiding and managing an ecosystem along a trajectory of recovery of its natural (characteristic or historic) structure, function, and composition (Franklin et al. 1981; Noss 1990; Noss et al. 2005). Two specific aspects of restoration are considered briefly in the following sections: (1) the enhancement of existing patches of native vegetation, and (2) the establishment of new patches of native vegetation.

Enhancement of Existing Patches

In many modified landscapes, patches of native vegetation remain but are structurally degraded (e.g., Saunders et al. 2003; see Chapter 10). Changing management regimes in these landscapes can sometimes lead to the natural recovery of structural complexity (Lindenmayer et al. 2003b). For example, in some grazing landscapes, a reduction in stocking rates will improve tree and shrub regeneration (Spooner et al. 2002; see Figure 18.1) and facilitate the restoration of conditions that are similar to ecosystem condition prior to landscape modification (Brooks and Merenlender 2001; Jansen and Robertson 2001).

In other examples, active management efforts are required to enhance vegetation structure and condition; for example, because natural regeneration has not taken place, because the seedbank has been depleted, or because spe-

Figure 18.1. Natural regeneration around a remnant tree at Table Top near Albury, southeastern Australia (photo by David Lindenmayer).

cies have been lost and/or ecosystem processes have changed. Examples include the ponderosa pine *(Pinus ponderosa)* forests of southwestern United States (Covington 2003) and the mid-boreal forests of Scandinavia (Linder and Östlund 1998). The reestablishment of appropriate fire regimes, deliberate tree girdling to create standing dead and down timber, and strategic thinning of dense regrowth forest stands may be appropriate restoration treatments in these and other similar kinds of ecosystems (Carey and Gill 1983; Carey et al. 1999a, b; Noss et al. 2006; also see Chapter 19).

The restoration of keystone structures (i.e., structural elements that are particularly important to many species; see Tews et al. 2004), should be particularly important in efforts to enhance existing patches of remnant vegetation. What constitutes keystone structures varies widely among ecosystems and may, for example, include ephemeral water bodies (Tews et al. 2004) or tree hollows (Gibbons and Lindenmayer 2002).

In addition to enhancing vegetation structure and condition, enlarging existing patches of native vegetation to a given size can be an important restoration strategy. In a forest example, McCarthy and Lindenmayer (1999) showed that the most effective way to increase the area of habitat for the greater glider *(Petauroides volans)*, a forest-dependent Australian arboreal marsupial, was to expand the size of old-growth patches by maintaining a harvest-free zone of up to several hundred meters around them. In the Australian State of Tasmania, Brereton et al. (1997) recommended that any grassy forest and woodland containing white gum *(Eucalyptus viminalis)* within a 3 km radius of a known group of the endangered bird the forty-spotted pardalote *(Pardalotus quadragintus)* should be exempt from clearing to promote habitat connectivity for the species and facilitate future population expansions.

Establishment of New Patches

In some landscapes, vegetation restoration involves the complete reestablishment of native vegetation cover following past human activities such as mining or land clearing (Martin et al. 2004). Active replanting and revegetation can be a very long-term process, especially if important habitat features like large cavity trees take a long time to develop (Saunders et al. 1993a). Vegetation restoration in areas largely devoid of native vegetation is substantially more difficult than enhancing existing vegetation cover because, for example, edaphic conditions may have changed or competition from exotic plants (such as exotic grasses) is intense (Keenan et al. 1997). Additional considerations include which species should be restored, where seeds for these species will come from, whether weed control is required, whether the soil is in a reasonable condition, and a wide range of other issues (Bennett et al. 2000). Such active vegetation restoration is a large and complex topic, and a detailed treatment of it is well beyond the scope of this book. Clear objectives about the nature of the vegetation community to be restored are crucial in all restoration projects. Often, revegetation schemes will attempt to re-create similar types and patterns of vegetation cover that existed in the local area prior to human modification (Lambeck 1997; Bennett et al. 2000).

Strategy 2: Maintain and/or Restore a Matrix That Is Structurally Similar to Native Vegetation

Rationale

According to the patch–matrix–corridor model of modified landscapes (sensu Forman 1995), the matrix accounts for the largest proportion of the landscape and exerts a dominant influence on many important ecological processes (Chapter 3). In general, a matrix characterized by a vegetation structure that is similar to patches of native vegetation is likely to have numerous benefits for native species. Three key benefits are (1) the provision of habitat for some native species, (2) enhanced landscape connectivity, and (3) reduced edge effects (Chapters 11, 14). Matrix management can target the biophysical nature of the matrix per se, and the spatial arrangement of land uses in the landscape.

Managing the Vertical Structure of the Matrix

The vertical structure of the matrix has a fundamental effect on species distribution patterns. In both forestry and agricultural landscapes, several studies from throughout the world have shown that land use practices that produce structural characteristics in the matrix similar to those found in retained areas of native vegetation support the highest numbers of native species (Thompson et al. 2003; Dunn 2004; Figure 18.2). For example, shade coffee plantations that retain native tree species in the overstory tend to support more native species than plantations where the coffee is grown in full sun (Siebert 2002; Petit and Petit 2003).

The vertical structure of the matrix also exerts a strong influence on landscape connectivity (Wiens 1997; Hokit et al. 1999; see Figure 18.2). Strategic management, such as retaining trees during logging of forest landscapes or modifying the landscape surrounding vegetation remnants, may facilitate animal and plant dispersal (Forest Ecosystem Management Assessment Team 1993). Taxa that can disperse through a sympathetically managed matrix will be less dependent on wildlife corridors (Rosenberg et al. 1997; Mabry and Barrett 2002), or the width of wildlife corridors needed to facilitate movement may be reduced (Forman 1995; Lindenmayer 1998). Fewer dispersal events may then be required to "rescue" populations in vegetation remnants (Stacey and Taper 1992; Mills and Allendorf 1996). Conversely, if conditions in the matrix are hostile, then larger corridors

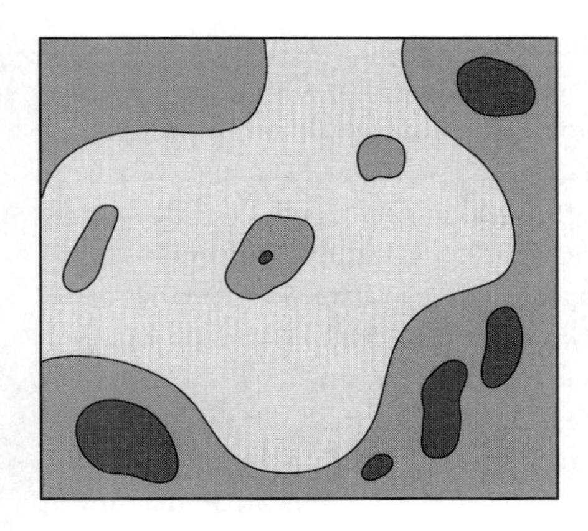

Land-use intensity	Low	Medium	High
Habitat suitability	Many species	Some species	Few species
Habitat connectivity	High for many species	High for some species	Low for many species
Edge effects	Needs protection	Constitutes buffer	Origin of edge effects

Figure 18.2. Schematic representation of a landscape with different land use intensities. Different land use intensities fulfill different roles with respect to the provision of habitat, habitat connectivity, and edge effects.

linking larger retained patches may be required (Taylor et al. 1993). Notably, management of the matrix may provide better habitat connectivity than wildlife corridors for those taxa that apparently disperse randomly in landscapes (Belthoff and Ritchison 1989; Date et al. 1996).

A third argument for managing the vertical structure of the matrix relates to edge effects (see Figure 18.2). As discussed in Chapter 11, the magnitude of edge effects is often strongly associated with the level of contrast in physical, structural, and other conditions between vegetation remnants and the surrounding matrix. It follows then that management strategies in the matrix to limit the level of contrast between remnant vegetation and the surrounding landscape should reduce the magnitude of some types of edge effects that might otherwise develop (Rodríguez et al. 2001; Tubelis et al. 2004).

One of the best examples of matrix management comes from the Douglas fir *(Pseudostuga menziesii)* forests in the Pacific Northwest of the United States, particularly as it relates to the conservation of the northern spotted owl *(Strix occidentalis caurina)*, although the provision of suitable vegetation cover for a wide range of other species also became prominent in the forest management planning process (Forest Ecosystem Management Assessment Team 1993). A change in landscape pattern as a result of extensive clearcutting of the Douglas fir forests of northwestern United States was widely considered to be the major process threatening populations of the northern spotted owl (Forsman et al. 1984; Yaffee 1994). Conservation efforts for the species initially involved the establishment of conservation areas on government land. However, it was recognized that the areas between the conservation areas were also extremely important. Thomas et al. (1990) recommended the 50-11-40 rule, which ensured that at least 50% of the matrix should be in forest stands that were at least 11 inches (27.5 cm) in diameter at breast height and had a 40% canopy cover. Subsequent management reports and guidelines further increased the amount of reserved land but also strengthened prescriptions for the retention of components of original forest cover in areas subject to timber harvesting (Tuchmann et al. 1996; Franklin et al. 1997). These matrix management strategies "soften the matrix" (Franklin 1993) and in so doing, they (1) maintain populations of many species in off-reserve forests, (2) improve the structural condition of regenerating stands after logging, (3) increase the likelihood that species can disperse between formally reserved areas, and (4) reduce edge effects at the landscape scale.

Box 18.1. Management of Spatial Patterns to Reduce Edge Effects in Forests

A useful example of an approach to reduce edge effects comes from the Chequamegon National Forest in Wisconsin, USA (Parker 1997), where the selection of appropriate spatial and temporal patterns of harvest units was central to the maintenance of interior forest conditions. The need for larger blocks of old-growth forest first emerged because of concerns over edge effects associated with ungulate herbivory on understory plants, which can extend up to 8000 m from patch boundaries (Alverson et al. 1994). Management regimes more comparable to natural conditions have included a shift from managed patch sizes of 20 ha to 5000 ha, thereby simulating historical, fire-regenerated patch sizes (Parker 1997).

Managing the Spatial Pattern of the Matrix

The spatial arrangement of land use types can have significant effects on the prevalence of edge effects (Chapter 11). It follows that one approach to mitigating negative edge effects is to manage spatial patterns of landscape vegetation cover to limit the length of human-created boundaries between patch types (Box 18.1). Such approaches are perhaps best developed in forested landscapes subject to logging and associated road building for the transport of wood.

In many jurisdictions, areas subject to clearfell timber harvesting are widely dispersed across forest landscapes with the potential to create spatial patterns characterized by large amounts of edge and limited interior area (Franklin and Forman 1987; Figure 18.3). This can place some vegetation types and species at risk that are dependent on interior microclimates or are susceptible to processes typical of edge environments such as extensive wind damage (Moore 1977; Savill 1983) and weed invasion (Yates et al. 2004). In these cases, planning the spatial arrangement of harvest units so they are aggregated rather than dispersed has the potential to limit the extent and magnitude of negative edge effects.

What constitutes appropriate sizes of cutover units may be guided by the patch sizes that are produced through natural disturbance regimes (Perera et al. 2004), although very large harvested areas are often politically and socially unacceptable (Hunter 1993; Haila et al. 1993). Thompson and Angelstam (1999) provide a useful example of the woodland caribou *(Rangifer tarandus)* from Canadian forests. The species requires large tracts of old forest that cover hundreds of square kilometers. Therefore, harvest units (that eventually regrow to an old-growth stage) need to be sufficiently large to match the patch sizes created by stand-replacing natural disturbances. If this does not occur, herds of the woodland caribou may become subdivided and susceptible to edge effects such as increased predation (Thompson and Angelstam 1999).

An important caveat associated with the aggregation of harvest units in forests to mitigate edge effects is the concurrent need for other activities within cutover units, such as stand structural retention to ensure internal structural complexity (see earlier discussion) and patch heterogeneity (see Strategy 5).

Strategy 3: Maintain and/or Restore Buffers around Sensitive Areas

Rationale

Edge effects are changes to the abiotic and biotic environment that occur at the boundaries between relatively unmodified vegetation and highly modified areas. Edge effects can have major negative impacts on native ecosystems (Chapter 11). At the landscape scale, edge effects may be effectively mitigated by minimizing the structural contrast between the matrix and patches of native vegetation (see Strategy 2). A complementary strategy at a more localized scale is to specifically create buffers around ecologically sensitive areas.

Figure 18.3. Dispersed cutting patterns created in the Douglas-fir forests of the Pacific Northwest of the USA (photo by Jerry Franklin).

Examples of Buffers

There are many examples demonstrating that remnants of premodification ecosystems may not maintain their species composition because of edge effects (reviewed by Kremsater and Bunnell 1999). Buffers are areas that surround and protect sensitive areas to better conserve the species within them (Hylander et al. 2004). Typical aims of buffers may be to limit the impacts of a disturbance regime on native ecosystems, or to maximize native species richness within a protected area (Baker 1992; Spackman and Hughes 1995). The concept is particularly well developed in the context of mitigating edge effects from logging on riparian and aquatic zones, where there are important benefits not only to native species but also to water quality and soil stability (e.g., Cockle and Richardson 2003; Quinn et al. 2004; Sweeney et al. 2004).

What constitutes a suitable buffer will depend on many factors such as the type of process that generates an edge effect (see Chapter 11), and the taxa or other attributes that need protection (Box 18.2). For example, the width of buffer strips to protect riparian and aquatic areas from pesticides and to reduce in-stream invertebrate mortality needs to exceed 50 m in Australian eucalypt plantations (Barton and Davies 1993). In contrast, buffers may need to be several hundred meters wide to effectively mitigate changed wind patterns (Harris 1984; Saunders et al. 1991). Recommended buffer widths can vary significantly between regions, even for the same group of species (Kinley and Newhouse 1997; Hagar 1999). For example, in eastern United States, buffer widths of 100 m have been recommended to maintain species richness and abundance of migratory neotropical songbirds in vegetation remnants (Keller et al. 1993; Hodges and Krementz 1996; Kilgo et al. 1998). Darveau et al. (1995) recommended buffers of at least 60 m to maintain bird species richness in logged Canadian boreal forest,

Box 18.2. Buffer Systems for Amphibians and Reptiles in Wetlands and Riparian Areas

Semlitsch and Bodie (2003) conducted a review of the use of wetlands, riparian areas, and surrounding terrestrial areas by amphibians and reptiles, primarily in North America. Given that many species use either or both aquatic areas and adjacent terrestrial land and can move long distances between such areas, Semlitsch and Bodie (2003) made broad recommendations for the sizes of buffers needed to maintain "core habitat" for amphibians and reptiles where surrounding land uses such as agriculture, forestry, and urbanization occurred. They recommended three terrestrial zones adjacent to aquatic and wetland areas—a terrestrial zone immediately adjacent to the aquatic area, a second zone that encompasses the core areas used by semiaquatic species, and a third zone that buffers the second from edge effects (Figure 18.4). The sum of the three zones was crudely estimated to extend 350 meters from the margin of a wetland or aquatic area and is more than 10 times wider than the 15 to 30 meter buffers used in many U.S. states (Semlitsch and Bodie 2003).

Notably, although the estimates made by Semlitsch and Bodie (2003) are crude, they are based on current ecological knowledge of the species to be protected. Hence the buffer width estimates provide a rational basis for reviewing existing prescriptions for aquatic ecosystem protection.

whereas buffers almost three times this width were needed to maintain bird species richness along streams in Vermont (northeastern United States) (Spackman and Hughes 1995).

Many other approaches to the development of buffer zones have been proposed. Mladenhoff et al. (1994) described 100 m restoration zones around remnant old-growth patches in second-growth matrix lands designed to buffer and reduce edge effects. In many forest landscapes, unlogged or selectively harvested stands can have a positive buffering effect for adjacent sensitive areas (Hylander et al. 2002). Similarly, planning of the spatial alignment of harvest units can mitigate impacts of abiotic effects such as wind damage (Lindenmayer et al. 1997; Berry 2001). Aquatic zones can be buffered by staggering logging operations to ensure that both sides of a riparian buffer are not logged at the same time (Recher et al. 1987; Hylander et al. 2002). All of these approaches have adjacency rules (sensu Davis et al. 2001) that specify waiting periods before an area neighboring a regenerating harvest unit can be cut. They also acknowledge that the importance of the buffering effect of retained areas will diminish as the surrounding disturbed area begins to recover.

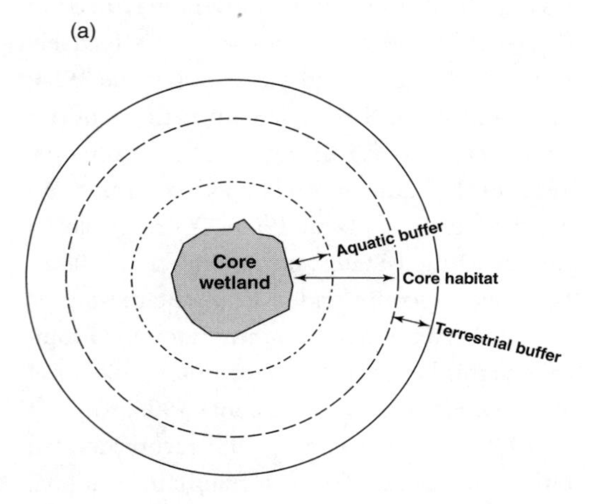

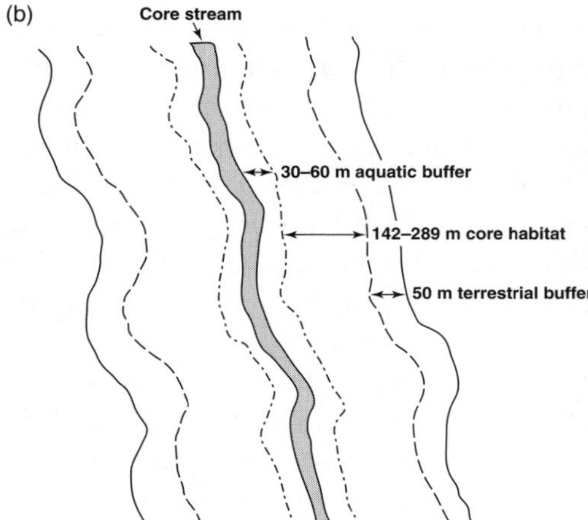

Figure 18.4. Proposed buffer zones for wetlands and streams (redrawn from Semlitsch, R.D. and Bodie, J.R. 2003: *Conservation Biology*, **17**, 1219–1228, Biological criteria for buffer zones around wetlands and riparian habitat for amphibians and reptiles, with permission from Blackwell Publishing).

To illustrate the potential value of buffering in a forest management context, Harris (1984) showed how an old-growth patch bounded by a recently clearfelled area may need to be an order of magnitude larger than one surrounded by mature forest to achieve the same area of interior forest habitat (Figure 18.5). In the case of fire risks and edge effects, mature forest buffers may reduce the chance of a fire in an old-growth patch (Harris 1984) because the probability of ignition and spread declines with increasing age in some forest types (Agee and Huff 1987).

Notably, although buffers are most widely used in forestry landscapes, recent work suggests they may be equally useful in agricultural landscapes. For example, streamside vegetation set aside as riparian buffers also provides important habitat for bats in Central American farming landscapes (Galindo-Gonzalez and Sosa 2003). Similarly, for eastern Australian farms, McIntyre et al. (2000) suggested that core wildlife areas covering 10% of a given farm be embedded within larger woodland areas managed at moderate intensity.

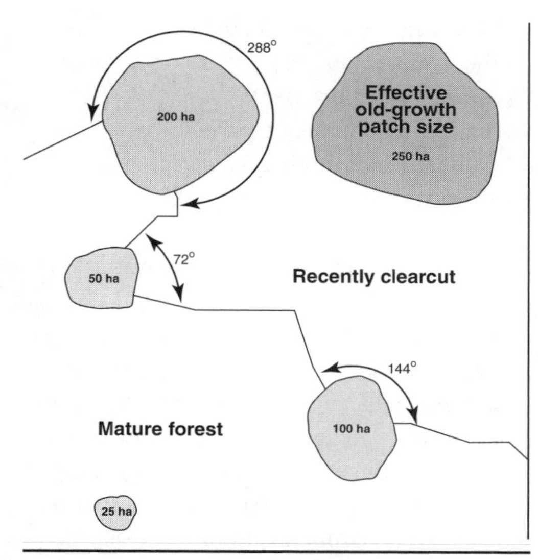

Figure 18.5. The buffering effect of surrounding matrix conditions on stands of old-growth forest. The diagram shows the size of an area of old-growth forest needed to maintain interior conditions in a matrix dominated by recently cut forests (250 ha in size) contrasted with one surrounded by mature forest (25 ha in size) (redrawn from Harris, L.D., Maser, C. and McKee, A. 1982: *Transactions of the North American Wildlife Natural Resources Conference*, **47**, 374–392, Patterns of old-growth harvest and implications for Cascades wildlife, with permssion from the Wildlife Management Institute).

Large-Scale Buffer Zones

Buffer zones can also be implemented at scales larger than simply along particular streams or surrounding patches of remnant revegetation (Gallent and Kim 2001). Noss and Harris (1986) outlined a strategy termed multiple-use-modules (MUMs) consisting of concentric management zones in the matrix buffering a core reserved area (see also Schonewald-Cox and Buechner 1990). Similarly, in biosphere reserves of the United Nations Educational, Scientific, and Cultural Organization (UNESCO), large multiple-use zones are designed to buffer more sensitive areas, where the primary focus is nature conservation (Shriar 2001; Nautiyal et al. 2003). Similar kinds of buffering approaches have been applied around reserves in many parts of the world, including Korea (Gallent and Kim 2001), Indonesia (Salafasky 1993), Nicaragua (Smith 2003), and Switzerland (Soutter and Musy 1993).

A Semiformal Approach to Buffer Design

Kelly and Rotenberry (1993) formalized a general approach for buffer zone design through a set of interrelated questions:

- What external forces or processes are likely to have an impact on the protected entity (species, community, or resources)?
- To what extent are the external forces likely to penetrate the boundary of the sensitive site and result in negative impacts?

- Can these forces be ranked in terms of their impacts to support a priority list of buffering requirements?
- Are the potentially negative forces amenable to hypothesis testing?
- How can data be gathered to test these hypotheses?
- How can the external forces be mitigated?

Burgman and Ferguson (1995) considered several of these issues in assessing threats to cool temperate rainforest fragments in the southeastern Australian state of Victoria. They recommended better planning and mapping of forest landscapes; adoption of a system of buffers, especially around rainforest areas; modified logging practices in stands adjacent to rainforest; planning road construction to avoid sensitive areas; and new research to fill knowledge gaps regarding the sensitivity of rainforest communities to human activities.

Strategy 4: Maintain and/or Restore Corridors and Stepping Stones

Rationale

Landscape connectivity (as defined in Chapter 12) can promote biological conservation because it may contribute positively to habitat connectivity for a range of individual species, and because it may connect ecological processes at multiple spatial scales (i.e., facilitate ecological connectivity) (see Chapters 6, 12). In addition to matrix management (see Strategy 2), creating corridors that link patches of native vegetation, and providing stepping stones of native vegetation throughout the landscape, are effective ways to enhance landscape connectivity.

Corridors

The establishment of wildlife corridors has been one of the most controversial topics in conservation biology (Simberloff et al. 1992) and their value will depend on the objectives of landscape management, the ecology of the species targeted for conservation, and the kinds of landscape changes corridors are intended to mitigate. It may not be possible to design corridor systems that conserve all species vulnerable to the effects of landscape alteration. Perhaps most importantly, the reservation, rehabilitation, or maintenance of corridors may be relatively expensive. An uncritical insistence on corridors may divert financial resources from other, potentially more appropriate or less costly conservation strategies such as setting aside larger, consolidated protected areas (Simberloff et al. 1992) and making the matrix more permeable for movement (Franklin et al. 1997; Rosenberg et al. 1997).

Given that there might be some negative impacts associated with wildlife corridors, key questions in corridor design and establishment include:

- Which species move between vegetation patches without corridors (e.g., Mabry and Barrett 2002) and which species depend on corridors and to what degree (Beier and Noss 1998; Berggren et al. 2002)?
- Is a corridor to function solely as a conduit for movement or also to provide suitable habitat?
- How suitable are areas being connected by the corridor as habitat for the species of interest?
- What kinds of movement are corridors designed to facilitate—small-scaled day to day movements or larger-scaled seasonal migration (see Chapter 6)?
- How is corridor use influenced by the conditions and human activities in the surrounding landscape (Rosenberg et al. 1997)?

- What is the spatial scale and extent of landscape change and how well matched is corridor establishment to that scale (Tilman et al. 1997)?

These questions can be fundamentally important for a range of reasons. For example, a range of studies have shown that habitat specialists, edge-sensitive biota, or taxa that disperse poorly may not be able to use most kinds of corridors as linear habitat or for movement, whereas strong dispersers, edge-tolerant taxa, and generalists may not need them to survive (Box 18.3). The species targeted for conservation by using wildlife corridors therefore needs careful appraisal (Hannon and Schmiegelow 2002). Similarly, if the spatial scale of landscape alteration is large, then corridor functionality may become substantially impaired (Tilman et al. 1997).

Given the kinds of issues just outlined here, in some cases, particular parts of landscapes might be useful ones to target for the establishment or restoration of wildlife corridors. These include known dispersal routes of animals (even if they do not support high-quality habitat for a particular species) or areas where reservation or restoration efforts may provide a high conservation return for effort such as watercourses and associated riparian vegetation (Fisher and Goldney 1997) or roadside verges, which can support significant connected areas of native vegetation (Fortin and Arnold 1997; Forman et al. 2002).

Finally, although much work considers the restoration of corridors to reestablish links in landscapes already subject to extensive subdivision, a more effective strategy will, in almost all cases, be to retain landscape connectivity prior to landscape modification rather than attempting to re-create it once it has been lost. This requires landscape planning prior to modification, which is valuable, but often not possible, particularly where long-term impacts on landscapes such as land clearing have a history spanning many decades, centuries, or even millennia.

Stepping Stones

Although wildlife corridors can be valuable for some taxa, for others habitat connectivity may be better provided by the maintenance or establishment of "stepping stones" or dispersed patches of vegetation. The provision of stepping stones may be a good strategy for mobile species (like butterflies and bats) that can move through the matrix (Lumsden et al. 1994; Schultz 1998; Law et al. 1999) but are unable to disperse long distances between isolated patches (Bennett 1998). A range of studies have highlighted the importance of stepping stones in contributing to landscape connectivity, including those of birds (Potter 1990; Fischer and Lindenmayer 2002a, b) and butterflies (e.g., Nève et al. 1996; see Chapter 12). Stepping stones may also be valuable for plants (Collingham and Huntley 2000), particularly as sources of pollen to maintain gene flow over large distances between scattered plant populations (Young et al. 1996; Date et al. 1991).

Small Patches as Valuable Complements to Large Patches

Stepping stones and corridors can provide additional habitat for those species that use native vegetation but which are not area-sensitive. Hence, even small areas of native vegetation can be valuable. Table 9.2 in Chapter 9 contains many examples highlighting the conservation value of small patches in a wide range of ecosystems, including forests, woodlands, grassland, and wetlands. In some ecosystems that have experienced extensive and intensive human modification, small patches are all that remains, and they warrant protection or careful management because of their relictual importance (Saunders et al. 1987; Schwartz and van Mantgem 1997; Box 18.4). Thus, although such areas are not as effective as larger patches for conservation, they are nevertheless valuable, and special conser-

Box 18.3. Wildlife Corridors and Arboreal Marsupials

Wildlife corridors are used to mitigate the impacts of timber harvesting on forest fauna in wood production forests in many states of Australia. Typically, they are unlogged strips in riparian areas that maintain water quality as well as play a role in wildlife conservation.

Lindenmayer et al. (1993) surveyed arboreal marsupials in 49 retained linear strips that varied in width from 40 m to 250 m. Each strip was surrounded by young, recently cut forest that was unsuitable for the target species. The aim of the study was to determine if the strips were being used by various species of possums and gliders. Two questions were examined:

1. Is the number of corridors occupied by different species of arboreal marsupials similar to that expected, based on the suitability of the forest in the strips for these animals?
2. Are there any features of the strips (such as their width or length) that made them more likely to be used by arboreal marsupials?

The habitat requirements of several species of arboreal marsupials within intact forest were known from earlier studies (Lindenmayer et al. 1991b), providing the basis for estimating the suitability of the habitat for arboreal marsupials in each of the retained strips. Four out of five species, including the two most common ones, were observed with a frequency that did not differ substantially from expectations, based on the quality of the habitat. In the case of Leadbeater's possum *(Gymnobelideus leadbeateri)*, 17 sites supported suitable habitat, but it was recorded from just one. Therefore, even though suitable forest occurred in many sites, other factors appeared to preclude Leadbeater's possum utilizing the retained linear strips.

Animals with a complex social system (e.g., those that live in colonies) and which consume widely dispersed food like large flightless insects were relatively rare in the retained strips. It seems likely that the narrow linear habitat strips make it difficult for these species to harvest food efficiently and undertake some aspects of group social behavior (Recher et al. 1987). The species that were most commonly observed in wildlife corridors have a diet consisting of readily available food (such as leaves) and have a relatively simple social system—they are solitary or live in pairs. These findings indicate that, although setting aside networks of wildlife corridors may be a valuable strategy for some arboreal marsupials, they may not be effective for the conservation of all species.

Some features of the strips were related to their use by arboreal marsupials (question 2 above). Significantly more arboreal marsupials occurred in strips that supported numerous trees with hollows, probably because these trees provided nest sites for them. In addition, the position of the strips in a forest landscape influenced their use. More animals were found in corridors that spanned forests on different parts of the topographic sequence (e.g., linked gullies to ridges) than in corridors confined to only a gully or a midslope. The reasons for this result may be related to the need for some animals to move through different parts of the forest to harvest a range of food types (Claridge and Lindenmayer 1994). The results highlight the importance of a good understanding of the biology and ecology of those species thought to be protected by establishing networks of wildlife corridors (Harris and Scheck 1991).

vation efforts are required for them to ensure they are not lost simply because they are small (Schwartz 1997). For example, more than half the amount of many types of native temperate woodland (e.g., white box [*Eucalyptus albens*] and yellow box [*Eucalyptus melliodora*]) in grazing landscapes in southeastern Australia occurs in patches smaller than 1 ha (Prober and Thiele 1995; Gibbons and Boak 2002). Yet, the cumulative conservation value of many small patches scattered throughout a landscape can be substantial, and up to three-quarters of native bird species may use patches <1 ha in some way (Fischer and Lindenmayer 2002c).

Box 18.4. The Value of Small Patches—Key Woodland Habitats in Sweden

Scandinavia has experienced prolonged human impacts on its forests, and many elements of its biota are threatened (Berg et al. 1994; Virkkala et al. 1994; Angelstam and Pettersson 1997). More than 200 species are classified as endangered and 3000 regarded as vulnerable in Sweden alone (National Board of Forestry 1995, 1996). There are now considerable efforts to better manage forest landscapes to conserve native species, but conventional reserve-based approaches are limited because much of the Swedish forest estate is privately owned, and only minimal amounts are in national parks and nature reserves. A primary element in conservation efforts in Swedish forests has been the Key Woodland Habitats initiative (Figure 18.6). Most areas designated as Key Woodland Habitats are small—typically 0.5 to 5 ha (Aune et al. 2005), and they predominantly occur on private land. Nevertheless, they are extremely important for conservation because (1) they support (or are predicted to support) threatened and endangered species (Gustafsson et al. 1999; Jonsson and Jonsell 1999), and/or (2) they are characterized by structural features rare elsewhere in forest landscapes, such as large quantities of dead wood, old large-diameter trees, and multiaged stands (Gustafsson 2000). In 2000 it was estimated there were 60000 to 80000 Key Woodland Habitats totalling ~1% of the Swedish forest estate (Gustafsson 2000; Aune et al. 2005). Although these areas will not support viable populations of some species (e.g., see Ericsson et al. 2005), they clearly play an extremely valuable and complementary role to national parks and nature reserves.

Strategy 5: Maintain and/or Restore Landscape Heterogeneity and Capture Environmental Gradients

Rationale

Other things being equal, heterogeneous human-modified landscapes tend to support more native species than homogeneous human-modified landscapes (Chapter 14). Landscape heterogeneity may be enhanced by establishing or maintaining patches of native vegetation of a range of sizes, shapes, and levels of isolation; and by maintaining a range of different land use types and land use intensities between them. In many human-modified landscapes, land use intensity is highest in the most productive parts of the landscape, and the largest remaining patches of native vegetation are restricted to the least productive areas (Chapter 2). This bias is undesirable because different species occupy different locations along environmental gradients. That is, some species only occur in the most productive environments or only occur in high densities in productive environments (Pressey 1995; Scott et al. 2001). For a wide range of species to be able to persist in modified landscapes, all parts of a given environmental gradient should not be subject to the same high-intensity form of human land use. In particular, the least intensively used areas should not be confined to relatively unproductive parts of the landscape, such as rocky ridges with thin or low-fertility soils.

Natural Heterogeneity as a Guide for Landscape Management

Ecosystems are naturally heterogeneous. For example, disturbance regimes may create heterogeneous land cover, such as different successional stages in different locations follow-

Figure 18.6. A Swedish Key Woodland Habitat (photo by Lena Gustafsson).

ing a wildfire (Whelan et al. 2001). In addition, landscapes are characterized by natural environmental gradients (e.g., in topography, climate, or soil type and soil depth; see Austin and Smith 1989). Such natural heterogeneity can be used as a guide to manage human-modified landscapes (Lindenmayer and Franklin 2002). Different species inhabit different environmental conditions in natural landscapes. Hence providing a range of environmental conditions that resemble the natural mix of conditions as closely as possible is likely to benefit the native suite of taxa in any given landscape.

This management strategy highlights the importance of overcoming often strong biases in landscape modification where the most productive areas are those most severely altered (Chapter 2). Some species can only persist in productive locations, and they may not be able to benefit from retained vegetation in less productive areas such as rocky ridges (Fischer et al. 2005b). Creating landscape heterogeneity across environmental gradients recognizes the range of adaptations of different species and communities to different natural disturbance types and intensities, and to different environmental conditions (such as soil moisture or temperature gradients).

Although the maintenance of landscape heterogeneity is best guided by an understanding of natural heterogeneity in a given landscape, some landscapes have long since lost their natural disturbance regimes and natural patterns of heterogeneity. Cultural farming landscapes in Europe exemplify a situation where this is the case. Notably, landscape heterogeneity tends to benefit native species richness in this situation, even if it is not based on natural heterogeneity patterns that once prevailed. For example, Benton et al. (2003) argued that heterogeneity within and between fields and landscapes was essential for maintaining the remaining species in

Table 18.1. Summary of pattern-oriented management strategies to mitigate the negative effects of human landscape modification on species and assemblages of species

Management strategy	*Purpose*
1. Maintain and/or restore large and structurally complex patches of native vegetation	• Provide core habitat for species that depend on native vegetation
2. Maintain and/or restore a matrix that is structurally similar to native vegetation	• Provide habitat for many species throughout the landscape • Provide habitat connectivity for many species throughout the entire landscape • Reduce structural contrast between modified and unmodified areas, thereby reducing edge effects throughout the landscape
3. Maintain and/or restore buffers around sensitive areas	• Protect sensitive habitats like aquatic systems • Reduce edge effects at a specific location
4. Maintain and/or restore corridors and stepping stones	• Provide habitat connectivity for species that cannot move through the matrix • Provide additional habitat for species that use native vegetation but are not area-sensitive
5. Maintain and/or restore landscape heterogeneity and capture environmental gradients	• Provide a diversity of habitats that are useful to a range of different species • Distribute different land-use intensities across natural gradients in climate, topography, and primary productivity

European agricultural landscapes (see also Attain and deLucio 2001; Tscharntke et al. 2005). Hence, as a general guiding principle, heterogeneity is preferable to intensive land use resulting in homogeneity, even if patterns of heterogeneity are not directly based on a detailed understanding of the natural landscape heterogeneity that once prevailed in a given area.

Summary

This chapter focuses on managing aspects of landscape pattern (particularly vegetation cover) to mitigate the negative impacts of landscape modification on biota. Five broad and general management strategies are briefly outlined that are likely to be widely applicable in many kinds of landscapes (Table 18.1).

Although each of these landscape pattern–based strategies is important, some individual species will not be adequately conserved by them and some key threatening processes will remain unmitigated despite their application. Additional approaches will be required, and they are addressed in Chapter 19.

Links to Other Chapters

This chapter is a distillation of landscape pattern–based management strategies designed to mitigate some of the threats to biota that arise from human modification of landscapes, discussed in detail in Parts I through III of this book. The need to maintain and restore large and structurally complex patches of native vegetation (Strategy 1) is designed to offset the problems of habitat loss and habitat degradation for individual species (Chapters 4, 5), and vegetation loss and vegetation deterioration for assemblages of species (Chapters 9, 10). It is also an approach to counter the negative impacts of small patch size (Chapter 9) and the associated edge effects that can arise as a consequence of boundary to interior ratios in small areas of remaining native vegetation (Chapter 11). Strategy 2 in this chapter concerns maintaining or creating structurally complex conditions in the matrix and, like other strategies, it resonates with the themes of several preceding chapters. Matrix management is a critical approach to reducing vegetation loss and deterioration (Chapters 9 and 10), reducing edge effects (Chapter 11), contributing to landscape connectivity (Chapter 12), and maintaining landscape heterogeneity (Chapter 14). Strategy 3 on buffers relates to many aspects of altered patterns of landscape change including vegetation loss (Chapter 9), vegetation deterioration (Chapter 10), and, particularly, edge effects (Chapter 11). The need to maintain or create corridors and stepping stones (Strategy 4) is a pattern-based management response to limit the loss of landscape connectivity (Chapter 12), and it also can be important for maintaining habitat connectivity for individual species (Chapter 6). The maintenance of landscape heterogeneity and capturing of environmental gradients (Strategy 5) builds on themes discussed in detail in Chapters 2 and 14.

Further Reading

The scientific literature contains a vast amount of material on reserves. Good entries to this massive topic are Noss and Cooperrider (1994) and Margules and Pressey (2000). The topic of restoration is an equally enormous one—entire journals such as *Restoration Ecology* and *Ecological Restoration* are dedicated to it. Detailed discussions on vegetation restoration programs are also provided in texts such as those by Saunders et al. (1993a), Bennett et al. (2000), and Hobbs and Yates (2000). The importance of matrix management is a feature of articles or books by Franklin (1993), DeGraaf and Miller (1996), Franklin et al. (1997), Hunter (1999), Craig et al. (2000), and Lindenmayer and Franklin (2002, 2003). Noss and Harris (1986), Burgman and Ferguson (1995), and Semlitsch and Bodie (2003) discuss some practical approaches for the design of buffers to better protect particular elements of the biota from edge effects and landscape alteration. The paper by Kelly and Rotenberry (1993) is an excellent discussion of some of the general considerations that underpin the design of buffer systems. Many books and papers discuss approaches to maintaining or creating landscape connectivity. Some of them include Bennett (1990b, 1998), Saunders and Hobbs (1991), Hobbs (1992), Lindenmayer (1994, 1998), Forman (1995), Beier and Noss (1998), and Haddad et al. (2003). Discussions of the importance of landscape heterogeneity appear in texts by Forman (1995) and Lindenmayer and Franklin (2002) as well as in the series of papers on countryside biogeography in Costa Rica by Luck and Daily (2003) and Mayfield and Daily (2005). Environmental gradients are discussed in depth by Austin and Smith (1989) and Austin (2002).

CHAPTER 19

Managing Individual Species and Ecological Processes to Mitigate the Decline of Species and Assemblages

Managing landscape pattern can be an efficient first step to mitigate the negative effects of landscape modification on native species and assemblages of species. This is because, even in the absence of detailed ecological knowledge, some types of landscape patterns are likely to benefit simultaneously a large number of native species (Chapter 18). However, although pattern-based mitigation strategies are a useful starting point for conservation in modified landscapes, they cannot guarantee that all important species or ecological processes are adequately catered to. For this reason, in this chapter we suggest five management strategies that are directly targeted at particular species or ecological processes. For each strategy, we briefly discuss its scientific basis with reference to earlier parts of the book; and we provide some examples where these strategies have been applied. We end this chapter with a brief discussion of ecosystem trajectories; that is, the changes of ecosystem patterns and processes through time and into the future.

Strategy 1: Maintain Key Species Interactions and Functional Diversity

Rationale

Landscape modification alters the composition of ecological communities. This can lead to changes in species interactions (Chapter 7). Some types of species interactions are particularly important for ecosystem functioning; examples are predation by large mammals, pollination, and seed dispersal. Species involved in such interactions are sometimes called keystone species, and their continued survival is important to avoid cascading effects of landscape change (Chapter 15). More generally, maintaining a diversity of species within different functional groups provides a safeguard for continued effective ecosystem functioning and is likely to enhance the ability of ecosystems to recover in response to disturbance (Chapter 15).

Successful attempts to maintain key species interactions invariably involve a good understanding of the ecology of the suites of species in question and the ecological processes that link interspecies interactions. This was illustrated by the conservation of the large blue butterfly (*Maculina arion*) in Britain (see the discussion in Chapter 7 in which key species interactions between butterflies, ants, rabbit grazing, disturbance regimes, and the presence of open vegetation types were carefully documented as part of species recovery and the maintenance of key ecosystem processes; Elmes and Thomas 1992). The case studies of two keystone species discussed in this chapter—American bison (*Bos bison*) and red-naped sapsucker (*Sphyrapicus nuchalis*)—further emphasize the importance of understanding species interactions and links with ecosystem processes.

Keystone Species

Keystone species are species that have a disproportionate effect on ecosystem function relative to their abundance (Power et al. 1996; Chapter 15). The American bison in North American grasslands is a good example of a keystone species (Knapp et al. 1999; Figure 19.1). Once abundant, populations of the species were severely reduced in the 1800s. The species has been extensively studied on the Konza Prairie, a native tallgrass prairie in northeastern Kansas, USA. In this area, the presence of bison fundamentally alters ecological processes and contributes to spatially heterogeneous and species-rich vegetation communities. For example, bison preferentially feed on grasses, thereby allowing forbs to flourish. Areas grazed by bison have the highest plant species diversity, the highest spatial heterogeneity, and the highest nitrogen availability in the prairie. In combination with appropriate fire regimes, bison grazing therefore is considered key to conserving and restoring ecosystem function and promoting population recovery in other species within tallgrass prairie ecosystems (Knapp et al. 1999).

Some ecosystems have multiple keystone species, each of which is fundamentally important to maintain ecosystem functioning. For example, the red-naped sapsucker in a subalpine ecosystem in Colorado fulfils two keystone roles (Daily et al. 1993). First, it creates nest cavities in aspen *(Populus tremuloides)* trees, which are used by many secondary cavity-nesting species that do not themselves excavate nest hollows. Second, the red-naped sapsucker drills holes into some shrubby willow species (*Salix* spp.). The sap thus released is a valuable food source for the red-naped sapsucker but also for a range of other native bird species. The loss of the sapsucker would fundamentally alter species composition; hence the sapsucker is a keystone species. Notably, other species are also important to maintain ecosystem functioning in this community. For example, not all aspen trees are chosen by the sapsucker for excavating cavities. Rather, the sapsucker can create cavities only in trees that are infected with a heartwood fungus. This fungus, therefore, is as important as the sapsucker to maintain ecosystem functioning—it is also a keystone species (Daily et al. 1993).

The examples of the American bison and the red-naped sapsucker demonstrate that keystone species relationships can be complex. However, to maintain functioning ecosystems it is important that potential threats to keystone species are identified and mitigated so that the presence and abundance of keystone species are maintained (Soulé et al. 2005).

Response Diversity

A superficial examination of the keystone species concept may suggest that species that are not keystone species are not needed to maintain ecosystem function. Indeed, by definition, the loss of nonkeystone species will not result in immediate and disproportionate changes to ecosystem function. Despite this, nonkeystone species are still important even if they do not appear to be dominant in the ecosystem (Lyons et al. 2005). For example, Walker et al. (1999) argued that many functionally dominant species have minor species as "functional equivalents"—species that are less common but fulfill ecological functions similar to those of the dominant species. Using data from grassland ecosystems, Walker et al. (1999) showed that the loss of certain dominant species often led to an increase in less common but functionally similar species. Uncommon and nonkeystone species thus provided an important "insurance role." When dominant equivalents were lost, uncommon and nonkeystone species could compensate to some extent for the loss of dominant taxa.

Elmqvist et al. (2003) generalized this notion and argued that a diversity of species within a given functional group increased the likelihood of high "response diversity"—in other words, there is a greater chance of some species responding positively to an exogenous disturbance to compensate for other, similar species that responded negatively to the disturbance. Hence, although nonkeystone species do not have the same immediate importance

Figure 19.1. Wild American bison *(Bos bison)* at Yellowstone National Park, Wyoming (photo by David Lindenmayer).

to maintain functioning ecosystems as do keystone species, they are nevertheless important because of their "insurance role" and the contribution they make to ecosystem resilience (also see Chapter 15).

Strategy 2: Maintain or Apply Appropriate Disturbance Regimes

Rationale

Human landscape modification often results in a change to historical disturbance regimes (e.g., through logging, grazing, intensive agriculture, or by altering fire regimes). Landscape-scale disturbances can substantially alter vegetation structure (Chapter 10) and may trigger cascades that cause fundamental and potentially irreversible changes to ecosystems (Chapter 15). Understanding the impacts that particular disturbance regimes have on ecosystem functioning therefore is a useful basis from which to choose the most appropriate disturbance regimes. In general, disturbance regimes that attempt to mirror historical disturbance regimes for a given landscape are likely to be a useful starting point for management (Hunter 1993; Korpilahti and Kuuluvainen 2002; Lindenmayer and McCarthy 2002; Perera et al. 2004).

Fire—an Example of a Commonly Altered Disturbance Regime

Fire can be a major form of natural and human disturbance in many landscapes and regions around the world (Agee 1993; Rülcker et al. 1994; Cary et al. 2003), and many organisms are well adapted to it (Dyrness et al. 1986; Bunnell 1995). However, fire regimes have been altered dramatically in areas where human landscape modification has occurred (Bradstock et al. 2002; Covington 2003), and some species and assemblages can be threatened as a result (Woinarski and Recher 1997; Noss et al. 2006). For example, in Australia, more species of na-

Figure 19.2. Burned habitat formerly occupied by the eastern bristlebird at Jervis Bay prior to a major wildfire in December 2003 (below: eastern bristlebird habitat sign; photos by David Lindenmayer).

tive birds are threatened by altered fire regimes than by any other threatening process except land clearing (Woinarski 1999; Garnett and Crowley 2000). Issues of the impacts of disturbances by fire and associated effects on biota are complex. This is because, in some landscapes, the absence of fire creates problems like a lack of regeneration of particular plant species (Zackrisson 1977), whereas in others, fires can be too frequent (Gill et al. 1999) or too intense (Covington et al. 1997).

Management practices using only pattern-based approaches may not meet desired conservation goals because they may fail to address the problems associated with altered disturbance regimes as key threatening processes. Indeed, if applied without adequate consideration, pattern-oriented management strategies may even exacerbate problems. The establishment of wildlife corridors, for example, may facilitate the inappropriate spread of fire between otherwise isolated areas. Thus highly focused management strategies are often needed to ensure the appropriate application of fire regimes in modified landscapes. One example is that of the endangered heathland eastern bristlebird (*Dasyornis brachypterus*) in eastern Australia, which has a highly restricted distribution, poor recolonization abilities, and small population sizes that are sensitive to frequent fire (Baker 1998; Figure 19.2). The species disperses poorly and appears to be vulnerable to the effects of large-scale fires that do not leave unburned refugia (Pyke et al. 1995; Baker 1997). Fire exclusion is the most appropriate conservation strategy for this species, but this may not be possible, given the proximity of human infrastructure to its habitat. Because of the need for fuel management in these areas, Baker (1997) recommended strategic slashing of vegetation. In other cases where prescribed burning is essential, the direction of fire fronts should be planned to provide escape routes for individual bristlebirds, given their limited movement ability (Baker 1997).

Another example of the problems created by altered fire regimes comes from the ponderosa pine (*Pinus ponderosa*) forests of southwestern United States. Here, past fire suppression together with livestock grazing (which reduces the fine fuels necessary to carry frequent, low-severity fires), old-growth logging, road building, and other human activities have created a forest condition highly susceptible to uncharacteristic stand-replacing fire, drought, and insect attack (Covington et al. 1997; Friederici 2003; Moore et al. 2004). High-severity fires in southwestern ponderosa pine forests appear to be increasing in frequency and spatial extent after decades of fire suppression (Covington 2003). A range of targeted management strategies will be needed to address issues associated with altered fire regimes as a key threatening process in ponderosa pine forests. These include fire treatments such as thinning and prescribed burning to limit fire severity, spatial variation of management treatments to ensure a variety of stand conditions including appropriate habitat structures for high profile species such as the Mexican spotted owl (*Strix occidentalis lucida*), removal of roads to curtail the spread of flammable exotic species, and the creation of reference areas that act as benchmarks to assess the impacts of management actions (Noss et al. 2006; Figure 19.3).

Notably, fire is by no means the only disturbance regime that can be altered by human activities. Other disturbances can be equally important in some areas. An example is the hydrological flow patterns in the Macquarie Marshes in eastern Australia. Here, upstream damming of the Macquarie River for the cotton industry has led to a massive decline of flooding events, which are needed to sustain key ecosystem processes and various elements of the biota in the marshes. As is often the case with other disturbance regimes, flooding regimes that more closely resemble historical flooding events would most likely result in improved conservation outcomes (Caldwallader 1986).

Strategy 3: Maintain Species of Particular Concern

Rationale

Managing landscape pattern will effectively protect the habitat of many species (Chapter 18). However, different species perceive the same landscape pattern in different ways, and differently from humans (Chapter 3; Part II). Hence a careful assessment is needed to identify species of conservation concern that are not adequately protected by managing landscape pattern alone. Complementary and highly focused conservation strategies can then be employed for such species. Here we describe several examples of highly focused conservation strategies, and two more detailed case studies of highly endangered species.

Figure 19.3. Fire restoration treatment in the ponderosa pine *(Pinus ponderosa)* forests of southwestern USA (photo by Reed Noss).

Examples of Highly Focused Conservation Strategies

Many different strategies can feature in highly focused conservation efforts. Among others, four of them include:

- The control of diseases and parasites. For example, the application of insecticides to the nests of the echo parakeet *(Psittacula egeus)* on Mauritius has controlled populations of tropical nest flies *(Passerimyia heterochaeta)* and increased fledging success rates (Jones and Duffy 1993; Greenwood 1996).
- Captive breeding, reintroduction, and translocation. There are numerous examples of species that have been closely managed or held in captivity to reestablish natural populations following population crashes and imminent extinction (IUCN 1998; Lyles 2002). One is Shevolski's horse *(Equus callabus przewalski)*, which was reduced to 12 animals in Prague Zoo in 1912, and then recovered to a current population of about 700. Another is the Arabian oryx *(Oryx leucoryx)*, which was reduced to 8 individuals in London Zoo but recovered to more than 400 individuals, although poaching of reestablished wild

populations has reemerged (Ostrowski et al. 1998).

- Control of predators or highly competitive species. Introduced animals and plants have been implicated in the extinction or decline of many species (Whittaker 1998; Simberloff 2000; Primack 2001), and there are numerous documented cases where population recovery has occurred when numbers of exotic predators or highly competitive species have been controlled or eradicated (e.g., Miller and Mullette 1985; O'Dowd et al. 2003). Exotic predator control has been pivotal to the success of endangered bird recovery programs in many parts of the world, with New Zealand probably being one of the most spectacular and successful examples (reviewed by Saunders 1994). Native taxa sometimes need to be controlled as part of focused management for a particular species. For example, in the United States, populations of the brown-heaed cowbird *(Molothrus ater)*, which is a brood parasite, need to be controlled to promote the recovery of Kirtland's warbler *(Dendroica kirtlandii)*. Populations of Kirtland's warbler have increased approximately fivefold since cowbird populations have been controlled (Solomon 1998).
- The focused protection of key areas critical for the breeding and survival of a species. An example is the winter habitat of the monarch butterfly *(Danaus plexippus)* in western North America. This species congregates at high densities in relatively small overwintering areas (Malcolm and Zalucki 1993), only some of which are protected from logging. Other examples are the localized overwintering areas that are visited repeatedly by a number of species of large herbivores in North America (Thompson and Angelstam 1999).

Box 19.1. Purpose-Built Structures—a Method to Enhance Habitat Connectivity for Particular Species of Conservation Concern

Managing landscape pattern can effectively enhance landscape connectivity; for example, by establishing a "soft" matrix, corridors, and stepping stones (Chapter 18). However, landscape connectivity may not necessarily translate into habitat connectivity for all species—some species, for instance, do not use corridors (Chapter 12). For such species, highly targeted, purpose-built structures can sometimes be used to enhance habitat connectivity. Such purpose-built structures are often useful where a species' habitat is dissected by human infrastructure such as roads (Forman et al. 2002) or railways (Rodríguez et al. 1996).

Figure 19.4 shows an overpass in Banff National Park in Alberta, Canada, that facilitates the safe movements of large mammals from one side of the Trans-Canada Highway to the other. Similarly, a rope bridge in New South Wales, Australia, is promoting habitat connectivity among populations of arboreal marsupials such as the sugar glider *(Petaurus breviceps)* and common brushtail possum *(Trichosurus vulpecula)*. Another example is that of a population of the mountain pygmy possum *(Buramys parvus)* in southeastern Australia, whose habitat connectivity was severely disrupted by the development of a recreational ski run. Mansergh and Scotts (1989) demonstrated that when two previously subdivided populations of the species were linked via a specially constructed boulder field, the social structure and breeding system of the mountain pygmy possum reverted to that characteristic of a continuous population.

In the case of each of these strategies (and numerous others that have not been mentioned; e.g., see Box 19.1), a critical aspect of focused conservation work is a rigorous diagnosis of the underlying causes of decline of a species and then targeted efforts to mitigate the impacts of the threatening process or interrelated set of threatening processes (Caughley and Gunn 1996). In the following section, a small number of case studies outlining highly focused conservation strategies are presented. Many other examples can be found throughout the conservation biology and general ecological

Figure 19.4. Specifically built grassed overpass in Banff National Park, Alberta (Canada) to reduce collisions between large mammals and motor vehicles, and thereby improve habitat connectivity for wildlife populations (photo by David Lindenmayer).

literatures (e.g., see Caughley and Gunn 1996; Thompson and Angelstam 1999; Sutherland 2000; Hunter 2002; Groom et al. 2005).

The Seychelles Magpie-Robin

The Seychelles magpie-robin *(Copsychus sechellarum)* is a celebrated case of focused conservation management. Its management highlights not only the need for targeted efforts to conserve a particular species but also both the complementary roles of pattern- and process-based work and the recognition that multiple threatening processes need to be tackled as part of successful recovery. The species formerly occurred on eight islands in the Seychelles group in the Indian Ocean, but feral cats *(Felis catus)* eradicated them from all but Frigate Island (Wilson and Wilson 1978; Watson et al. 1992). Cat eradication on two occasions stimulated the recovery of the single remaining population of the species. However, even though cat numbers had been controlled during the second recovery event, monitoring and survey work indicated that populations were markedly lower than they had been following earlier cat eradication work. By the 1980s, survival of young per adult pair began to decline, indicating that other factors were limiting the population. Careful field study of the Seychelles magpie-robin then established the habitat requirements of the species and revealed that key foraging areas were bare earth and leaf litter, often associated with gardens tended by humans. Altered land use patterns associated with a decline in the number of gardens reduced the amount of suitable habitat for the Seychelles magpie-robin, thereby capping population numbers. In addition, there was a shortage of nesting sites in hollow trees. Those that were available were remote from suitable foraging areas—leaving unguarded nests more vulnerable to predation. Finally, a density-dependence effect was identified in which young birds in low-density populations moved to vacant areas and began breeding, but those in high-density populations remained as nonbreeding individuals in the natal territory.

The work on the Seychelles magpie-robin has been important because it highlights the links between understanding the patterns of distribution of a species and the set of ecological processes giving rise to observed spatial distribution patterns. Using this information, focused management actions for the Seychelles magpie-robin have been complex and have involved ongoing predator control, habitat manipulation, the provision of supplemental food, and the establishment of populations of birds on other islands in the Seychelles group (Gretton et al. 1991; Komdeur 1996).

The Western Swamp Tortoise

Another example of a threatened species that requires highly focused management strategies is the western swamp tortoise *(Pseudomydura umbrina)* from southwestern Australia. The species was discovered in 1839, but no more specimens were seen until 1953. It is the only member of its genus (Cogger 1995) and may possibly be the only living representative of the subfamily Pseudemydurinae (Kuchling et al. 1992). It is one of Australia's most endangered reptiles, and it has a highly restricted distribution (covering about 100–150 km^2) near Perth (Cogger 1995). Two swamps, one of 65 ha and the other 155 ha, were reserved. However, the western swamp tortoise disappeared from the larger reserve during the 1980s, and by the late 1980s only a few dozen animals remained in the smaller swamp (Kuchling et al. 1992). Habitat for the western swamp tortoise has been extensively modified for agriculture and human settlement (Kuchling et al. 1992). In addition, when swamps dry out, tortoises estivate in holes or under deep leaf litter where they are vulnerable to predation by the introduced red fox *(Vulpes vulpes)*. Populations also have very low reproductive rates. In summary, the species has suffered from habitat loss, habitat degradation, and increased predation pressure (Kuchling et al. 1992; Threatened Species Network 2003).

Given the array of factors threatening remaining populations of the western swamp tortoise, several recovery strategies are part of conservation efforts for the species. Feral predator control, reserve expansion, and habitat rehabilitation are three key ones (Kuchling et al. 1992; Burbidge and Kuchling 1994). Another has been a captive breeding program, which was initiated in 1988 and substantially increased the numbers of animals—more than 170 tortoises have been successfully reared in captivity (Threatened Species Network 2003). The population recovery has been encouraging. In the mid-1980s there were about 30 animals remaining, but by 2001 there were 110 animals (Threatened Species Network 2003). As with many threatened species, highly targeted management actions will continue to be a critical part of the species' recovery.

Caveats

An important caveat of maintaining particular species of concern is that conservation strategies developed for one species may be opposed to the requirements of some other species (Landres et al. 1988; Lindenmayer et al. 2000). There are many cases where this occurs. For example, conservation efforts for the endangered pupfish *(Cyprinodon diabolis)* in North America destroyed habitat for the Ash Meadows naucorid *(Ambrysus amargosus)*—an endangered waterbug (Polhemus 1993). Another case is the conservation of rare heathland birds in eastern Australia. Different species respond differently to heathland fires of different frequency, intensity, and patchiness (Whelan 1995; Woinarski 1999). For example, the response of the endangered eastern bristlebird is markedly different from that of another rare species—the ground parrot *(Pezoporus wallicus)*, which is a stronger flier that quickly recolonizes burnt areas and then occupies them for many years (see Woinarski 1999; Keith et al. 2002b). Hence a fire management strategy developed in any one location for one of these species will not be optimal for the other.

Strategy 4: Control Aggressive, Overabundant, and Invasive Species

Rationale

Landscape change tends to result in habitat loss for many species. However, it also tends to strongly favor a small number of native or introduced species (Chapter 3). Some species that benefit from anthropogenic landscape change can become overly abundant and can negatively affect other species via aggressive behavior, competition, or predation (Chapters 7, 10, 11). Controlling invasive or overabundant plant and animal species therefore is important to maintain functioning and diverse ecosystems. Several examples are provided in the next sections.

The Beaver in Tierra del Fuego: An Invasive Introduced Animal Species

The North American beaver *(Castor canadensis)* was introduced to Tierra del Fuego (in the far south of South America) in 1946 to create a fur industry (MacDonald 2001). The species lacks any significant natural or human predation and has spread to occupy essentially all suitable aquatic and riparian areas, including sites at the upper timberline. Estimated densities for North American beaver colonies are 6.6 to 8.5 colonies/km of stream in some parts of Tierra del Fuego (Franklin in Lindenmayer and Franklin 2002)—three to four times higher than populations in the natural part of the species' distribution (MacDonald 2001). Except for some of the smallest (first order) tributaries, essentially all aquatic ecosystems have been significantly modified. Key physical and chemical properties of streams, wetlands, and other aquatic areas have been comprehensively altered (Figure 19.5), leading to large-scale landscape modification and flow-on negative impacts on many species and assemblages closely associated with aquatic zones.

Conservation strategies for many elements of the biota within landscapes in Tierra del Fuego, particularly those associated with aquatic areas, will ultimately depend on the active control of large populations of the North American beaver. The case of the North American beaver is far from an unusual one where populations of an exotic species can be so large or have such profound impacts that there are massive flow-on impacts on other species, ecological processes, and patterns of landscape cover (Mooney and Hobbs 2000). Invasive or overabundant species can become threatening processes per se, which require major control efforts to ensure the persistence of many other elements of the biota (Clavero and Garcia-Berthou 2005; Department of Environment and Heritage 2005).

The Brown-Headed Cowbird: An Overabundant Native Animal Species

Problems of overabundant taxa are not limited to exotic species. Native species may also become so abundant in modified environments that they can have a significant negative impact on many other species. The brown-headed cowbird *(Molothrus ater)* in North America is a useful example. This species is well known as an edge-favoring nest parasite in many areas (see Chapter 11). The species can destroy the eggs and instigate nest abandonment (Elliot 1999; Smith et al. 2003). This not only significantly reduces the number of offspring produced by host birds but also skews the sex ratio of offspring of young that are successfully raised (Zanette et al. 2005). To counter these problems, removal of brown-headed cowbirds is increasingly being applied to recover declining populations of host species (Morrison et al. 1999; Smith et al. 2000)—as already discussed for Kirtland's warbler *(Dendroica kirtlandii)*.

Figure 19.5. Riparian and aquatic areas in Tierra del Fuego that have been extensively damaged by the North American Beaver (photo by David Lindenmayer).

Weed Invasion and Control

No vegetation type is immune from weed invasion, and the degree of invasion depends on edaphic and microclimatic conditions, vegetation structure, the type and frequency of natural and artificial disturbance (such as grazing; see Chapter 10), and the proximity of source populations of weed species (Fox and Fox 1986; Cronk and Fuller 1995). In a wide-ranging study of plant invasions worldwide, Lonsdale (1999) found other important variables related to the proportion of exotic species in a vegetation community. Although the number of exotic plants varies widely between different parts of the world, the proportion is higher outside nature reserves, within plant communities that are species-rich, and on islands. For example, the island of Hawaii has more introduced plants than native ones (860 versus 850) (Eldredge and Miller 1995).

The most effective strategy to control invasive weed species is to prevent them from arriving in a new environment in the first place (Low 1999; Myers et al. 2000). Quarantine regulations in many countries aim to prevent the importation of environmental weeds. Unfortunately, such quarantine services are too under-resourced to perform adequately the functions they are assigned in relation to the prevention of potential weed species and other invasive organisms. This is true both for wealthy countries like the United States and Australia, and for a wide range of other countries around the world (Tyndale-Biscoe 1997; Simberloff et al. 2005). These problems are being exacerbated by increased trade associated with the integration and globalization of the economies of trading nations (Everett 2000).

When an invasive and damaging weed species is first discovered, early action may eliminate problems arising from weed invasion before it becomes logistically and financially impractical to do so (McNeeley et al. 2003). These recommendations are relevant not just to weeds but to all invasive organisms.

There are numerous weed control methods, but it is beyond the scope of this book to delve into them. Some include herbicide application, fire (such as burning after flowering and before seed set), removal of seedlings by hand, ringbarking, cutting at ground level, and the removal of stems and roots. Another method to control pest plants is biological control. This involves the use of a predatory or parasitic organism to control a weed (Harley and Forno 1992). However, few environmental weeds have been successfully managed with biological control agents, and some, such as those with large seed reserves in the soil, may not be suited to such control measures (Briese 2000). Moreover, in some cases, the introduction of biological control agents has led to substantial environmental damage (Low 1999). The appropriate method for controlling weeds will depend on the extent of the problem, available resources, the biology and habitat requirements of the target species, and the potential impacts on nontarget species. In other cases, limiting human infrastructure, such as the establishment of roads into areas supporting native vegetation, can be an important part of weed control because vehicles can be a significant source of weed seeds (Wace 1977; Forman et al. 2002).

Strategy 5: Minimize Ecosystem-Specific Threatening Processes

Rationale

This book has focused on the direct and indirect impacts of landscape change on both individual species and assemblages of species. However, landscape change is not the only factor threatening biodiversity (Chapter 1). A range of additional threatening processes can be equally important or even more important in some landscapes. The identification and control of potential threats that are not directly related to landscape change are critical in modified landscapes. For example, human hunting is a major threatening process for many species in Europe (Revilla et al. 2001) as well as Central America (Redford 1992), Africa (Bennett 2000), and Asia (Rabinowitz 1995). The ibex *(Capra ibex ibex)* provides a useful example; it was exterminated from France prior to the development of a system of National Parks where hunting was precluded (Grodinski and Stüwe 1987; Skonhoft et al. 2002). Conservation of species threatened by human hunting is dependent on tackling this problem directly rather than pursuing pattern-oriented mitigation strategies, such as maintaining vegetation cover and landscape connectivity or buffering edge effects.

Many other examples exist of threatening processes that are not directly related to landscape modification but that are important to consider and mitigate to protect species and ecosystems in some situations. Examples include the presence of chemicals in the food web, which can adversely affect a range of species, including amphibians, reptiles, and birds (e.g., Hall and Henry 1992; Olsen et al. 1993; Gibbons et al. 2000; Oaks et al. 2004), or air pollution from mining (Letnic and Fox 1997; Read 1998). Similarly, human-enhanced global warming may pose a threat to many species (Thomas et al. 2004), especially ones already restricted to cool climates at high elevations (Brereton et al. 1995; Hughes 2003).

An awareness of such threatening processes and how they may impact on particular species or ecosystems is important to complement strategies designed to mitigate landscape modification impacts, with additional conservation protocols targeting other significant threatening processes.

Considering Ecosystem Trajectories

Ecosystems are not static. In addition to natural change, some types of ecosystem change reflect adaptations to changing environmental conditions, such as species' range shifts in response to global warming (e.g., Hughes 2003; Hampe and Petit 2005). Thus, not every type of ecosystem change is necessarily bad or requires mitigation. However, some types of ecosystem change are clearly undesirable from a wide range of perspectives. An example is environmental degradation in Australian farming areas (Ghassemi et al. 1995). Land degradation such as soil erosion and increasing levels of secondary salinity is already severe in some areas such as the wheatbelt in Western Australia (Department of Conservation and Land Management 2000; Wallace et al. 2003) and is continuing in many other parts of the continent (State of the Environment Advisory Council 1996). In this case, considering ecosystem trajectories is essential to avoid predictable but undesirable changes before they occur. For example, Dorrough and Moxham (2005) calculated a narrow 30-year window of opportunity during which large-scale tree regeneration will need to take place in the Australian state of Victoria if large-scale losses of tree cover are to be avoided. Extinction debts under such circumstances could be substantial, and Recher (1999) warned that unless agricultural practices change, every second species of Australian land bird may suffer extinction within this century.

Considering likely future trajectories of ecosystems is essential to gauge whether a particular set of mitigation strategies is likely to be successful, or if additional strategies are required. Scenario planning can be a useful tool in this context (Foran and Poldy 2002; Peterson et al. 2003a). It facilitates the explicit consideration of alternative futures (Santelmann et al. 2004) and can highlight inconsistencies between the future demands on ecosystem processes and their likely future supply (Foran and Poldy 2002; Maass et al. 2005). For example, work in several parts of the world has highlighted negative correlations between human population size or density and several groups of species (e.g., Naeem et al. 1999; Parks and Harcourt 2002; Luck et al. 2004). Human migration patterns that are resulting in increasing urbanization and the further development of human settlements on productive land (UNEP 1999; Population Action International 2000) will create additional future problems for efforts to conserve biodiversity in human-modified landscapes (Foran and Poldy 2002).

Summary

Although impact mitigation approaches based on managing landscape patterns are essential for managing modified landscapes and will benefit many species, in many circumstances a valuable additional conservation strategy will be to target management efforts at particular species and mitigate the impacts of particular threatening processes. This chapter outlines five general strategies that aim to maintain selected individual species and important ecological processes (Table 19.1).

1. Maintain key species interactions and maintain functional diversity.
2. Apply or maintain appropriate disturbance regimes.
3. Maintain species of particular concern.
4. Control aggressive, overabundant, and invasive species.
5. Minimize ecosystem-specific threatening processes.

Importantly, species-based and threatening process-based approaches should be implemented in concert with the landscape pattern–based strategies that were outlined in the previous chapter.

Table 19.1. Summary of process-oriented and species-oriented management strategies to mitigate the negative effects of human landscape modification on species and assemblages of species

Management strategy	*Purpose*
1. Maintain key species interactions and functional diversity	• Protect important ecosystem processes • Protect characteristic ecosystem structure and feedbacks
2. Apply or maintain appropriate disturbance regimes	• Encourage characteristic vegetation structure • Create characteristic spatial and temporal variability in vegetation patterns
3. Maintain species of particular concern	• Ensure the continued survival of rare or threatened species
4. Control aggressive, overabundant, and invasive species	• Reduce competition and predation by undesirable species that could negatively affect desirable species • Maintain characteristic species composition
5. Minimize ecosystem-specific threatening processes	• Identify problems that may affect biodiversity but are not directly related to landscape modification • Establish protocols to eliminate these problems

Links to Other Chapters

Chapter 18 outlined pattern-based mitigation approaches that act as an adjunct to the species and threatening process-based approaches that feature in this chapter. Chapter 8 discusses in more detail the threatening processes that can lead to the decline of individual species.

Further Reading

The book by Caughley and Gunn (1996) is an excellent discussion of the compelling reasons why threatening processes are the root cause of the decline of individual species and why it is essential they are appropriately tackled as part of developing effective programs for the recovery of individual species. The conservation biology literature contains numerous examples of management actions that were targeted at conserving a particular species. Thompson and Angelstam (1999) present many valuable case examples that emphasize why a focus on particular species is often a critical strategy in comprehensive conservation planning. The role of keystone species and response diversity are discussed by Soulé et al. (2003, 2005) and Elmqvist et al. (2003). The notion of using natural disturbance regimes to guide anthropogenic disturbance regimes is discussed by Hunter (1993), Attiwill (1994), and Perera et al. (2004).

CHAPTER 20

Guiding Principles for Mitigating the Decline of Species and Assemblages of Species

This chapter completes Part V and the themes associated with mitigating the negative impacts of landscape change on biota. It contains two short sections. First, we reemphasize the value of the 10 general management strategies that were generated collectively from the discussions in Chapters 18 and 19. The final section of this chapter is a brief exploration of some of the nonscientific aspects of impact mitigation in human-modified landscapes.

Guiding Principles That Consider Patterns and Processes

We believe it is important to recognize the complementary role of managing both spatial patterns and ecological processes and the need to embrace both single-species and multispecies approaches. That is, any comprehensive approach to biological conservation that attempts to counter the impacts of human-derived landscape alteration needs to explicitly manage for spatial patterns and ecological processes as well as embrace both single-species and multispecies approaches. This underpins the 10 broad management strategies that were discussed in Chapters 18 and 19 and which are summarized in Table 20.1. Explicitly recognizing the complementarity of managing landscape patterns and ecological processes, while managing for selected single species as well as species assemblages, is central to the thinking embodied in this book. The 10 principles outlined in Table 20.1 bring together some of the most widely accepted guiding principles in applied ecology, with the goal to provide a tangible but scientifically credible set of guiding principles for biological conservation in modified landscapes. The precise application of these principles in any given landscape or region will be dependent on many factors, such as the objectives of management, the type and intensity of threatening processes, and the key elements of the biota of conservation concern. Hence there are no simple prescriptive recipes about how large patches need to be, what levels of connectivity are appropriate, or how much heterogeneity is required that can be applied uncritically in all landscapes.

Putting It All Together: A Case Study from South America

There are few real-world management examples where a major and comprehensive effort has been made to link process-based and pattern-based strategies cohesively to mitigate the impacts of landscape alteration on native species and species assemblages. An outstanding example is the Rio Condor project in Tierra Del Fuego, Chile (Figure 20.1). The area encompassed by the project is part of a major timber production operation. This industrial forestry project incorporates a comprehensive program to mitigate the impacts of landscape change to conserve taxa at the regional, landscape, and forest-stand levels. The management approaches that have been implemented include (after Franklin in Lindenmayer and Franklin 2002):

- The creation of a system of large ecological reserves

Table 20.1. Guiding principles for mitigating the decline of species and species assemblages in modified landscapes*

Guiding principle
Pattern-based management strategies (see Chapter 18)
• Maintain and/or restore large and structurally complex patches of native vegetation
• Maintain and/or restore a matrix that is structurally similar to native vegetation
• Maintain and/or restore buffers around sensitive areas
• Maintain and/or restore corridors and stepping stones
• Maintain and/or restore landscape heterogeneity and capture environmental gradients
Species- and process-based management strategies (see Chapter 19)
• Maintain key species interactions and functional diversity
• Maintain or apply appropriate disturbance regimes
• Maintain species of particular concern
• Control aggressive, overabundant, and invasive species
• Minimize ecosystem-specific threatening processes

*Drawn from Chapter 18 and 19, respectively.

- The identification and protection of sensitive locations within off-reserve areas, including riparian and wetland buffers, sensitive sites (e.g., steep slopes and high-elevation forest), important biological sites, or local biodiversity hotspots
- The maintenance and enrichment of structural conditions in harvested areas by aggregated retention of components of original (prelogged) forest, and the conservation of coarse woody debris

These strategies have a range of key aims, including: (1) the maintenance of habitat for particular species across a range of spatial scales, (2) the maintenance of vegetation cover for suites of species, (3) the maintenance of structural complexity of forest stands to limit habitat degradation for individual species and vegetation deterioration for assemblages of taxa, (4) the prevention of habitat isolation for individual species and the maintenance of landscape connectivity for species assemblages, (5) the maintenance of landscape heterogeneity for species assemblages, and (6) careful consideration and subsequent buffering of edge effects (e.g., accelerated levels of windthrow at the boundaries of cutover units). Importantly, the Rio Condor project has also made provision for carefully targeted management of particular entities, such as the protection of nesting sites of the Andean condor *(Vultura gryphus)*. Finally, the implementation of a comprehensive environmental monitoring and research program to obtain new information has been a key part of the adaptive evolution of the project. There are also other important social and nonscientific dimensions to the Rio Condor project. There has been an environment permitting process (with a linked set of codes of practice and/or monitoring program) as well as the appointment of an ombudsman-like land-steward position to act as an independent arbiter and adviser on environmental matters. An interdisciplinary commission drawing on members from the Chilean Academy of Sciences operated between 1994 and 1999 to provide additional scientific and other oversight to the Rio Condor project (Franklin in Lindenmayer and Franklin 2002).

Postscript—Rio Condor

In late 2004, the ownership of the land managed under the Rio Condor project changed (Jerry Franklin, pers. comm.), and the area has been set aside as a reserve. Nevertheless, the approaches that were taken under the previous land stewardship illustrate a valuable combined set of approaches for integrating landscape management as part of attempts to limit the potentially negative impacts of landscape alteration on biota.

Beyond Science: Mitigating Landscape Change in the Real World

The primary focus of this book has been on the ecological theory and science associated with identifying and understanding the impacts of landscape modification on biota. In the earlier sections of Part V we have explored some of the ways these impacts might be mitigated—again from a largely scientific perspective. However, there are many significant human dimensions

Figure 20.1. Views of the Rio Condor landscape (photos by Jerry Franklin).

associated with landscape management and the resources and values (including the biodiversity values) that flow from landscapes. These social dimensions have profound effects on the ways in which landscapes and natural resources are managed and in many cases will greatly outweigh the importance of science in decision making (Functowicz and Ravetz 1991; Johnson and Herring 1999). Indeed, in the words of Ludwig et al. (1993, p. 17):

> Resource problems are, after all, human problems that are generated through economic and political systems which humans design.

Given that habitat loss and vegetation loss are the most serious threats to the persistence of much of the world's terrestrial biota (Primack 2001; Fahrig 2003), the control of human activities to limit the amount and rate of vegetation loss is critical. This has many important political and social dimensions. For example, approaches to reduce the rate of vegetation loss and its effects on native species may include setting aside additional formal and semiformal protected areas, purchasing private land for conservation, gazetting new environmental legislation, developing incentives to retain and manage native vegetation (e.g., compensation packages for landowners who elect to conserve vegetation), and creating disincentives to remove or degrade native vegetation (e.g., fines and other penalties and removing perverse subsidies) (e.g., Whitten et al. 2002). Other strategies with a strong human dimension include the better integration of conservation objectives with commodity production such as timber harvesting or farming and grazing enterprises. On this basis, for the remainder of this chapter, we briefly touch on some of the human and nonscientific dimensions of the ways in which the impacts of landscape modification may be mitigated. Topics discussed include reserves, off-reserve conservation, and the notion of landscape accounting.

Reserves

Reserves and protected areas represent one way to mitigate the impacts of landscape modification on native species and assemblages—they can be useful for protecting assemblages of taxa through limiting vegetation loss and deterioration, maintaining landscape connectivity, and maintaining areas that are largely free from degrading edge effects. Reserves can also be set aside for the conservation of single species and to reduce habitat loss and habitat degradation for them as well as limit the impacts of habitat isolation on populations of an individual taxon. Lindenmayer and Franklin (2002) argue that large ecological reserves should always play a key role in any credible conservation plan.

Reserves are not limited to public land, and conservation on private land is a major challenge in all parts of the world (Figgis 1999; Knight 1999; Hilty and Merenlender 2003). For example, in western Europe special conservation regulations for private land can be implemented in the public interest (Raff 2003; EU Habitat Directive 2005). In some jurisdictions, reserves have been purchased by private individuals or nongovernment organizations and are then set aside and/or managed primarily for conservation purposes. In Great Britain, the Royal Society for the Protection of Birds has accumulated substantial land holdings. Indeed, the importance of private land conservation is underscored by the fact that the amount of private land in many nations is significantly greater than the amount of public land (Knight 1999; Hilty and Merenlender 2003).

Although reserves are important for mitigating the impacts of landscape modification, they also have some important practical limitations (Box 20.1). These include:

- Limited available land to allocate to reserves and social and economic impediments to expansion and management

- Limited size
- Geographical isolation and therefore limited ability to maintain large-scale ecological connectivity
- Lack of representativeness
- Difficulties in capturing highly mobile taxa such as migratory or nomadic species (e.g., large carnivores)
- Difficulties in capturing taxa with fine scale or patchy distribution patterns
- Instability of abiotic and biotic conditions within reserves, including ongoing human impacts within reserves
- Potential impacts from intensification of exploitation of unreserved land once reserve systems are established

The limitations of formal reserve systems mean that complementary off-reserve approaches are critical for biological conservation. The following section briefly discusses some of these approaches.

Box 20.1. Reserves in the Real World

Setting aside reserves can be an important way of protecting areas for native species and limiting vegetation loss (Margules and Pressey 2000). However, in many countries, particularly rapidly developing countries with large and fast-growing human populations, gazetting reserves without considering the interests, behavior, and attitudes of local people can lead to conflicts and ultimately to degradation of the reserve (e.g., Western and Gichohi 1993; Parks and Harcourt 2002; Struhsaker et al. 2005). For example, Khirthar National Park is Pakistan's first and one of its most important national parks. It has several threatened species and some spectacular natural features. It is also the home of almost 100000 people who grow crops and graze animals; these people are a critical element in all conservation strategies (e.g., Yamada et al. 2004). In another example, Fuller et al. (2004) used satellite imagery to show that about 3 million ha of rainforest was cleared in the Indonesian province of Kalimantan between 1997/1998 and 2002. More than two-thirds of this was in protected areas or places that were proposed to be reserves (Fuller et al. 2004). These cases highlight that simply setting aside land does not guarantee the conservation of native species. This does not mean that reserves are not important, but rather that to reach conservation goals social, economic, and political factors also need to be considered.

Off-Reserve Conservation

Many kinds of measures can be used to mitigate the negative impacts of landscape modification—with varying levels of success in different jurisdictions. While scientific input to off-reserve conservation is fundamentally important, other academic disciplines such as environmental policy analysis or ecological economics provide appropriate frameworks for how best to put science into practice. A detailed discussion of the nonscientific aspects of off-reserve conservation is beyond both the scope of this book and the expertise of its authors. The following paragraphs are designed to give a flavor of the raft of possible approaches to implementing scientific principles for off-reserve conservation.

One approach to enhance off-reserve conservation is the development of codes of practice, which are used widely in many countries. Balancing multiple values of natural resources and/or natural environments lies at the core of most codes of practice. Codes of practice may cover such diverse activities as grazing by domestic livestock, mining, regulation of firewood cutting, prescriptions for logging (and associated works like road building and maintenance), and the extraction and application of irrigated water. The general thrust of codes of practice is not dissimilar to that which underpins approaches such as certification and some industries such as those linked with timber and paper production have embarked on certification systems that aim to enhance the prospects for ecological sustainability in forests (Wallis et al. 1997; Daily and Walker 2000). The concept of certification has become important with the increasing globalization of trade in that the approach attempts to ensure that forest management decisions in one country or by one generation should not impinge negatively on residents of another country or on future generations (Viana et al. 1996). Certification is not restricted to the forestry sector. It has been applied to marine resources (Marine Stewardship Council 2005), and its use in agricultural and livestock grazing industries is mooted as a future development.

Incentive schemes may also be used to enhance biological conservation outside reserves, especially on private land. Agri-environment schemes in the European Union are examples of incentive schemes. Traditionally, the European Union has subsidized its farmers for the production of agricultural output. Agri-environment schemes also subsidize farmers but differ from the traditional payments in that they subsidize environmental management measures rather than agricultural output. The schemes available differ substantially between different countries in the European Union (Kleijn and Sutherland 2003). Some of the schemes most frequently taken up by farmers in a range of countries are outlined in Table 20.2. The effectiveness of agri-environment schemes has been much debated (e.g., Fox 2004; Kleijn et al. 2004; Vickery et al. 2004). One potential problem with agri-environment schemes to date has been that they often apply at very small scales; that is, at the scale of an individual field or farm. Focusing management efforts at such a local scale may be ineffective when degrading processes occurr at larger spatial scales (Kleijn et al. 2004; Tscharntke et al. 2005). In addition, a lack of monitoring and evaluation has meant that the general effectiveness of agri-environment schemes to date is essentially unknown (Kleijn and Sutherland 2003).

Finally, regulation and legislation can be used to promote conservation activities. For example, in western Europe, the Habitat Directive is probably the strongest legislative tool for biodiversity conservation in the European Union (EU Habitat Directive 2005). Perhaps the most well known legal instrument for the protection of native species is the US Endangered Species Act 1973, which is a piece of umbrella legislation under which all species threatened with extinction can benefit from conservation actions such as habitat protection. More than 1200 species have been added to the list of endangered species since the act was passed in 1973, and activities that harm these taxa and their habitat are prohibited, including on private land (Primack 2002). The US Endangered Species Act has been used as a template for similar legislation enacted in other countries around the world. For example, in Australia, the federal government's Environment Protection and Biodiversity Conservation Act 1999 (EPBC) has provision to make lists of populations, subspecies, species, and ecological communities that are vulnerable, threatened, or endangered (Department of Environment and Heritage 2005). Private citizens and organizations that undertake practices which threaten or endanger listed entities can be prosecuted under the EPBC Act. Notably, both under the US Endangered Species Act and the Australian EPBC Act, although all endangered species may, in theory, be conserved, the reality is that low-profile taxa such as invertebrates, fungi, or mosses rarely appear on lists of threatened organisms (Primack 2002; Department of Environment and Heritage 2005), and are rarely protected (Hunter 2002). Moreover, the US Endangered Species Act and the Australian EPBC Act are only as strong as the political will of governments to invoke such legislation, and their powers may be weakened by politicians with anticonservation interests.

Integrating Protected Areas and Off-Reserve Conservation

In many cases it may not be possible to formally set aside large areas where the priority goal is conservation. Other integrated approaches are

Table 20.2. Examples of agri-environment schemes in the European Union that have a high uptake in selected countries*

Country	*Example of a scheme with high uptake*
Austria	Crop rotation stabilization
Belgium	Planting of cover crop between two crops
	Restricting stocking densities
Denmark	Maintenance of extensive grassland
	Organic farming
Germany	Maintenance of grassland ecosystems
Greece	Reduction of nitrogen pollution
	Conservation of endangered breeds
Italy	Reduction of fertilizer and pesticide inputs
	Maintenance of countryside and landscape
Netherlands	Management agreements for meadow birds and species-rich vegetation
Switzerland	Maintenance of low-intensity grazing meadows

*For more details, more countries, and additional examples see Kleijn and Sutherland (2003).

possible. One is the use of biosphere reserves. Land tenures in a given biosphere reserve may encompass private land (including urban and periurban areas), grazing leases, conservation reserves, and state forest, and there may accordingly be a range of different land uses, including irrigated crops and other forms of horticulture (e.g., Puszcza Kampinoska Biosphere Reserve near Warsaw in Poland; Andrzejewska 2003). The overarching aim of a biosphere reserve is to identify approaches to ecologically sustainable development while, at the same time, dealing with a myriad of major conservation and land management problems (Brunckhorst 1999; Shriar 2001). There are ~500 biosphere reserves in almost 100 nations worldwide.

Landscape "Accounting" and Improved Whole of Landscape Management

Many workers recognize that landscape change is a major factor contributing to the present extinction crisis worldwide (Primack 2001). Addressing this problem to ensure that landscapes provide for the requirements of assemblages as well as particular species will be pivotal for attempts to develop ecologically sustainable landscape management practices. However, part of the problem in addressing this issue is that in the vast majority of agricultural, forestry, and other modified areas, deep-seated landscape management problems are, in part, underpinned by a major overallocation of natural resources. Some serious thinking and a sensible vision are urgently needed about landscape futures in many jurisdictions around the world. There is a current propensity for some political leaders, resource economists, and policy makers to treat the environment like a "magic pudding" that can provide all things to all people, and every time another slice of it is taken away then it somehow magically grows back. The problems of agricultural landscapes are repeated many times the world over. There is not enough "environmental margin" in most agricultural areas to stem the major losses of native biota that are presently occurring. In these landscapes, large-scale soil and land degradation is taking place (UNEP 1999) and water is overallocated, while water flows for environmental purposes are limited and declining. The land is also overallocated; competition with the production of crops, domestic livestock grazing, and firewood cutting means there is presently not enough land in agricultural zones where, for example, sufficient areas of native forest or woodland can be adequately conserved and regenerate naturally, and where substantial vegetation restoration plantings can be established. Despite this, agricultural lobby groups and policy makers in many parts of the world are pushing for greater areas of crops, higher populations of domestic livestock, or greater timber and pulp yields. This situation has resulted in massive losses of biodiversity and has the potential to "undermine the capacity of ecosystems to sustain food production, maintain freshwater and forest resources, regulate climate and air quality, and ameliorate infectious diseases" (Foley et al. 2005).

Despite such warnings, there is rarely any serious "landscape accounting" for these developments. That is, whether landscapes can realistically and ecologically support the continued expansion of these industries and, at the same time, maintain other key environmental values, ecosystem services, and viable populations of native plants and animals. Ultimately, many kinds of landscape management problems cannot be tackled seriously until levels of resource allocation, and type and level of resource use are better planned. Landscapes are not magic puddings. Careful thought is needed to articulate what we want them to look like in the future—and then better planning is needed to meet these visions (see Chapter 19 for a brief discussion of scenario planning).

Summary

The previous two chapters in Part V each featured a set of general management strategies for enhanced biological conservation in landscapes subject to human modification. The five management strategies in Chapter 18 were pattern-orientated ones targeting the size, structure, and spatial arrangement of land cover types. The corresponding five strategies outlined in Chapter 19 concerned the management of individual species and ecological processes. These two sets of strategies are highly complementary, and any comprehensive approach to biological conservation will need to consider both sets of strategies. That is, any comprehensive approach to biological conservation that attempts to counter the impacts of human-derived landscape alteration needs to explicitly recognize the complementary roles of managing spatial patterns and ecological processes, and also needs to embrace both single-species and multispecies approaches.

The primary focus of this book has been on the ecological theory and science associated with identifying and understanding the impacts of landscape modification on biota. However, it is fundamentally important to recognize the many significant human, socioeconomic, and political dimensions associated with landscape change. These nonscientific dimensions have profound effects on the ways in which landscapes and natural resources, including biodiversity, are managed.

Links to Other Chapters

This chapter links logically to the previous two (Chapters 18 and 19), which draw out general principles for managing altered landscapes subject to past and continuing human modification.

Further Reading

Davis et al. (2001), Ludwig et al. (2001), Clark (2002), and Dovers and Wild River (2003) provide wide-ranging discussions on why the human dimension and associated policy process are fundamental to decision making in natural resource management, including in landscapes subject to extensive human modification. Hilty and Merenlender (2003) outline many of the key issues associated with the need for conservation on private land. Margules and Pressey (2000) provide a useful overview of reserve design issues. Lindenmayer and Franklin (2002) discuss the limitations of reserves and highlight the critical need for off-reserve conservation strategies, including the detailed case study of Rio Condor. Daily and Ehrlich (1999) and Ludwig et al. (2001) discuss the role of engaging with academic disciplines outside of science, especially in a conservation policy context. An accessible introduction to ecological economics is provided by Armsworth and Roughgarden (2001). Agri-environment schemes in the European Union were reviewed by Kleijn and Sutherland (2003). Soulé et al. (2005) provide an interesting discussion of the relationship between conservation science and legislation.

PART VI
Synthesis

This part concludes the book. It contains a single chapter, which provides a brief synthesis of the book, highlights its main conclusions, and outlines future priorities for research and management (Figure VI.1).

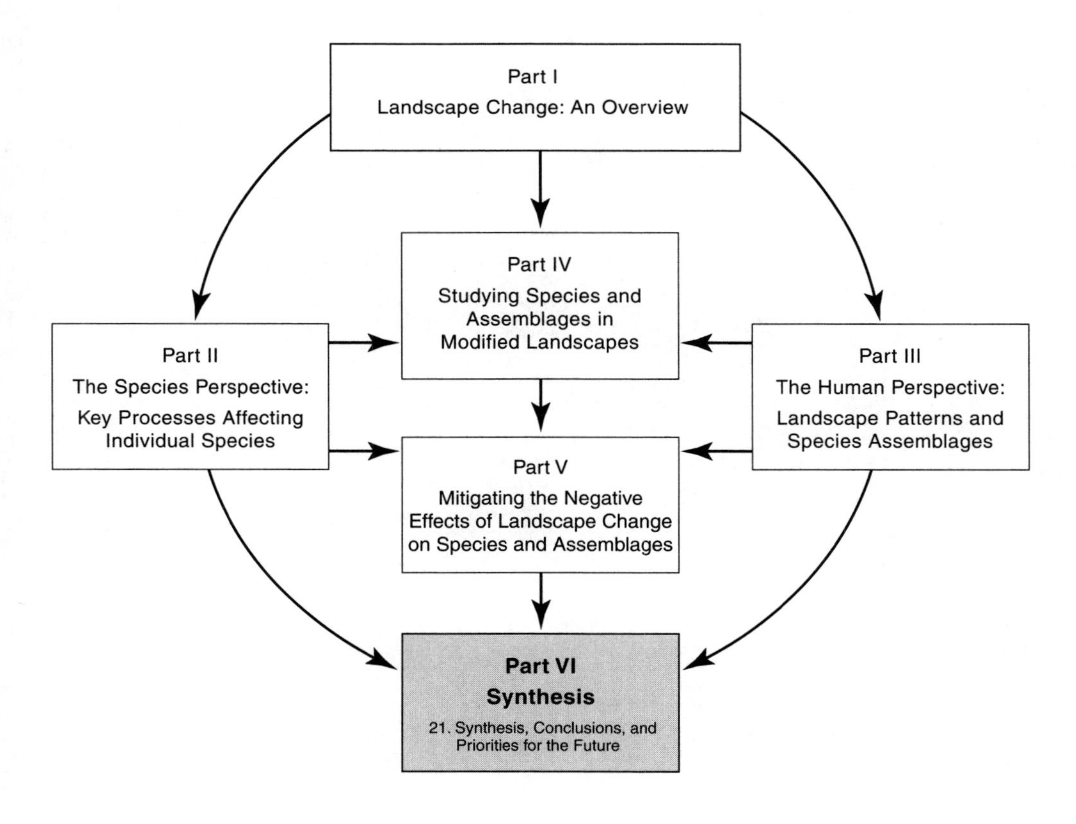

Figure VI.1. Topic interaction diagram showing links between the different parts of the book, and the topics covered in Part VI.

CHAPTER 21
Synthesis, Conclusions, and Priorities for the Future

This book has attempted to take the reader on a journey through a series of topics associated with how landscapes change, the processes associated with such changes, the ecological changes that result from landscape alteration, the ways that landscape change can be studied, and how the effects of landscape change on biota may be mitigated. Despite the large body of work completed to date, many areas of ecological and conservation theory that can help describe, forecast, and mitigate ecological changes have yet to be well developed (e.g., Simberloff 1992; Belyea and Lancaster 1999). Future research and landscape management challenges are therefore considerable. In this final chapter, we provide some concluding remarks on the themes of landscape change. We briefly summarize how we have sought to tackle Bunnell's (1999a) "fragmentation panchreston," and then outline some themes for future research. These themes arise from our review of the literature for this book, and they provide an indication of key knowledge gaps that need to be addressed for effective conservation management in modified landscapes in the future.

Tackling the "Fragmentation Panchreston"

Bunnell (1999a) suggested that the term "fragmentation" has become a panchreston because it has become a catch-all phrase that means different things to different people. Thus the term "fragmentation" itself, as well as associated terms in the literature on modified landscapes, like "thresholds" or "assembly rules," have become vague entities due to inconsistencies in the definitions used by different authors (Belyea and Lancaster 1999; Fahrig 2003). In addition, we felt there was considerable confusion about the links between various themes frequently associated with landscape modification, such as habitat loss, vegetation clearing, connectivity, nested subset theory, edge effects, and other topics. This confusion, in turn, has created inefficiencies in both ecological research and applied conservation management.

The objective of this book was to tackle the fragmentation panchreston. We have tried to define key terms carefully and use them consistently throughout the book. In addition, we set out and applied an explicit conceptual framework to highlight the links between different topics. In this context, one of the ways we have attempted to tackle the panchreston has been to explicitly identify when we are dealing with particular processes, and alternatively, when we focus on landscape patterns. Similarly, we distinguished where the focus was on single-species responses from where the aim was to explore the response of groups of species and assemblages. On this basis, our approach to tackling the fragmentation panchreston can be envisaged as a series of logically linked steps that progress from the careful definition of the specific problem to be tackled within the broad domain of "fragmentation topics," through hypothesis formulation against a background of a particular type of conceptual landscape model, selection of appropriate study methods, and application of mitigation methods matched to

Table 21.1. Sequence of steps to deal with the fragmentation panchreston

Step	*Description*
1. Domain conceptualization	Recognition that fragmentation is composed of multiple processes and multiple patterns and can influence both individual species and assemblages
2. Problem definition	Definition of the key problem—which processes and/or patterns are the key one(s) to study
3. Response definition	Definition of the appropriate response to be studied (i.e., individual species, species assemblage, landscape patterns, processes)
4. Landscape model selection	Appropriate selection of underlying landscape model to guide the design of a study, review, or experiment
5. Hypothesis formulation	Formulate appropriate hypotheses or alternative models to be tested or reviewed relative to problem definition, response definition, and underlying conceptual landscape model
6. Experimental design or definition of procedure of desktop studies	Implementation of appropriate experimental design or protocol based on steps 1 through 5 above
7. Implementation	Application of appropriate field and/or desktop study methods and analytical techniques
8. Outcomes	Analysis and interpretation
9. Impact mitigation	Develop appropriate mitigation strategies matched to address the identified undesirable impacts of landscape change
10. Monitoring and evaluation	Use insights obtained to guide future work

address the impacts of landscape change that have been quantified (Table 21.1).

The logical and sequential way in which we have attempted to tackle the fragmentation panchreston is important for a range of reasons. In particular, we believe that the area of study of the effects of landscape change on biodiversity is presently "bogged" and making relatively little tangible progress. This is because (1) much of the literature is focused on "story-telling" of particular case studies (see Shrader-Frechette and McCoy 1993) that describe pattern-based species-specific or landscape-specific findings that are difficult to generalize to other taxa or locations (in part because the underlying mechanisms giving rise to emergent patterns have not been examined or quantified); (2) many workers appear to be overwhelmed by the blizzard of ecological details that accompanies "fragmentation" studies, making it difficult to separate what is important from what is not; this, in turn, creates problems in defining key questions and formulating appropriate hypotheses to guide future work; (3) many workers appear to be captive to a particular conceptual framework such as a given model of landscape cover (e.g., the "island model"; see Chapter 3) and this constrains problem solving and the development of robust conservation management strategies (Haila 2002); and (4) terminology is often used too loosely, leading to further confusion.

Of course, we have been guilty of these kinds of errors in our past work on landscape change; which is one of the reasons why we wrote this book. Yet we do feel that to make better progress in studies of landscape change and the mitigation of its negative effects on biodiversity, a structured approach like that outlined in Table 21.1 may help to focus future work and avoid further confusion. Less ambiguity and confusion, in turn, should help speed the development of an improved understanding of the effects of landscape change on biota, and importantly, rapidly evolve more effective conservation and impact mitigation strategies.

Improved future work will benefit from key questions being better defined and framed. Which aspects of "fragmentation" are to be examined (e.g., habitat loss, habitat subdivision, edge effects)? What is the response being quantified—the presence or abundance of a particular species? the reproductive success of that species? or an aggregate measure like species richness? What patterns and processes associated with landscape change might give rise to the patterns of abundance, breeding success, or assemblage composition that are being documented, and how can they be quantified? What is the appropriate landscape model that forms the backdrop against which particular hypotheses might be tested? Which methods are the best ones to examine the kinds of problems that are most in need of testing? Given particular research outcomes, what might be the best management strategies to mitigate the key negative problems stemming from landscape alteration? Is mitigation successful? and can monitoring provide useful new insights for future work? Although these questions might sound trite, our personal experience in working in altered landscapes together with our detailed exploration of the literature that accompanied writing this book has indicated that these simple questions are rarely explicitly considered in studies of landscape change. Overcoming this problem in the future may help to dismantle Bunnell's (1999a) fragmentation panchreston (Box 21.1).

Although we have presented our concerns about the fragmentation panchreston, we recognize that there also has been important progress made in the past 3 decades. The disciplines of conservation biology and landscape ecology have grown considerably since their inception (Soulé 1985; Forman and Godron 1986), and the importance of issues such as spatial scale and landscape context is now well acknowledged in mainstream ecology. Similarly, the need to conduct rigorous conservation science and better preserve biodiversity is well appreciated by many governments around the world.

Box 21.1. Tackling the Fragmentation Panchreston and "Unfragmenting" a Rare Bird Species

The endangered eastern bristlebird *(Dasyornis brachypterus)* has been the focus of extensive field studies at Booderee National Park on the south coast of New South Wales (southeastern Australia) for many years (e.g., Baker 1997). Early work has established strong relationships between two patterns—the spatial distribution of the bird and the spatial distribution of patches of complex multilayered heath. Recent collaborative work has identified the process linking the two patterns—predation impacts of the exotic red fox *(Vulpes vulpes)*, which confined eastern bristlebirds to dense heathland thickets that were inaccessible to foxes. An intensive and extensive poison baiting program for foxes over the past 4 years has significantly reduced populations of these feral predators. As a result, populations of the eastern bristlebird have now expanded well beyond complex multilayered heaths and occupy all vegetation types in Booderee National Park except warm temperate rainforests. Hence what was formerly a "fragmented" population confined to patches of a highly specific habitat type has become an unfragmented population. The work on the eastern bristlebird highlights the critical importance of linking two spatial patterns with an underlying ecological process (in this case predation) to best target conservation management strategies.

Summary of Key Concepts and Themes

We have explicitly recognized a fundamental difference between (1) particular ecological processes that impact upon individual species, and (2) the human-perceived vegetation patterns that arise in modified landscapes, which can be correlated with the spatial distribution of individual species or species assemblages.

We acknowledge that patterns and processes cannot be unequivocally teased apart in all instances (Bunnell 1999a). However, a general awareness of the degree to which various relationships between variables are causal or correlative is important (Shipley 2000). For single species, the ecological processes that are causally related to their distribution patterns are often reasonably well known. The spatial distribution of an assemblage, on the other hand, is the emergent pattern of multiple species' individualistic responses to their biophysical environment (including species interactions). Relationships between assemblage-level measures of species occurrence (such as species richness) and landscape pattern therefore are often correlative rather than causal. Both types of relationships and insights—process-oriented ones and pattern-oriented ones—are useful and complementary (Part I). Linking processes and patterns is a fundamental challenge in landscape ecology (Hobbs 1997), and we hope that this book is a useful contribution to tackling this challenge.

We have used the terms "habitat loss," "habitat degradation," and "habitat isolation" to describe threatening processes that affect individual species (Table 21.2). Thus, under our terminology, habitat loss, habitat degradation, and habitat isolation are species-specific threats because habitat itself is a species-specific concept (Part II). Conversely, human-created patterns that typically occur in modified landscapes include lost or degraded native vegetation, recently created edge environments, lost landscape connectivity, and changes to landscape heterogeneity (Part III; Table 21.2). Any comprehensive approach to mitigating the negative effects of landscape change on biota should consider both landscape pattern and ecological processes (Part V; Table 21.2).

Future Research Priorities

Already, an enormous body of work has been completed on issues related to human landscape modification. As a result, many insights on landscape change and "fragmentation" are now largely unquestioned within the scientific community (Harrison and Bruna 1999). For example, we know that many species are being lost (Pimm et al. 1995; Gaston and Spicer 2004). We know that intensification of human land use tends to have negative effects on native species and species assemblages (Benton et al. 2003). We know that maintaining native vegetation cover is useful to conserve native species (Haila 2002). We also know that some simple mitigation measures like the maintenance of large and well-connected patches of native vegetation are more often useful for conservation than not (Noss and Beier 2000; Haila 2002). However, there are many other issues we know much less about. In the following section, we discuss four broad themes that researchers may wish to consider in the future: (1) suggested priorities for autoecological studies, (2) suggested priorities for ecosystem studies, (3) issues of spatial and temporal scale, and (4) challenges of integration within and beyond the science of ecology.

Challenges at the Individual Species Level

Understanding Habitat Loss and Habitat Degradation

Recent work by Fahrig (2003), Fazey et al. (2005a), and others has indicated that much work on the landscape in conservation biology has focused on topics such as habitat subdivision and habitat isolation, when in fact the underlying and driving influence has been habitat loss and vegetation loss. A key issue must be to devise studies that can separate the effects of habitat loss and habitat subdivision (Fahrig 2003) while at the same time recognizing that the processes of habitat loss and habitat subdivision are almost always confounded.

Understanding Dispersal and Recolonization

Dispersal is a fundamental process in population maintenance (Wolfenbarger 1946). In many modified landscapes, populations in some patches experience localized extinctions, which are then reversed by dispersal and re-

Table 21.2. A summary of key themes in this book

Themes	Part II: Threatening processes affecting individual species	Part III: Human-created changes to landscape pattern	Part V: Mitigation of the negative impacts of landscape alteration
How much?	Habitat loss	Loss of vegetation cover	Limit habitat loss, habitat restoration, limit vegetation loss, vegetation restoration
How big?	Reduction in habitat patch size	Reduction in vegetation patch size, increase in negative edge effects	Increase habitat area, increase vegetation patch size, buffer edge effects
How continuous?	Habitat isolation	Loss of landscape connectivity, increase in negative edge effects	Establishment of corridors, stepping stones, matrix management, buffer edge effects
How complex?	Habitat degradation	Vegetation deterioration or loss of vertical structural complexity Loss of horizontal landscape heterogeneity	Maintenance of characteristic vegetation structure, maintenance of landscape heterogeneity
Which species?	Emphasis on individual species	Emphasis on aggregate species responses (e.g., species richness, nested subset theory)	Focus on biodiversity in general, and selected individual species
Processes or patterns?	Emphasis on processes	Emphasis on patterns	Management of both landscape pattern and key ecological processes

colonization (Hanski 1994a, 1999a; Krohne 1997). Thus dispersal and recolonization are fundamental to the long-term population persistence of many species (Brown and Kodric-Brown 1977; Mills and Allendorf 1996; Palomares et al. 2000). Despite the importance of these key biological processes, there is limited knowledge about them per se, including in modified landscapes (Chepko-Sade and Halpin 1987; Stenseth and Lidicker 1992a; Dieckmann et al. 1999). For example, dispersal behavior can change dramatically in response to the spatial availability and quality of suitable habitat (Travis et al. 1995) as often occurs when landscapes are modified by humans (Norton et al. 1995). Similarly, there are few empirical studies worldwide that have examined how patches are recolonized after localized extinctions take place (but see Middleton and Merriam 1981; Palomares et al. 2000). Although it is widely believed that large patches close to a potential source of animals will be more likely to be occupied than small, more isolated ones (Fritz 1979; Hanski 1994a), few empirical field studies of this hypothesis have been undertaken (Johnson and Gaines 1990).

Information on dispersal and recolonization is critical for developing natural resource management strategies (Lubchenco et al. 1991) such as the establishment of wildlife corridors (Andreassen et al. 1996; Rosenberg et al. 1997; Bennett 1998) and other strategies for conserving wildlife in human-modified landscapes (e.g., matrix management; Lamberson et al. 1994; Lindenmayer and Franklin 2002).

Data on dispersal can be extremely difficult to obtain from some standard field techniques such as ring banding or radio tracking, and long-term studies are often required to gather high-quality data (e.g., Nelson 1993; Palomares et al. 2000). However, the use of these techniques may change animals' movement patterns and thwart attempts to obtain valid field results (Oakleaf et al. 1993; Baptista and Gaunt 1997). In addition, many dispersal events may be missed by methods such as trapping, and traditional methods are often unable to determine whether actual gene flow (through reproduction) has taken place. A better understanding of dispersal may be gained by intertwining demographic and genetic studies (e.g., Aars et al. 1998; Manel et al. 2003). We have employed such a "landscape genetics" approach using molecular genetic techniques in the Tumut fragmentation study for native species like the bush rat *(Rattus fuscipes)* and the greater glider *(Petauroides volans)* (Hewittson 1997; Peakall et al. 2003; Taylor et al. 2006). One of the values of these types of studies is in elucidating whether or not population models for heterogeneous landscapes are prone to substantial errors (Karieva et al. 1997; Hanski 1999b). The usefulness of such studies could be further magnified by connecting estimates of effective interpatch dispersal with measures of matrix condition and landscape cover (such as the presence of corridors or retained vegetation).

Challenges at the Ecosystem Level

Overcoming Taxonomic Biases

The vast majority of work on landscape alteration has focused on birds and mammals, although some invertebrate groups such as butterflies and beetles have also received reasonable attention. Remarkably few studies have examined the effects of habitat loss and habitat subdivision on plants (Hobbs and Yates 2003), particularly groups such as bryophytes (Zartman 2003). There is a need to broaden the scope of inference about the impacts of landscape change by studying a wider range of taxonomic groups such as mosses and lichens (Pharo et al. 2004) as well as invertebrate assemblages that make a disproportionately large contribution to overall levels of species richness in many ecosystems (Majer et al. 1994; Didham et al. 1996). There is also a need to complete more studies where the responses of different taxonomic groups in the same system are quantified to assess contrasting impacts of landscape change (e.g., Robinson et al. 1992; Lindenmayer et al. 2003c).

Studying Biodiversity outside "Patches"

Conditions in the matrix have massive ramifications for almost all aspects of species response to landscape change, including (1) spatial and temporal pattern of habitat suitability (Fischer et al. 2004a), (2) patterns of dispersal (Franklin 1993), (3) the magnitude of edge effects (Lehtinen et al. 2003), and (4) extinction proneness (Laurance 1991). Yet few ecological studies have been specifically designed to quantify the importance of the matrix for biota, and many others have overlooked the use of the matrix in influencing species' responses to landscape alteration (Simberloff et al. 1992). These deficiencies need to be addressed as part of better informed future work. For example, new generations of metapopulation models will need to include positive or negative conditions in the matrix to improve their predictive ability and take account of real landscape conditions (e.g., Ricketts 2001). This may include redefining patches within these models to include surrounding areas in the matrix that undergo temporal and/or spatial changes in their suitability for organisms, such as the regeneration of forests after logging. Sisk et al. (1997) provided an elegant example in which they modeled bird assemblages within California oak woodlands that were embedded in a matrix of either grassland or chapparal. There is also a place for the development of new types of models that treat landscapes as consisting of multiple patch types of different quality and which can change on a temporal basis.

An increased emphasis on the role and importance of the matrix for biota is important for other reasons. For example, although the bulk of the conservation biology literature has focused on relatively pristine environments (Fazey et al. 2005a), in fact much of the world's biota occurs outside reserve systems (Daily 2001; Lindenmayer and Franklin 2002). Similarly, the focus on reserves and large patches has led to conservation importance of intermediate and small-sized patches of native vegetation being undervalued or ignored altogether (Zuidema et al. 1996; Turner 1996; Fischer and Lindenmayer 2002c).

Cumulative Effects and Regime Shifts

Landscape modification may lead to synergistic or cumulative impacts on biota and ecosystem processes. These impacts can be difficult to quantify (Cocklin et al. 1992a), even with well-designed studies. For example, the long-term effects of changes in the condition of patches of remnant vegetation may remain undetected within the time frames of most impact studies (Saunders et al. 2003). Traditional forms of landscape analysis also may not detect cumulative effects at a landscape scale. The potential for negative patch and landscape-level cumulative effects, and the current paucity of methods for detecting and assessing them (Cocklin et al. 1992b; Burris and Canter 1997) make studies of cumulative impacts a vital area for further work (Paine et al. 1998).

An additional key challenge for future work will be to identify unsustainable ecosystem trajectories (Manning et al. 2004b) and possibly irreversible regime shifts before they occur (Scheffer et al. 2001; Folke et al. 2004). Notably, as ecological degradation continues in a given landscape, mitigation often becomes increasingly difficult and costly (Milton et al. 1994). Tools such as scenario planning (Peterson et al. 2003a) are increasingly being used to explore and conceptualize future ecosystem trajectories. They can identify inconsistencies in the demand and supply of ecosystem processes, thereby highlighting unsustainable land use trends at a time when mitigation measures can be effective and efficient (Peterson et al. 2003b; Santelmann et al. 2004; Maass et al. 2005).

Issues of Scale

Scaling Up in Space

Species and ecosystem responses to landscape change will often need to be studied at multiple spatial scales. Species occurrence has often been studied at the patch level rather than the landscape level (McGarigal and McComb 1999; Fahrig 2003), leading to a possible mismatch between the scale of human land use and the scale at which biota are sampled (Manning et al. 2004a). Hence more studies genuinely focused at the landscape scale are required (e.g., Villard et al. 1999; Radford et al. 2005). It is important to recognize that not all parts of a landscape are created equal—typically the most productive areas are also the most perturbed (Chapter 2). Natural environmental gradients have been ignored by some workers who have compared perturbed high-productivity environments with unperturbed low-productivity environments. Where primary productivity and land use intensity are confounded, the lack of significant difference in a target response (e.g., species abundance) across the study area does not reliably indicate that there have been no negative impacts of landscape change.

Scaling Up in Time

The amount of time elapsed since landscape change commenced is a significant factor influencing the occurrence of some species in altered landscapes (e.g., Suckling 1982; Loyn 1987). This is expected, given concepts such as extinction debts among biota in isolated vegetation remnants (Diamond et al. 1987; Tilman et al. 1994; McCarthy et al. 1997). Long-term studies and/or prolonged periods of recurrent monitoring are often needed to quantify long-term changes (Debinksi and Holt 2000). Despite this, virtually every review of research programs highlights the dearth of long-term ecological studies of ecosystem, community, species, or population responses to landscape change. For example, even in a wealthy country such as Australia, there are only five long-term ecological research sites registered nationwide. The full magnitude of many of the effects of human modification of landscapes will only be determined by well-supported long-term studies, which will, in turn, provide the essential data to develop more effective mitigation strategies.

There are many challenges for maintaining long-term studies, such as the maintenance of core funding, an often limited number of publications during many phases of the work, and determining when a long-term study is needed, how long that study needs to be maintained, and when it is reasonable to terminate the study. Of course, not all work needs to be long-term to be useful—some behavioral and cross-sectional studies can deliver useful findings relatively quickly.

Designing Natural Experiments

Part IV summarized some of the tools that can be used to study the effects of landscape change on individual species and species assemblages, including experiments, observational studies, and modeling. All have some strengths but also some limitations. We believe that in many cases, small-scale work such as traditional experiments and microcosm studies have limited value to derive larger-scaled land management guidelines (Noss and Beier 2000; but see Srivastava et al. 2004). However, with careful attention to experimental design, it is possible to adopt approaches typical of smaller scale "true" experiments for use at larger spatial scales. "Natural experiments" (Chapter 16) can use "real world" land use practices or management actions as "treatments" in research and produce robust results of considerable value for landscape management (Dunning et al. 1995). We believe that an increasing use of well-designed natural experiments which take advantage of existing landscape conditions or conditions that are about to change could add considerable new knowledge on how to better conserve native species and species assemblages in landscapes subject to human modification.

The Challenge of Integration

Integrating Disparate Areas within Ecology

One of the main reasons for writing this book was to establish explicit links between often disparate areas in ecology and conservation biology. We believe that approaches which explicitly recognize the complementarity of managing landscape patterns and ecological processes deserve further attention. Similarly, there are several largely unexplored links between currently disparate areas in ecology that deserve further research attention. Useful insights may be obtained by strengthening the links between (1) work conducted at different spatial scales and levels of ecological organization (Forman 1964; Diamond 1973); (2) comparing results from different taxonomic groups in the same system (Robinson et al. 1992; Gascon et al. 1999; Lindenmayer et al. 2003c); (3) work on species interactions (like trophic interactions or mutualisms) and landscape pattern (e.g., Cordeiro and Howe 2003; Bodin et al. 2006); and (4) work on ecosystem processes, resilience, biota, and commodity production (e.g., Elmqvist et al. 2003; Ricketts et al. 2004).

Integrating Academic Disciplines

Managing Earth's ecosystems is an interdisciplinary challenge (Daily and Ehrlich 1999). Discussion has increased over the last few years about how to effectively integrate multiple academic disciplines (e.g., Armsworth and Roughgarden 2001; Campbell 2005). Bringing together academics from different disciplines is challenging because different disciplines have different ways of communicating and different frameworks to ensure internal intellectual rigor. These different disciplines also make different implicit assumptions about the world. Despite these difficulties, more collaborative projects involving experts from a range of academic disciplines are urgently required. This is particularly important in human-modified landscapes where many stakeholders represent a wide range of interests and values, and many potentially competing land uses (e.g., Maass et al. 2005).

Integrating Theory and Practice

An unfortunate dichotomy has developed in many areas of ecology whereby theory and applied science often act as almost separate disciplines (Fazey 2005). The effects of landscape alteration have not been immune from such a divide. However, this dichotomy needs to be addressed to determine where and when theory is useful for conservation and management and where and when it is not (Fazey and McQuie 2005). This is not necessarily straightforward because a well-developed piece of theory that is found to apply well to some landscapes or to particular species will often fail when tested in new circumstances (Simberloff 1992). Two approaches could be useful in this regard. First, there appears to be a need for more meta-analyses that assess the generality of particular phenomena across different landscapes or taxonomic groups. However, in the words of Southwood (1966), a challenge with such work will be to overcome the "[d]ifferences in methodologies [that] can invalidate comparisons between studies." Nevertheless, careful evaluation of multiple case studies may help generate empirical generalizations that provide guidance in the absence of case study–specific data (Shrader-Frechette and McCoy 1993).

A second approach uses systematic reviews to rigorously assess the evidence supporting a particular piece of theory (Fazey et al. 2005b). As outlined in Chapter 17, a well-developed framework has been created for systematic reviews in medical research, and it is coordinated by a body called the Cochrane Collaboration. In addition to overseeing the review process, the Cochrane Collaboration also plays a major role in disseminating research findings to practitioners. Fazey et al. (2005b) argued that many elements of the Cochrane Collaboration's approach could be usefully embraced by conservation organizations.

More generally, the need for stakeholder participation is increasingly well-recognized in the applied ecological literature (Saunders 1996; Berkes 2004; Higgs 2005). Few land managers read the scientific literature, and it takes many years for new ideas and empirical evidence to "trickle down" to the level where improved management practices are implemented. Work is needed on the research adoption process (Walker 1998):

1. Where it has been successful, where it has failed, and why
2. The organizational and other structures that promote or impair capitalizing on new research (see Kirkpatrick 1998)
3. The interplay between research, policy, and management

Too many scientists appear to operate under a "strategy of hope"; that is, simply hoping that their work will be useful for management professionals and policy makers but doing nothing to further that goal (Hamel and Prahalad 1989). Therefore, more scientists need to engage in direct communication of scientific results to stakeholders, including resource managers, policy professionals, and politicians.

Concluding Remarks

Landscape change and "habitat fragmentation" are young but now massive topics in conservation biology that encompass many diverse and interrelated processes, landscape patterns, and biotic responses at a range of scales. As outlined at the start of this book, the literature on these topics is now so large that it is virtually impossible for any one person or group of people to keep up with it. Indeed, Belovsky et al. (2004) believe that many current workers are unaware of many valuable past studies, leading to the "reinvention of wheels" and retesting of so-called silver bullets that failed previously when subject to rigorous empirical assessment (e.g., indicator species approaches; Landres et al. 1988; Temple and Wiens 1989; Niemi et al. 1997; Lindenmayer et al. 2000). We contend that there are no silver bullets for biological conservation (cf. Myers et al. 2000; Box 21.2). Especially in human-modified landscapes, biological conservation is not a tidy scientific project (Ewers 2005), and a mix of strategies will be needed to halt the current extinction crisis.

For scientists, a possible way forward is to set two broad research priorities. The first is to review, integrate, and synthesize existing knowledge in a way that is scientifically defensible, robust, and of interest to scientists, policy makers, and land managers. Such syntheses are of great value—they tend to be widely read and widely cited, and they provide easy access to a body of literature that may otherwise seem overwhelming. Notably, concise synthesis articles are now key components of several leading-edge ecological journals, including *Ecology Letters, Frontiers in Ecology and the Environment*, and *Trends in Ecology and Evolution*. Arguably, there is plenty of room for more syntheses of existing information, and the production of such work should be an important future priority for scientists.

The second broad research priority is to produce highly focused empirical work that answers a particular question of either practical or theoretical significance. Such work should carefully consider current knowledge gaps (e.g., see our own subjective list above) and, ideally, will create mechanistic links between patterns and processes and contribute to ecological theory as well as outline the practical implications of the work. The explicit formulation of practical implications has become a key requirement of several highly regarded applied ecological journals like *Frontiers in Ecology and the Environment, Biological Conservation*, and the *Journal of Applied Ecology*. Together, focused empirical work and syntheses of existing information may help to provide clearer "insights into system dynamics," which are still lacking in many cases (Bissonette and Storch 2002).

Notably, different researchers have different skills and interests. This diversity can be a valuable strength of applied ecology. Focused work can fill important knowledge gaps and can provide valuable inputs for integration attempts and reviews. A key challenge lies in recognizing and capitalizing on the wide range of interrelated topics that are relevant to conservation management in human-modified landscapes. This is critical because human-modified landscapes will be where the ultimate future of much of Earth's terrestrial biota will be decided.

Summary

The first part of this chapter revisits Bunnell's (1999a) fragmentation panchreston (i.e., the confusion that has arisen by the term "fragmentation" meaning many different things to different people). We have tried to tackle the panchreston by separating key threatening processes associated with landscape modification and their effects on individual species, from the human-perceived patterns of vegetation cover that characterize altered landscapes, which are often correlated with the occurrence of suites of taxa and entire assemblages. We differentiate habitat loss, habitat degradation, and habitat isolation as processes threatening individual species from the human-created patterns of lost or degraded native vegetation, lost landscape connectivity, and edge effects that can be associated with changes in species assemblages. Such a differentiation is dependent on the precise and consistent use of terminology, which is pivotal to (1) avoiding confusion between workers, (2) ensuring that the complex interacting factors associated with landscape alteration can be appropriately teased apart, and (3) helping to determine when it is appropriate to focus on ecological processes and individual species versus human-created patterns and aggregate measures such as species richness. Notably, studies of (and mitigation strategies for) processes and the responses of individual species are not mutually exclusive to those targeting landscape patterns and species assemblages. Rather, approaches focusing on processes or patterns, respectively, are highly complementary, and the development of informed mitigation strategies will often require both types of approaches.

The final part of this chapter outlines some future research issues that could be fruitful in developing better insights in the impacts of landscape alteration on biota and how to mitigate such impacts. We set out some of the future challenges at the individual species and ecosystem levels. We also discuss scale issues and the need for integration across disparate areas of ecology, across different academic disciplines, and across the range of stakeholders involved in the management of human-modified landscapes.

Box 21.2. Why There Are No Magic Bullets in Landscape Management and Biological Conservation

In many sections of this book, we present important caveats about the uncritical application of concepts, methods, equations, and general tools. Whether the topic is species-loss equations or habitat suitability indices we have outlined the reasons why these approaches are not "magic bullets" and why their uncritical application could actually deliver poor outcomes for landscape management and biological conservation. A key issue is that landscape management revolves around what Ludwig et al. (2001) term "wicked environmental problems" that are characterized by (1) multiple legitimate human perspectives, (2) large amounts of uncertainty, (3) no definitive formulation, (4) no stopping rules to determine when a problem has been appropriately addressed, and (5) no test for a solution—in part because the outcome will often depend on how the problem is framed and by whom (Maddox 2000).

About the Authors

David B. Lindenmayer is a professor at the Centre for Resource and Environmental Studies at the Australian National University in Canberra, Australia. He has published widely on wildlife ecology and conservation biology. Two of his recent books include *Conserving Forest Biodiversity* (with Jerry Franklin, Island Press, 2002) and *Practical Conservation Biology* (with Mark Burgman, CSIRO Publishing, 2005).

Joern Fischer is a postdoctoral fellow at the Centre for Resource and Environmental Studies at the Australian National University in Canberra, Australia. His main interest is conservation in human-modified landscapes. He has published both empirical and conceptual work in a range of international journals.

References

Aanderaa, R., Rolstad, J., and Sognen, S.M. 1996. *Biological Diversity in Forests*. Norges Skogeierforbund og A/S Landbruksforlaget. Oslo, Norway.

Aars, J. and Ims, R.A.. 1999. The effect of habitat corridors on rates of transfer and interbreeding between vole demes. *Ecology,* **80**, 1648–1655.

Aars, J., Ims, R.A., Liu, H-P., Mulvey, M. and Smith, M.H. 1998. Bank voles in linear habitats show restricted gene flow as revealed by mitochondrial DNA (mtDNA). *Molecular Ecology,* **7**, 1383–1389.

Aars, J., Johannesen, E. and Ims, R.A. 1999. Demographic consequences of movements in subdivided root vole populations. *Oikos,* **85**, 204–216.

Abensperg-Traun, M. and Smith, G.T. 2000. How small is too small for small animals? Four terrestrial arthropod species in different-sized remnants in agricultural Western Australia. *Biodiversity and Conservation,* **8**, 709–726.

Åberg, J., Swenson, J.E. and Angelstam, P. 1995. The effect of matrix on the occurrence of Hazel grouse *(Bonasa bonasia)* in isolated habitat fragments. *Oecologia,* **103**, 265–269.

Adams, W.M. 2004. *Against Extinction: The Story of Conservation*. Earthscan, London.

Agee, J.K. 1993. *Fire Ecology of the Pacific Northwest Forests*. Island Press, Washington, D.C.

Agee, J.K. 1999. Fire effects on landscape fragmentation in interior west forests. Pp. 43–60 in *Forest Fragmentation: Wildlife Management Implications*. J.A. Rochelle, L.A. Lehmann, and J. Wisniewski, eds. Brill, Leiden, Germany.

Agee, J.K. and Huff, M.H. 1987. Fuel succession in a western hemlock/douglas fir forest. *Canadian Journal of Forest Research,* **17**, 697–704.

Agger, P. and Brandt, J. 1988. Dynamics of small biotopes in Danish agricultural landscapes. *Landscape Ecology,* **1**, 227–240.

Akaike, H. 1973. Information theory and an extension of the maximum likelihood principle. Pp. 267–281 in *International Symposium on Information Theory*. 2nd ed. B.N. Petran and F. Csaki, eds. Akademiai Kiado, Budapest.

Akçakaya, H.R. and Ferson, S. 1992. *RAMAS/Space*. Applied Biomathematics, Setauket, New York.

Alatalo, R.V., Lundberg, A. and Bjorklund, M. 1982. Can the song of male birds attract other males? An experiment with the pied flycatcher *Ficedula hypoleuca*. *Bird Behaviour,* **4**, 42–45.

Allee, W.C. 1931. *Animal Aggregations*. University of Chicago Press, Chicago.

Allee, W.C., Emerson, A.E., Park, O., Park, T. and Schmidt, K.P. 1949. *Principles of Animal Ecology*. Saunders, Philadelphia.

Allen, T.F.H. and T. B. Starr. 1988. *Hierarchy: Perspectives for Ecological Complexity*. The University of Chicago Press, Chicago.

Allen, T.F.H. and T.W. Hoekstra. 1992. *Toward a Unified Ecology*. Columbia University Press, New York.

Alverson, W.S., Kuhlmann W. and Waller, D.M. 1994. *Wild Forests Conservation Biology and Public Policy*. Island Press, Washington, DC, USA.

Amarasekare, P. 1994. Spatial population structure in the banner-tailed kangaroo rat, *Dipodomys spectabilis*. *Oecologia,* **100**, 166–176.

Ambuel, B. and Temple, S. 1983. Area-dependent changes in the bird communities and vegetation of southern Wisconsin forests. *Ecology,* **64**, 1057–1068.

Anderson, L. and Burgin, S. 2002. Influence of woodland remnant edges on small skinks (Richmond, New South Wales). *Austral Ecology,* **27**, 630–637.

Andreassen, H.P. and Ims, R. 2001. Dispersal in patchy vole populations: Role of patch configuration, density dependence and demography. *Ecology,* **82**, 2911–2926.

Andreassen, H.P., Halle, S. and Ims, R. 1996. Optimal width of movement corridors for root voles: Not too narrow and not too wide. *Journal of Applied Ecology,* **33**, 63–70.

Andrén, H. 1992. Corvid density and nest predation in relation to forest fragmentation: a landscape perspective. *Ecology,* **73**, 794–804.

Andrén, H. 1994. Effects of habitat fragmentation on birds and mammals in landscapes with different proportions of suitable habitat: a review. *Oikos,* **71**, 355–366.

Andrén, H. 1997. Habitat fragmentation and changes in biodiversity. *Ecological Bulletin,* **46**, 171–181.

Andrén, H. 1999. Habitat fragmentation, the random sample hypothesis and critical thresholds. *Oikos,* **84**, 306–308.

Andrén, H. and Angelstam, P. 1988. Elevated predation rates as an edge effect in habitat islands: experimental evidence. *Ecology,* **69**, 544–547.

Andrewartha, H.G. and Birch, L.C. 1954. *The Distribution and Abundance of Animals.* University of Chicago Press, Chicago.

Andrzejewska, A. 2003. Physiography and nature monitoring in the Kampinoski National Park and its buffer zone. *Ecohydrology and Hydrobiology,* **3**, 247–254.

Angelstam, P. 1996. The ghost of forest past: natural disturbance regimes as a basis for reconstruction for biologically diverse forests in Europe. Pp. 287–337 in *Conservation of Faunal Diversity in Forested Landscapes.* R.M. DeGraaf and R.I. Miller, eds. Chapman and Hall, London.

Angelstam, P. and Pettersson, B. 1997. Principles of Swedish forestry biodiversity management. *Ecological Bulletin,* **46**, 91–203.

Angermeier, P.I. 1995. Ecological attributes of extinction-prone species: loss of freshwater fishes of Virginia. *Conservation Biology,* **9**, 143–158.

Antongiovanni, M. and Metzger, J.P. 2005. Influence of matrix habitats on the occurrence of insectivorous bird species in Amazonian forest fragments. *Biological Conservation,* **122**, 441–451.

Armbruster, P. and Lande, R. 1993. A population viability analysis for African elephant *(Luxodonta africana):* How big should reserves be? *Conservation Biology,* **7**, 602–610.

Armesto, J.J., Rozzi, R., Smith-Ramirez, C. and Arroyo, M.T. 1998. Conservation targets in South American temperate forests. *Science,* **282**, 1271–1272.

Armsworth, P.R. and Roughgarden, J.E. 2001. An invitation to ecological economics. *Trends in Ecology and Evolution,* **16**, 229–234.

Arnold, G.W. and Weeldenburg, J.R. 1998. The effects of isolation, habitat fragmentation and degradation by livestock grazing on the use by birds of patches of gimlet *Eucalyptus salubris* woodland in the wheatbelt of Western Australia. *Pacific Conservation Biology,* **4**, 155–163.

Arnold, G.W., Steven, D.E. and Weeldenburg, J.R. 1993. Influences of remnant size, spacing pattern and connectivity on population boundaries and demography in Euros *Macropus robustus* living in a fragmented landscape. *Biological Conservation,* **64**, 219–230.

Arrhenius, O. 1921. Species and area. *Journal of Ecology,* **9**, 95–99.

Ås, S. 1999. Invasion of matrix species in small habitat patches. *Conservation Ecology,* **3**, 1–15.

Askins, R.A. and Philbrick, M.J. 1987. Effects of changes in regional forest abundance on the decline and recovery of a forest bird community. *Wilson Bulletin,* **99**, 7–21.

Askins, R.A., Philbrick, M.J. and Sugeno, D.S. 1987. Relationships between the regional abundance of forest and the composition of bird communities. *Biological Conservation,* **39**, 129–152.

Atmar, W. and Patterson, B.D. 1993. The measure of order and disorder in the distribution of species in fragmented habitat. *Oecologia,* **96**, 373–382.

Atmar, W. and Patterson, B.D. 1995. *The Nestedness Temperature Calculator: A Visual Basic Program, Including 294 Presence–Absence Matrices.* AICS Research Incorporate and The Field Museum, Chicago.

Attain, J.A., and deLucio, J.V. 2001. The role of landscape structure in species distribution of birds, amphibians, reptiles and lepidopterans in Mediterranean landscapes. *Landscape Ecology,* **16**, 147–159.

Attiwill, P.M. 1994. Ecological disturbance and the conservative management of eucalypt forests in Australia. *Forest Ecology and Management,* **63**, 301–346.

August, P.V. 1983. The role of habitat complexity and heterogeneity in structuring tropical mammal communities. *Ecology,* **64**, 1495–1507.

Aune, K., Jonsson, B.G. and Moen, J. 2005. Isolation and edge effects among woodland key habitats in Sweden: is forest policy promoting fragmentation? *Biological Conservation,* **124**, 89–95.

Austin, M.P. 1999. A silent clash of paradigms: some inconsistencies in community ecology. *Oikos,* **86**, 170–178.

Austin, M.P. 2002. Spatial prediction of species distribution: an interface between ecological theory and statistical modelling. *Ecological Modelling*, **157**, 101–118.

Austin, M.P. and Smith, T.M. 1989. A new model for the continuum concept. *Vegetatio*, **83**, 35–47.

Austin, M.P., Meyers, J.A. and Doherty, M.D. 1994a. *Data Capability, Sub-project 3. Modelling of Landscape Patterns and Processes using Biological Data*. Division of Wildlife and Ecology, CSIRO, Canberra.

Austin, M.P., Nicholls, A.O., Doherty, M.D. and Meyers, J.A. 1994b. Determining species response functions to an environmental gradient by means of a ß-function. *Journal of Vegetation Science*, **5**, 215–228.

Australian Agriculture, Fisheries and Forestry 2003. *At a Glance 2003*. Australian Agriculture, Fisheries and Forestry, Commonwealth of Australia, Canberra.

Avery, R.A. 1979. *Lizards: A Study in Thermoregulation*. Edward Arnold Publishers, London.

Baker, J. 1997. The decline, response to fire, status and management of the eastern bristlebird. *Pacific Conservation Biology*, **3**, 235–243.

Baker, J. 1998. *Ecotones and Fire and the Conservation of the Eastern Bristlebird*. PhD thesis, University of Wollongong, Wollongong.

Baker, W.L. 1992. The landscape ecology of large disturbances in the design and management of nature reserves. *Landscape Ecology*, **7**, 181–194.

Bakker, K.K., Naugle, D.E. and Higgins, K.F. 2002. Incorporating landscape attributes into models for migratory grassland conservation. *Conservation Biology*, **16**, 1638–1646.

Balmford, A. 1996. Extinction filters and current resilience: the significance of past selection pressures for conservation biology. *Trends in Ecology and Evolution*, **11**, 193–196.

Banks, S.C., Lawson, S.J., Finlayson, G.R., Lindenmayer, D.B., Ward, S.J. and Taylor, A.C. 2004. The effects of habitat fragmentation on demography and genetic variation in a marsupial carnivore. *Biological Conservation*, **122**, 581–597.

Baptista, L.F. and Gaunt, S.L.L. 1997. Bioacoustics as a tool in conservation studies. Pp. 212–213 in *Behavioral Approaches to Conservation in the Wild*. J.R. Clemmons and R. Buchholz, eds. Cambridge University Press, Cambridge.

Barbour, M.S. and Litvaitis, J.A. 1993. Niche dimensions of New England cottontails in relation to habitat patch size. *Oecologia*, **95**, 321–327.

Barnett, J.L., How, R.A. and Humphreys, W.F. 1978. The use of habitat components by small mammals in eastern Australia. *Australian Journal of Ecology*, **3**, 277–285.

Barrett, G., Peles, J.D. and Harper, S.J. 1995. Reflections on the use of experimental landscapes in mammalian ecology. Pp. 157–174 in *Landscape Approaches in Mammalian Ecology and Conservation*. W. Lidicker, ed. University of Minnesota Press, Minneapolis, Minnesota.

Barrett, G.W. 2000. Birds on farms: ecological management for agricultural sustainability. Supplement to *Wingspan*, **10(4)**. Birds Australia, Hawthorn, Victoria.

Barrett, G.W., Ford, H.A. and Recher. H.F. 1994. Conservation of woodland birds in a fragmented rural landscape. *Pacific Conservation Biology*, **1**, 245–256.

Barrett, G.W., Silcocks, A., Barry, S., Cunningham, R. and Poulter, R. 2003. *The New Atlas of Australian Birds*. Birds Australia, Melbourne.

Bart, J. and Forsman, E.D. 1992. Dependence on northern spotted owl *Strix occidentalis caurina* on old growth forests in western U.S.A. *Biological Conservation*, **62**, 95–100.

Barton, D.R., Taylor, W.D. and Biette, R.M.. 1985. Dimensions of riparian buffer strips required to maintain ba habitat in southern Ontario streams. *North American Journal of Fisheries Management*, **5**, 364–378.

Barton, J.L. and Davies, P.E. 1993. Buffer strips and streamwater contamination by atrazine and pyrethenoids aerially applied to *Eucalyptus nitens* plantations. *Australian Forestry*, **56**, 201–210.

Batary, P. and Baldi, A. 2004. Evidence of an edge effect on avian nest success. *Conservation Biology*, **18**, 389–400.

Baur, A., and Baur, B. 1992. Effect of corridor width on animal dispersal: a simulation study. *Global Ecology and Biogeography Letters*, **2**, 52–56.

Bawa, K.S. 1990. Plant-pollinator interactions in tropical rain forests. *Annual Review of Ecology and Systematics*, **21**, 399–422.

Bayne, E.M. and Hobson, K.A. 1997. Comparing the effects of landscape fragmentation by forestry and agriculture on predation of artificial nests. *Conservation Biology*, **11**, 1418–1429.

Bayne, E.M. and Hobson, K.A. 1998. The effects of habitat fragmentation by forestry and agriculture on the abundance of small mammals in the southern boreal mixedwood forest. *Canadian Journal of Zoology*, **76**, 62–69.

Beattie, A. and Erlich, P.R. 2001. *Wild Solutions*. Melbourne University Press, Melbourne.

Beer, J. and Fox, M. 1997. Conserving vegetation in fragmented rural landscapes: edge effects and eucalypt forest remnants. Pp. 313–318 in *Conservation Outside Nature Reserves*. P. Hale and D. Lamb, eds. University of Queensland, Brisbane.

Beier, P. 1993. Determining minimum habitat areas and habitat corridors for cougars. *Conservation Biology*, **7**, 94–108.

Beier, P. 1995. Dispersal of juvenile cougars in fragmented habitat. *Journal of Wildlife Management*, **59**, 228–237.

Beier, P. and Noss, R. 1998. Do habitat corridors provide connectivity? *Conservation Biology*, **12**, 1241–1252.

Beissinger, S.R. 1986. Demography, environmental uncertainty, and the evolution of mate desertion in the snail kite. *Ecology*, **67**, 1445–1459.

Beissinger, S.R. and McCullough, D.R. 2002. *Population Viability Analysis*. University of Chicago Press, Chicago.

Beissinger, S.R. and Westphal, M.I. 1998. On the use of demographic models of population viability in endangered species management. *Journal of Wildlife Management*, **62**, 821–841.

Bell, D.J., and Oliver, W.L. 1992. Northern Indian tall grasslands: management and species conservation with special reference to fire. Pp. 109–123 in *Tropical Ecosystems: Ecology and Management*. K.P. Singh and J.S. Singh, eds. Wiley, New Dehli.

Bell, D.J., Oliver, W.L. and Ghose, R.K. 1990. Rabbits, hares and pikas, status survey and conservation action plan. Pp. 109–123 in Lagomorph Specialist Group, International Union for the Conservation of Nature, Switzerland, Gland.

Bellinger, R.G., Ravlin, F.W. and McManus, M.L. 1989. Forest edge effects and their influence on the gypsy moth (Lepidoptera: Lymantriidae) egg mass distribution. *Environmental Entomology*, **18**, 840–843.

Belovsky, G.E., Botkin, D.B., Crowl, T., Cummins, K., Franklin, J., Hunter, M., Joern, J., Lindenmayer, D.B., MacMahon, J, Margules, C. and Scott, M. 2004. Ten suggestions to strengthen the science of ecology. *BioScience*, **54**, 345–351.

Belsky, A.J. 1994. Influences of trees on savanna productivity: tests of shade, nutrients, and tree–grass competition. *Ecology*, **75**, 922–932.

Belthoff, J.R. and Ritchison, G. 1989. Natal dispersal of eastern screech-owls. *Condor*, **91**, 254–265.

Belyea, L.R. and Lancaster, J. 1999. Assembly rules within a contingent ecology. *Oikos*, **86**, 402–416.

Bender, D.J. and Fahrig, L. 2005. Matrix structure obscures the relationship between interpatch movement and patch size and isolation. *Ecology*, **86**, 1023–1033.

Bender, D.J., Contreras, T.A. and Fahrig, L. 1998. Habitat loss and population decline: a meta-analysis of the patch size effect. *Ecology*, **79**, 517–529.

Bender, D.J., Tischendorf, L. and Fahrig, L. 2003. Evaluation of patch isolation metrics for predicting animal movement in binary landscapes. *Landscape Ecology*, **18**, 17–39.

Bengtsson, J., Angelstam, P., Elmqvist, T., Emanuelsson, U., Folke, C., Ihse, M., Moberg, F. and Nyström, M. 2003. Reserves, resilience and dynamic landscapes. *Ambio*, **32**, 389–396.

Bengtsson, J., Ahnstrom, J. and Weibull, A.C. 2005. The effects of organic agriculture on biodiversity and abundance: a meta-analysis. *Journal of Applied Ecology*, **42**, 261–269.

Benkman, C.W. 1993. Logging, conifers, and the conservation of crossbills. *Conservation Biology*, 5, 115–119.

Bennett, A.F. 1987. Conservation of mammals within a fragmented forest environment: the contributions of insular biogeography and autecology. Pp. 41–52 in *Nature Conservation: The Role of Remnants of Native Vegetation*. D.A. Saunders, G.W. Arnold, A.A. Burbidge and A.J. Hopkins, eds. Surrey Beatty, Sydney.

Bennett, A.F. 1990a. Land use, forest fragmentation and the mammalian fauna at Naringal, south-western Victoria. *Australian Wildlife Research*, **17**, 325–347.

Bennett, A.F. 1990b. *Habitat Corridors: Their Role in Wildlife Management and Conservation*. Department of Conservation and Environment, Melbourne.

Bennett, A.F. 1991. Roads, roadsides and wildlife conservation: a review. Pp. 99–117 in *Nature Conservation 2: The Role of Corridors*. D.A. Saunders and R.J. Hobbs, eds. Surrey Beatty, Sydney.

Bennett, A.F. 1998. *Linkages in the Landscape: The Role of Corridors and Connectivity in Wildlife Conservation*. IUCN, Gland, Switzerland.

Bennett, A.F. 2003. Habitat fragmentation. Pp. 440–445 in *Ecology: An Australian Perspective*. P. Attiwill and B. Wilson, eds. Oxford University Press, Melbourne.

Bennett, A.F., and Baxter, B. 1989. Diet of the long-nosed potoroo, *Potorous tridactylus* (Marsupialia: Potoroidae), in southwestern Victoria. *Wildlife Research*, **16**, 267–285.

Bennett, A.F. and Ford,. L.A. 1997. Land use, habitat change and the conservation of birds in fragmented rural environments: a landscape perspective from the Northern Plains, Victoria, Australia. *Pacific Conservation Biology*, **3**, 244–261.

Bennett, A.F., Lumsden, L.F., Alexander, J.S.A., Duncan, P.E., Johnson, P.G., Robertson, P. and Silveira, C.E. 1991 Habitat use by arboreal marsupials along an environmental gradient in northeastern Victoria. *Wildlife Research*, **18**, 125–146.

Bennett, A.F., Henein, K. and Merriam, G. 1994. Determinants of corridor quality: chipmunks and fencerows in a farmland mosaic. *Biological Conservation*, **68**, 155–165.

Bennett, A.F., Kimber, S. and Ryan, P. 2000. *Revegetation and Wildlife: A Guide to Enhancing Revegetated Habitats for Wildlife Conservation in Rural Environments*. Bushcare Research Report 2/00. Bushcare National Research and Development Program Research Report. Environment Australia, Canberra.

Bennett, E.L. 2000. Timber certification: where is the voice of the biologist? *Conservation Biology*, **14**, 921–923.

Bennett, S., 1997. Colonising cane toads *(Bufo marinus)* cause population declines in native predators: reliable anecdotal information and management implications. *Pacific Conservation Biology*, **3**, 65–72.

Benson, J. 1999. *Setting the Scene: The Native Vegetation of New South Wales*. Native Vegetation Advisory Council, Royal Botanic Gardens, Sydney.

Benton, T.G., Vickery, J.A. and Wilson, J.D. 2003. Farmland biodiversity: is habitat heterogeneity the key? *Trends in Ecology and Evolution*, **18**, 182–188.

Berendonk, T.U. and Bonsall, M.B. 2002. The phantom midge and a comparison of metapopulation structures. *Ecology*, **83**, 116–128.

Berg, A., Ehnstrom, B., Gustaffson, L., Hallingback, T., Jonsell, M. and Weslien, J. 1994. Threatened plant, animal and fungus species in Swedish forests: distribution and habitat associations. *Conservation Biology*, **8**, 718–731.

Berg, A., Nilsson, S.G. and Bostrom, U. 1992. Predation on artificial wader nests on large and small bogs along a south–north gradient. *Ornis Scandinavia*, **23**, 13–16.

Berger, J. 1990. Persistence of different-sized populations: an empirical assessment of rapid extinctions in bighorn sheep. *Conservation Biology*, **4**, 91–98.

Berger, J., Stacey, P.B., Bellis, L. and Johnson, M.P. 2001. A mammalian predator–prey imbalance: grizzly bear and wolf extinction effect avian neotropical migrants. *Ecological Applications*, **11**, 947–960.

Berggren, A., Birath, B. and Kindvall, O. 2002. Effects of corridors and habitat edges on dispersal behaviour, movement rates and movement angles in Roesel's bush-cricket *(Metriopetra roeseli)*. *Conservation Biology*, **16**, 1562–1569.

Berglund, H. and Jonsson, B.G. 2003. Nested plant and fungal communities: the importance of area and habitat quality in maximizing species capture in boreal old-growth forests. *Biological Conservation*, **112**, 319–328.

Berglund, H. and Jonsson, B.G. 2005. Verifying an extinction debt among lichens and fungi in northern Swedish boreal forests. *Conservation Biology*, **19**, 338–348.

Berkes, F. 2004. Rethinking community-based conservation. *Conservation Biology*, **18**, 621–630.

Berry, L. 2001. Edge effects on the distribution and abundance of birds in a southern Victorian forest. *Wildlife Research*, **28**, 239–245.

Berry, O., Tocher, M.D., Gleeson, D.M. and Sarre, S.D. 2005. Effect of vegetation matrix on animal dispersal: genetic evidence from a study of endangered skinks. *Conservation Biology*, **19**, 855–864.

Best, L.B., Bergin, T.M. and Freemark, K.E. 2001. Influence of landscape composition on bird use of rowcrop fields. *Journal of Wildlife Management*, **65**, 442–449.

Bezkorowajnyj, P.G., Gordon, A.M. and McBride, R.A. 1993. The effect of cattle foot traffic on soil compaction in a silvo-pastoral system. *Agroforestry Systems*, **21**, 1–10.

Bierregaard, R.O. and Stouffer, P.C. 1997. Understorey birds and dynamic habitat mosaics in Amazonian rainforests. Pp. 138–153 in *Tropical Forest Remnants: Ecology, Management and Conservation*.

W.F. Laurance and R.O. Bierregaard, eds. University of Chicago Press, Chicago.

Bierregaard, R.O., Lovejoy, T.E., Kapos, V., Santos, A. and Hutchings, R.W. 1992. The biological dynamics of tropical rainforest fragments. *BioScience,* **42**, 859–866.

Bierregaard, R.O. Gascon, C., Lovejoy, T.E., and Mesquita, R., eds. 2001. *Lessons from Amazonia: The Ecology and Conservation of a Fragmented Forest.* Yale University Press, New Haven, Connecticut.

Billington, H.L. 1991. Effect of population size on genetic variation in a dioecious conifer. *Conservation Biology,* **5**, 115–119.

BirdLife International. 2000. *Threatened Birds of The World.* BirdLife International, Lynx Editions, London.

Bissonette, J.A. and Storch, I. 2002. Fragmentation: is the message clear? *Conservation Ecology,* **6**, 14. http://www.consecol.org/vol6/iss2/art14.

Blackburn, T.M. and Gaston, K.J. 2003. Introduction: why macroecology? Pp. 1–4 in *Macroecology: Concepts and Consequences.* T.M. Blackburn and K.J. Gaston, eds. Blackwell Publishing, Oxford.

Blake, J.G. 1983. The trophic structure of bird communities in forest patches in east-central Illinois. *Wilson Bulletin,* **95**, 416–430.

Blakers, M., Davies, S.J. and Reilly, P.N. 1984. *The Atlas of Australian Birds.* Royal Australasian Ornithithological Union, Melbourne University Press, Melbourne.

Blazquez, M.C. 1996. Activity and habitat use in a population of *Ameiva ameiva* in Southeastern Colombia. *Biotropica,* **28**, 714–719.

Block, W.M. and Brennan, L.A. 1993. The habitat concept in ornithology: theory and applications. Pp. 35–91 in *Current Ornithology,* Vol. 11. D.M. Power, ed. Plenium Press, New York.

Blouin, M.S. and Connor, E.F. 1985. Is there a best shape for nature reserves? *Biological Conservation,* **32**, 277–288.

Bock, C.E., Smith, H.M. and Bock, J.H. 1990. The effect of livestock grazing upon abundance of the lizard, *Sceloporus scalaris,* in southeastern Arizona. *Journal of Herpetology,* **24**, 445–446.

Bodin, Ö., Tengö, M., Norman, A., Lundberg, J. and Elmqvist, T. 2006. The value of small size: loss of forest patches and threshold effects on ecosystem services in southern Madagascar. *Ecological Applications,* **16**, 440–451.

Boecklen, W.J. 1997. Nestedness, biogeographic theory, and the design of nature reserves. *Oecologia,* **112**, 123–142.

Bondrup-Nielsen, S. 1983. Density estimation as a function of livetrapping grid and home range size. *Canadian Journal of Zoology,* **61**, 2361–2365.

Boone, R.B. and Hunter, M.L. 1996. Using diffusion models to simulate the effects of land use on grizzly bear dispersal in the Rocky Mountains. *Landscape Ecology,* **11**, 51–64.

Borghesio, L. and F. Giannetti. 2005. Habitat degradation threatens the survival of the Ethiopian bush crow *Zavattariornis stresemanni. Oryx,* **39**, 44–49.

Boulinier, T., Nichols, J.D., Hines, J.E., Sauer, J.R., Flather, C.H. and Pollock, K.H. 2001. Forest fragmentation and bird community dynamics: inference at regional scales. *Ecology,* **82**, 1159–1169.

Bowne, D.R., Peles, J.D. and Barrett, G.W. 1999. Effects of landscape spatial structure on movement patterns of the hispid cotton rat *(Sigmodon hispidus). Landscape Ecology,* **14**, 53–65.

Boyce, M.S. 1992. Population viability analysis. *Annual Review of Ecology and Systematics,* **23**, 481–506.

Boyce, S.G. 1995. *Landscape Forestry.* Wiley, New York.

Bradford, D.F., Neale, A.C., Nash, M.S., Sada, D.W. and Jaeger, J.R. 2003. Habitat patch occupancy by toads *(Bufo punctatus)* in a naturally fragmented desert landscape. *Ecology,* **84**, 1012–1023.

Bradshaw, F.J. 1992. Quantifying edge effect and patch size for multiple-use silviculture: a discussion paper. *Forest Ecology and Management,* **48**, 249–264.

Bradstock, R.A., Williams, J.E. and Gill, A.M., eds. 2002. *Flammable Australia. The Fire Regimes and Biodiversity of a Continent.* Cambridge University Press, Melbourne.

Braithwaite, L.W. 1984. The identification of conservation areas for possums and gliders in the Eden woodpulp concession district. Pp. 501–508 in *Possums and Gliders.* A.P. Smith and I.D. Hume, eds. Surrey Beatty and Sons, Sydney, Australia.

Braithwaite, L.W., Belbin, L., Ive, J. and Austin, M.P. 1993. Land use allocation and biological conservation in the Batemans Bay forests of New South Wales. *Australian Forestry,* **56**, 4–21.

Braithwaite, W. 2004. Do current forestry practices threaten forest fauna? A perspective. Pp. 513–536 in *Conservation of Australia's Forest Fauna*. 2nd ed. D. Lunney, ed. Royal Zoological Society of New South Wales, Sydney.

Braschler, B. and Baur, B. 2005. Experimental small-scale grassland fragmentation alters competitive interactions among ant species. *Oecologia*, **143**, 291–300.

Brashares, J.S. 2003. Ecological, behavioural, and life-history correlates of mammal extinctions in West Africa. *Conservation Biology*, **17**, 733–743.

Breininger, D.R. and Carter, G.M. 2003. Territory quality transitions and source–sink dynamics in a Florida scrub-jay population. *Ecological Applications*, **13**, 516–529.

Breininger, D.R., Toland, B., Oddy, D., and Legare, M.L. 2006 Landcover characterizations and Florida scrub-jay *(Aphelocoma coerulescens)* population dynamics. *Biological Conservation*, **128**, 169–181.

Brereton, R., Bennett, S. and Mansergh, I. 1995. Enhanced greenhouse climate change and its potential effect on selected fauna of southeastern Australia: a trend analysis. *Biological Conservation*, **72**, 339–354.

Brereton, R.N., Bryant, S.C. and Rowell, M. 1997. *Habitat Modelling of the Forty-Spotted Pardalote and Recommendations for Management*. Report to the Tasmanian Regional Forest Agreement. Environment and Heritage Technical Committee, Hobart, Tasmania.

Briese, D.T. 2000. Classical biological control. Pp. 139–160 in *Australian Weed Management Systems*. B. Sindel, ed. R.G. and F.J. Richardson, Melbourne.

Briggs, S. and Taws, N. 2003. Impacts of salinity on biodiversity: clear understanding or muddy confusion? *Australian Journal of Botany*, **51**, 609–617.

Bright, P.W. 1993. Habitat fragmentation: problems and predictions for British mammals. *Mammal Review*, **23**, 101–111.

Bright, P.W. 1998. Behaviour of specialist species in habitat corridors: arboreal doormice avoid corridor gaps. *Animal Behaviour*, **56**, 1485–1490.

Brittingham, M.C. and Temple, S.A. 1983. Have cowbirds caused songbirds to decline? *BioScience*, **33**, 31–35.

Brokaw, N.V. and Lent, R.A. 1999. Vertical structure. Pp. 373–399 in *Managing Biodiversity in Forest Ecosystems*. M. Hunter III, ed. Cambridge University Press, Cambridge.

Bromham, L., Cardillo, M., Bennett, A.F. and Elgar, M.A. 1999. Effects of stock grazing on the ground invertebrate fauna of woodland remnants. *Austral Ecology*, **24**, 199–207.

Brook, B.W. 2000. Pessimistic and optimistic bias in population viability analysis. *Conservation Biology*, **14**, 564–566.

Brook, B.W. and Bowman, D.M. 2004. The uncertain blitzkrieg of Pleistocene megafauna. *Journal of Biogeography*, **31**, 517–523.

Brook, B.W., O'Grady, J.J., Burgman, M.A., Akçakaya, H.R. and Frankham, R. 2000. Predictive accuracy of population viability analysis in conservation biology. *Nature*, **404**, 385–387.

Brook, B.W., Burgman, M.A., Akçakaya, H.R., O'Grady, J.J. and Frankham, R. 2002. Critiques of PVA ask the wrong questions: throwing out the heuristic baby with the numerical bath water. *Conservation Biology*, **16**, 262–263.

Brook, B.W., Sodhi, N.S. and Ng, P.K. 2003. Catastrophic extinctions follow deforestation in Singapore. *Nature*, **424**, 420–423.

Brooker, L. 2002. The application of focal species knowledge to landscape design in agricultural lands. *Landscape and Urban Planning*, **60**, 185–210.

Brooker, L. and Brooker, M. 2002. Dispersal and population dynamics of the blue-breasted fairy-wren, *Malurus pulcherrimus*, in fragmented habitat in the Western Australian wheatbelt. *Wildlife Research*, **29**, 225–233.

Brooker, M. and Brooker, L. 2001. Breeding biology, reproductive success and survival of blue-breasted fairy-wrens in fragmented habitat in the Western Australian wheatbelt. *Wildlife Research*, **28**, 205–214.

Brooks, C.N. and Merenlender, A.M. 2001. Determining the pattern of oak woodland regeneration for a cleared watershed in northwest California: a necessary first step for restoration. *Restoration Ecology*, **9**, 1–12.

Brooks, T. and Balmford, A. 1996. Atlantic forest extinctions. *Nature*, **380**, 115.

Brooks, T.M., Pimm, S.L. and Oyugi, J.O. 1999. Time lag between deforestation and bird extinction in tropical forest fragments. *Conservation Biology,* **13**, 1140–1150.

Brooks, T.M., Mittermeier, R.A., Mittermeier, C.G., da Fonseca, G.A.B., Rylands, A.B., Konstant, W.R., Flick, P., Pilgrim, J., Oldfield, S., Magin, G. and Hilton-Taylor, C. 2002. Habitat loss and extinction in the hotspots of biodiversity. *Conservation Biology,* **16**, 909–923.

Brothers, T.S. and Spingarn, A. 1992. Forest fragmentation and alien plant invasion of central Indiana old-growth forests. *Conservation Biology,* **6**, 91–100.

Brotons, L., Desrochers, A. and Turcotte, Y. 2001. Food hoarding behaviour of black-capped chickadees *(Poecile atripillus)* in relation to forest edges. *Oikos,* **95**, 511–519.

Brown, K. and Hutchings, R.W. 1997. Disturbance, fragmentation and the dynamics of diversity in Amazonian forest butterflies. Pp. 138–153 in *Tropical Forest Remnants. Ecology, Management and Conservation.* W.F. Laurance and R.O. Bierregaard, eds. University of Chicago Press, Chicago.

Brown, G.P. and Weatherhead, P.J. 2000. Thermal ecology and sexual size dimorphism in northern water snakes, *Nerodia sipedon. Ecological Monographs,* **70**, 311–330.

Brown, G.W. 2001. The influence of habitat disturbance on reptiles in a box–ironbark eucalypt forest of southeastern Australia. *Biodiversity and Conservation,* **10**, 161–176.

Brown, G.W., Nelson, J.L. and Cherry, K.A.. 1997. The influence of habitat structure on insectivorous bat activity in montane ash forests of the Central Highlands of Victoria. *Australian Forestry,* 60, 138–146.

Brown, J.H. and Kodric-Brown, A. 1977. Turnover rates in insular biogeography: effect of immigration on extinction. Ecology, **58**, 445–449.

Brown, K.A. and Gurevitch, J. 2004. Long-term impacts of logging on forest diversity in Madagascar. *Proceedings of the National Academy of Sciences,* **101**, 6045–6049.

Brown, M.J. 1985. Benign neglect and active management in Tasmania's forests: a dynamic balance or ecological collapse? *Forest Ecology and Management,* **85**, 279–289.

Brualdi, R.A. and J.G. Sanderson. 1999. Nested species subsets, gaps, and discrepancy. *Oecologia,* **119**, 256–264.

Bruna, E.M., Vasconcelos, H.L. and Heredia, S. 2005. The effect of habitat fragmentation on communities of mutualists: Amazonian ants and their host plants. *Biological Conservation,* **124**, 209–216.

Brunckhorst, D. 1999. *Models to Integrate Sustainable Conservation and Resource Use: Bioregional Reserves beyond Bookmark.* Nature Conservation Council Annual Conference on Integrated Natural Resource Management, University of Sydney, March 1999.

Bulinski, J. 1999. A survey of mammalian browsing damage in Tasmanian eucalypt plantations. *Australian Forestry,* **62**, 59–65.

Bullock, J.M., Kenward, R.E., Hails, R.S. and Webb, N.R. 2002. *Dispersal Ecology.* Cambridge University Press, Cambridge.

Bunnell, F. 1995. Forest-dwelling fauna and natural fire regimes in British Columbia: patterns and implications for conservation. *Conservation Biology,* **9**, 636–644.

Bunnell, F. 1998. Evading paralysis by complexity when establishing operational goals for biodiversity. *Journal of Sustainable Forestry,* **7**, 145–164.

Bunnell, F. 1999a. Foreword. Let's kill a panchreston: giving fragmentation a meaning. Pp. vii–xiii in *Forest Wildlife and Fragmentation: Management Implications.* J. Rochelle, L.A. Lehmann and J. Wisniewski, eds. Brill, Leiden, Germany.

Bunnell, F. 1999b. What habitat is an island? Pp. 1–31 in *Forest Wildlife and Fragmentation. Management Implications.* J. Rochelle, L.A. Lehmann and J. Wisniewski, eds. Brill, Leiden, Germany.

Bunnell, F., Dunsworth, G., Huggard, D. and Kremsater, L. 2003. *Learning to Sustain Biological Diversity on Weyerhauser's Coastal Tenure.* Weyerhauser Company, Vancouver, British Columbia.

Burbidge, A.A. and Kuchling, G. 1994. *Western Swamp Tortoise Recovery Plan.* Wildlife Management Program No. 11. Department of Conservation and Land Management, Perth.

Burbidge, A.A. and McKenzie, N.L. 1989. Patterns in modern decline of Western Australia's vertebrate fauna: causes and conservation implications. *Biological Conservation,* **50**, 143–198.

Burbrink, F.T., Phillips, C.A. and Heske, E.J. 1998. A riparian zone in southern Illinois as a potential dispersal corridor for reptiles and amphibians. *Biological Conservation,* **86**, 107–115.

Burdon J.J. and Chilvers, G.A. 1994. Demographic changes and the development of competition in a native eucalypt forest invaded by exotic pines. *Oecologia,* **97**, 419–423.

Burgess, R.L., and Sharpe, D.M. 1981. Introduction. Pp. 1–5 in *Forest Island Dynamics in Man-Dominated Landscapes.* R.L. Burgess and D.M. Sharpe, eds. Springer-Verlag, New York.

Burgman, M., Ferson, S. and Akçakaya, H.R. 1993. *Risk Assessment in Conservation Biology.* Chapman and Hall, New York.

Burgman, M.A. 2000. Population viability analysis for bird conservation: prediction, heuristics, monitoring and psychology. *Emu,* **100**, 347–353.

Burgman, M.A. and Ferguson, I.S. 1995. *Rainforest in Victoria: A Review of the Scientific Basis of Current and Proposed Protection Measures.* Report to Victorian Department of Conservation and Natural Resources, Forest Services Technical Reports 95-4. Melbourne, Victoria, Australia.

Burgman, M.A. and Lindenmayer, D.B. 1998. *Conservation Biology for the Australian Environment.* Surrey Beatty and Sons, Chipping Norton, New South Wales.

Burgman, M.A., Akçakaya, H.R. and Loew, S.S. 1988. The use of extinction models for species conservation. *Biological Conservation,* **48**, 9–25.

Burgman, M.A., Breininger, D.R., Duncan, B.W. and Ferson, S. 2001. Setting reliability bounds on habitat suitability indices. *Ecological Applications,* **11**, 70–78.

Burgman, M.A., Lindenmayer, D.B. and Elith, J. 2005. Managing landscapes for conservation under uncertainty. *Ecology,* **86**, 2007–2017.

Burke, D.M. and Nol, E. 1998. Influence of food abundance, nest-site habitat, and forest fragmentation on breeding Ovenbirds. *Auk,* **115**, 96–104.

Burkey, T.V. 1989. Extinction in nature reserves: the effect of fragmentation and the importance of migration between reserve fragments. *Oikos,* **55**, 75–81.

Burnett, S. 1992. Effects of a rainforest road on movements of small mammals: mechanisms and implications. *Wildlife Research,* **19**, 95–104.

Burnham, K.P. and Anderson, D.R. (1998). *Model Selection and Inference: A Practical Information-Theoretic Approach.* Springer-Verlag, New York.

Burris, R.K. and Canter, L.W. 1997. Cumulative impacts are not properly addressed in environmental assessments. *Environmental Impact Assessment Review,* **67**, 5–18.

Burrows, G.E. 1999. A survey of 25 remnant vegetation sites in the southwest slopes, New South Wales. *Cunninghamia,* **6**, 283–299.

Burton, P.J., Messier, C., Smith, D.W., and Adamowicz, W.L. eds. 2003. *Towards Sustainable Management of the Boreal Forest.* NRC Research Press, Ottawa, Canada.

Busack, S.D. and Bury, R.B. 1974. Some effects of off-road vehicles and sheep grazing on lizard populations in the Mojave desert. *Biological Conservation,* **6**, 179–183.

Cadenasso, M.L. and Pickett, S.T. 2001. Effect of edge structure on the flux of species into forest interiors. *Conservation Biology,* **15**, 91–97.

Caldwallader, P.L. 1986. Flow regulation in the Murray River system and its effect on the native fish fauna. Pp. 115–133 in *Stream Protection: The Management of Rivers for Instream Uses.* I.C. Campbell, ed. Water Studies Institute, Chisholm Institute of Technology, Melbourne.

Cale, P. 1999. *The Spatial Dynamics of the White-Browed Babbler in a Fragmented Agricultural Landscape.* PhD thesis, University of New England, Armidale, NSW.

Cale, P. 2003. The influence of social behaviour, dispersal and landscape fragmentation on population structure in a sedentary bird. *Biological Conservation,* **109**, 237–248.

Cale, P.G. and Hobbs, R.J. 1994. Landscape heterogeneity indices: problems of scale and applicability, with particular reference to animal habitat description. *Pacific Conservation Biology,* **1**, 183–193.

Campbell, L.M. 2005. Overcoming obstacles to interdisciplinary research. *Conservation Biology,* **19**, 574–577.

Cardillo, M. and Bromham, L. 2001. Body size and risk of extinction in Australian mammals. *Conservation Biology,* **15**, 1435–1440.

Carey, A.B. and J.D. Gill. 1983. Direct habitat improvement: some recent advances. Pp. 80–87 in *Snag Habitat Management: Proceedings of the Symposium.* U.S. Forest Service General Technical Report RM-99.

Carey, A.B., Kershner, J., Biswell, B. and Dominguez de Toledo, L. 1999a. Ecological scale and forest development: squirrels, dietary fungi, and vascular plants in managed and unmanaged forests. *Wildlife Monographs,* **142, 1–71.**

Carey, A.B., Lippke, B.R. and Sessions, J. 1999b. Intentional systems management: managing forests for biodiversity. *Journal of Sustainable Forestry*, **9**, 83–125.

Carlson, A. and Edenhamn, P. 2000. Extinction dynamics and the regional persistence of a tree frog metapopulation. *Proceedings of the Royal Society of London Series B*, **267**, 1311–1313.

Carlton, J.T., Vermeij, G.J., Lindberg, D.R., Carlton, D.A. and Dudley, E.C. 1991. The first historical extinction of a marine invertebrate in an ocean basin: the demise of the eelgrass limpet *Lottia avleus*. *Biological Bulletin*, **180**, 72–80.

Caro, T.M. 2001. Species richness and abundance of small mammals inside and outside an African national park. *Biological Conservation*, **98**, 251–257.

Caro, T. 2004. Preliminary assessment of the flagship species concept at a small scale. *Animal Conservation*, **7**, 63–70.

Carpenter, S., Chisholm, S.W., Krebs, C.J., Schindler, C.J. and Wright, R.F. 1995. Ecosystems experiments. *Science*, **269**, 324–327.

Carr, C.W., Yugovic, J.V. and Robinson, K.E. 1992. *Environmental Weed Invasions in Victoria: Conservation and Management Implications*. Department of Conservation and Environment and Ecological Horticulture, Melbourne.

Carroll, C., Noss, R.E., Paquet, P.C. and Schumaker, N.H. 2003. Extinction debt of protected areas in developing landscapes. *Conservation Biology*, **18**, 1110–1120.

Carroll, C., Noss, R.E., Paquet, P.C. and Schumaker, N.H. 2004. Use of population viability analysis selection algorithms in regional conservation plans. *Ecological Applications*, **13**, 1773–1789.

Carter, J., Ackleh, A.S., Leonard, B.P. and Wang, H.B. 1999. Giant panda (*Ailuropoda melanoleuca*) population dynamics and bamboo (subfamily Bambusoideae) life history: a structured population approach to examining carrying capacity when the prey are semelparous. *Ecological Modelling*, **123**, 207–223.

Cary, G., Lindenmayer, D.B. and Dovers, S., eds. 2003. *Australia Burning. Fire Ecology, Policy and Management Issues*. CSIRO Publishing, Melbourne.

Cascante, A., Quesada, M., Lobo, J.J. and Fuchs, E.A. 2002. Effects of dry tropical forest fragmentation on the reproductive success and genetic structure of the tree *Samanea saman*. *Conservation Biology*, **16**, 137–147.

Catchpole, C.K. and Slater, P.J. 1995. *Bird Song. Biological Themes and Variations*. Cambridge University Press, Cambridge.

Catling, P.C., Coops, N.C. and Burt, R.J. 2001. The distribution and abundance of ground-dwelling mammals in relation to time since wildlife and vegetation structure in southeastern Australia. *Wildlife Research*, **28**, 555–564.

Catterall, C.P., Green, R.J. and Jones, D.N. 1991. Habitat use by birds across a forest–suburb interface in Brisbane: implications for corridors. Pp. 247–258 in *Nature Conservation 2: The Role of Corridors*. D.A. Saunders and R.J. Hobbs, eds. Surrey Beatty and Sons, Chipping Norton, NSW.

Caughley, G. 1994. Directions in conservation biology. *Journal of Animal Ecology*, **63**, 215–244.

Caughley, G.C. and Gunn, A. 1996. *Conservation Biology in Theory and Practice*. Blackwell Science, Cambridge, Massachusetts.

Chalfoun, A.D., Thompson, F.R. and Ratnaswamy, M.J. 2002. Nest predators and fragmentation: a review and meta-analysis. *Conservation Biology*, **16**, 306–318.

Chen, J. 1991. *Edge Effects: Microclimatic Pattern and Biological Responses in Old-Growth Douglas Fir Forests*. PhD thesis, University of Washington, Seattle.

Chen, J., Franklin, J.F. and Spies, T.A. 1990. Microclimatic pattern and basic biological responses at the clearcut edges of old-growth Douglas fir stands. *Northwest Environmental Journal*, **6**, 424–425.

Chen, J., Franklin, J.F. and Spies, T.A. 1992. Vegetation responses to edge environments in old-growth Douglas-fir forests. *Ecological Applications*, **2**, 387–396.

Chen, J., Franklin, J.F. and Spies, T.A. 1995. Growing season microclimate gradients from clearcut edges into old-growth Douglas fir forests. *Ecological Applications*, **5**, 74–86.

Chepko-Sade, B.D. and Halpin, Z.T. 1987. *Mammalian Dispersal Patterns*. University of Chicago, Chicago.

Chesson, P. 2001. Metapopulations. Pp. 161–176 in *Encyclopaedia of Biodiversity*. S.A. Levin, ed. Academic Press, San Diego.

Chettri, N., Sharma, E. and Deb, D.C. 2001. Bird community structure along a trekking corridor of Sikkim Himalaya: a conservation perspective. *Biological Conservation*, **102**, 1–16.

Chettri, N., Sharma, E., Deb, D.C. and Sundriyal, R.C. 2002. Impact of firewood extraction on tree structure, regeneration and woody biomass productivity in a trekking corridor of the Sikkim Himalaya. *Mountain Research and Development*, **22**, 150–158.

Cincotta, R.P., Wisenwski, J. and Engelman, R. 2000. Human population in the biodiversity hotspots. *Nature*, **404**, 990–992.

Claridge, A.W. 1993. *Hypogeal Fungi as a Food Resource for Wildlife in the Managed Forests of South-Eastern Australia*. PhD thesis, Australian National University, Canberra, Australia.

Claridge, A.W. and Lindenmayer, D.B. 1994. The need for a more sophisticated approach toward wildlife corridor design in the multiple-use forests of southeastern Australia: the case for mammals. *Pacific Conservation Biology*, **1**, 301–307.

Clark, R.G. and Shutler, D. 1999. Avian habitat selection: pattern from process in nest-site use by ducks? *Ecology*, **80**, 272–287.

Clark, T.W. 2002. *The Policy Process. A Practical Guide for Natural Resource Professionals*. Yale University Press, New Haven, Connecticut.

Clark, T.W., Warneke, R.M. and George, G.G. 1990. Management and conservation of small populations. Pp. 1–18 in *Management and Conservation of Small Populations*. T.W. Clark and J.H. Seebeck, eds. Conference Proceedings, Melbourne, September 26–27, 1989. Chicago Biological Society, Chicago.

Clarke, G.M. and Young, A.G., eds. 2000. *Genetics, Demography and Viability of Fragmented Populations*. Cambridge University Press, Cambridge.

Clavero, M. and Garcia-Berthou, E. 2005. Invasive species are a leading cause of animal extinctions. *Trends in Ecology and Evolution*, **20**, 110–110.

Clevenger, A.P. and Waltho, N. 2000. Factors influencing the effectiveness of wildlife underpasses in Banff National Park, Alberta, Canada. *Conservation Biology*, **14**, 47–56.

Cockburn, A. 1991. *An Introduction to Evolutionary Ecology*. Blackwell Scientific Publications, Melbourne.

Cockburn, A., Scott, M.P. and Scotts, D.J. 1985. Inbreeding avoidance and male-biased natal dispersal in *Antechinus* spp. (Marsupialia, Dasyuridae). *Animal Behaviour*, **33**, 908–915.

Cockle, K.L., and Richardson, J.S. 2003. Do riparian buffer strips mitigate the impacts of clearcutting on small mammals? *Biological Conservation*, **113**, 133–140.

Cocklin C., Parker S. and Hay J. 1992a. Notes on cumulative environmental-change. I. Concepts and issues. *Journal of Environmental Management*, **35**, 31–49.

Cocklin C., Parker S. and Hay J. 1992b. Notes on cumulative environmental-change. II. A contribution to methodology. *Journal of Environmental Management*, **35**, 51–67.

Coffman, C.J., Nichols, J.D. and Pollock, K.H. 2001. Population dynamics of *Microtus pennsylvanicus* in corridor-linked patches. *Oikos*, **93**, 3–21.

Cogger, H. 1995. *Reptiles and Amphibians of Australia*. Reed Books Australia, Port Melbourne, Victoria.

Cole, E.K., Pope, M.D. and Anthony, R.G. 1997. Effects of road management on movement and survival of Roosevelt elk. *Journal Wildlife Management*, **61**, 1115–1126.

Collett, D.A. 1991. *Modelling Binary Data*. Chapman and Hall, London, England.

Collinge, S.K. and Forman, R.T.T. 1998. A conceptual model of land conversion processes: predictions and evidence from a microlandscape experiment with grassland insects. *Oikos*, **82**, 66–84.

Collingham Y.C. and Huntley B. 2000. Impacts of habitat fragmentation and patch size upon migration rates. *Ecological Applications*, **10**, 131–144.

Colwell, R.K. 1973. Competition and coexistence in a simple tropical community. *American Naturalist*, **107**, 737–760.

Commonwealth of Australia 2001. *National Land and Water Audit 2001: Australian Dryland Salinity Assessment 2000*. Commonwealth of Australia, Canberra.

Commonwealth of Australia 2002. *Australia's Natural Resources 1997–2002 and Beyond*. National Land and Water Resources Audit, Commonwealth of Australia, Canberra.

Connell, J.H. 1978. Diversity in tropical rainforests and coral reefs. *Science*, **199**, 1302–1310.

Conner, R.N. 1988. Wildlife populations: minimally viable or ecologically functional? *Wildlife Society Bulletin*, **16**, 80–84.

Connor, E.F. and McCoy, E.D. 1979. The statistics and biology of the species–area relationship. *American Naturalist,* **113**, 791–833.

Connor, E.F. and Simberloff, D. 1979. The assembly of species communities: chance or competition. *Ecology,* **60**, 1132–1140.

Connor, R.N. and Rudolph, D.C. 1989. *Red-Cockaded Woodpecker Colony Status and Trends on the Angelina, Davy Crockett and Sabine National Forests.* Research Paper SO-250, U.S.D.A. Forest Service Southern Forest Experimental Station, New Orleans, Louisana.

Cook, W.M., Lane, K.T., Foster, B.L. and Holt, R.D. 2002. Island theory, matrix effects and species richness. *Ecology Letters,* **5**, 619–623.

Cook, W.M., Anderson, R.M. and Schweiger, E.W. 2004. Is the matrix really inhospitable? Vole runway distribution in an experimentally fragmented landscape. *Oikos,* **104**, 5–14.

Cooper, C.B. and Walters, J.R. 2002. Experimental evidence of disrupted dispersal causing decline of an Australian passerine in fragmented habitat. *Conservation Biology,* **16**, 471–478.

Cooper, C.B., Walters, J.R. and Ford, H. 2002. Effects of remnant size and connectivity on the response of brown treecreepers to habitat fragmentation. *Emu,* **102**, 249–256.

Coppolillo, P., Gomez, H., Maisels, F. and Wallace, R. 2004. Selection criteria for suites of landscape species as a basis for site-based conservation. *Biological Conservation,* **115**, 419–430.

Cordeiro, N.J. and Howe, H.F. 2003. Forest fragmentation severs mutualism between seed dispersers and an endemic African tree. *Proceedings of the National Academy of Sciences of the United States of America,* **100**, 14052–14056.

Cornelius, C., Cofre, H. and Marquet, P.A. 2000. Effects of habitat fragmentation on bird species in a relict temperate forest in semiarid Chile. *Conservation Biology,* **14**, 534–543.

Costanza, R., Darge, R., Degroot, R., Farber, S., Grasso, M., Hannon, B., Limburg, K., Naeem, S., O'Neill, R.V., Paruelo, J., Raskin, R.G., Sutton, P. and Vandenbelt, M. 1997. The value of the world's ecosystem services and natural capital. *Nature,* **387**, 253–260.

Coulson, T., Mace, G.M., Hudson, E. and Possingham, H. 2001. The use and abuse of population viability analysis. *Trends in Ecological Evolution,* **16**, 219–221.

Covington, W.W. 2003. The evolutionary and historical context. Pp. 26–47 in *Ecological Restoration of Southwestern Ponderosa Pine Forests.* P. Friederici, ed. Island Press, Washington, D.C.

Covington, W.W., Fulé, P.Z., Moore, M.M., Hart, S.C., Kolb, T.E., Mast, J.N., Sackett, S.S. and Wagner, M.R. 1997. Restoring ecological health in ponderosa pine forests of the Southwest. *Journal of Forestry,* **95**, 23–29.

Cox, P.A. and Elmquist, T. 2000. Pollinator extinction in the Pacific islands. *Conservation Biology,* **14**, 1237–1239.

Cox, P.A., Elmquist, T., Pierson, E.D. and Rainey, E. 1991. Flying foxes as strong interactors in south Pacific island ecosystems: a conservation hypothesis. *Biological Conservation,* **115**, 419–430.

Craig, J.L., Saunders, D.A. and Mitchell, N. 2000. *Conservation in Production Environments: Managing the Matrix.* Surrey Beatty and Sons, Chipping Norton, Australia

Crawley, M.J. 1993. *GLIM for Ecologists.* Blackwell Scientific Publications, Oxford.

Crawley, M.J., and Harral, J.E. 2001. Scale dependence in plant biodiversity. *Science,* **291**, 864–868.

Cresswell, W.J. and Smith, G.C. 1992. The effects of temporally autocorrelated data on methods of home range analysis. Pp. 272–284 in *Wildlife Telemetry: Remote Monitoring and Tracking of Animals.* I.G. Priede and S.M. Swift, eds. Ellis Horwood Ltd, New York, USA.

Croizat, L.C. 1960. *Principia Botanica: or, Beginnings of Botany.* Codicator, Weldon and Wesley, Hutchin, England.

Crome, F.H. 1976. Some observations on the biology of the Cassowary in northern Queensland. *Emu,* **76**, 8–14.

Crome, F.H. 1994. Tropical rainforest fragmentation: some conceptual and methodological issues. Pp. 61–76 in *Conservation Biology in Australia and Oceania.* C. Moritz and J. Kikkawa, eds. Surrey Beatty, Chipping Norton.

Crome, F.H. and Bentrupperbaumer, J. 1993. Special people, a special animal and a special vision: the first steps to restoring a fragmented tropical landscape. Pp. 267–279 in *Nature Conservation*

3: Reconstruction of Fragmented Ecosystems. D.A. Saunders, R.J. Hobbs and P.R. Ehrlich, eds. Surrey Beatty, Chipping Norton.

Crome, F.H. and Moore, L.A. 1990. The southern cassowary *(Casuarius casuarius)* in north Queensland. *Australian Wildlife Research*, **17**, 369–385.

Cronk, Q.C.B. and Fuller, J.L. 1995. *Plant Invaders*. Chapman and Hall, London.

Crooks, K.R., and Soulé, M.E. 1999. Mesopredator release and avifaunal extinctions in a fragmented system. *Nature*, **400**, 563–566.

Culver, D.C., Master, L.S., Christman, M.C. and Hobbs, H.H. 2000. Obligate cave fauna of the 48 contiguous United States. *Conservation Biology*, **14**, 386–401.

Cunningham, M. and Moritz, C. 1998. Genetic effects of forest fragmentation on a rainforest restricted lizard (Scincidae, *Gnypetoscincus queenslandiae*). *Biological Conservation*, **83**, 19–30.

Cunningham, R.B. and Lindenmayer, D.B. 2005. Modeling count data of rare species: Some statistical issues. *Ecology*, **86**, 1135–1142.

Cunningham, R.B., Lindenmayer, D.B. and Lindenmayer, B.D. 2004. Sound recording of bird vocalisations in forests. I. Relationships between bird vocalisations and point interval counts of bird numbers: a case study in statistical modelling. *Wildlife Research*, **31**, 579–585.

Cunningham, S.A. 2000. Depressed pollination in habitat fragments causes low fruit set. *Proceedings of the Royal Society of London, Series B*, **267**, 1149–1152.

Curtis, J.T. 1956. The modification of mid-latitude grasslands and forests by man. Pp. 721–736 in *Man's Role in Changing the Face of the Earth*. W.L. Thomas, ed. University of Chicago Press, Chicago.

Cutler, A. 1991. Nested faunas and extinction in fragmented habitats. *Conservation Biology*, **5**, 496–505.

Daily, G.C., ed. 1997. *Nature's Services: Societal Dependence on Natural Ecosystems*. Island Press, Washington, D.C.

Daily, G.C. 1999. Developing a scientific basis for managing Earth's life support systems. *Conservation Ecology*, **3**, 14. http://www.consecol.org/vol13/iss12/art14.

Daily, G.C. 2001. Ecological forecasts. *Nature*, **411**, 245.

Daily, G.C. and Ehrlich, P.R. 1999. Managing Earth's ecosystems: an interdisciplinary challenge. *Ecosystems*, **2**, 277–280.

Daily, G.C. and Walker, B.H. 2000. Seeking the great transition. *Nature*, **403**, 243–245.

Daily, G.C., Ceballos, G., Pacheco, J., Suzan, G. and Sanchez-Azofeifa, A. 2003. Countryside biogeography of neotropical mammals: conservation opportunities in agricultural landscapes of Costa Rica. *Conservation Biology*, **17**, 1814–1826.

Daily, G.C., Ehrlich, P.R. and Haddad, N.M. 1993. Double keystone bird in a keystone species complex. *Proceedings of the National Academy of Sciences of the United States of America*, **90**, 592–594.

Daily, G.C., Ehrlich, P.R. and Sanchez-Azofeifa, G.A. 2001. Countryside biogeography: use of human-dominated habitats by the avifauna of southern Costa Rica. *Ecological Applications*, **11**, 1–13.

Dale, V.H., Brown, S., Haeuber, R.A., Hobbs, N.T., Huntly, N., Naiman, R.J., Riebsame, W.E., Turner, M.G. and Valone, T.J. 2000. Ecological principles and guidelines for managing the use of land. *Ecological Applications*, **10**, 639–670.

Dangerfield, M., Pik, A., Britton, D., Holmes, A., Gillings, M., Oliver, I., Briscoe, D. and Beattie, A. 2003. Patterns of invertebrate biodiversity across a natural edge. *Austral Ecology*, **28**, 227–236.

Darlington, P.J.J. 1957. *Zoogeography*. John Wiley and Sons, New York.

Darveau, M., Beauchesne, P., Belanger, L., Hout, J. and Larue, P. 1995. Riparian forest strips as habitat for breeding birds in boreal forest. *Journal of Wildlife Management*, **59**, 67–78.

Darwin, C. 1859. *On the Origin of Species by Means of Natural Selection*. J. Murray, London.

Daszak, P., Berger, L., Cunningham, A.A., Hyatt, A.D., Green, D.E. and Speare, R. 2000. Emerging infectious diseases and amphibian population declines. *Emerging Infectious Diseases*, **5**, 735–748. http://www.cdc.gov/ncidod/eid/vol5no6/daszak.htm.

Date, E.M., Ford, H.A. and Recher, H.F. 1991. Frugivorous pigeons, stepping stones and weeds in northern NSW. Pp. 241–245 in *Nature Conservation 2: The Role of Corridors*. D.A. Saunders and R.J. Hobbs, eds. Surrey Beatty, Chipping Norton, NSW.

Date, E.M., Recher, H.F., Ford, H.A. and Stewart, D.A. 1996. The conservation and ecology of rainforest pigeons in northern New South Wales. *Pacific Conservation Biology,* **2**, 299–308.

Davies, K.F. and Margules, C.R. 1998. Effects of habitat fragmentation on carabid beetles: experimental evidence. *Journal of Animal Ecology,* **67**, 460–471.

Davies, K.F., Margules, C.R. and Lawrence, J.F. 2000. Which traits of species predict population declines in experimental forest fragments? *Ecology,* **81**, 1450–1461.

Davies, K.F., Gascon, C. and Margules, C.R. 2001a. Habitat fragmentation. Pp. 81–97 in *Conservation Biology: Research Priorities for the Next Decade.* M.E. Soulé and G.H. Orions, eds. Island Press, Washington, D.C.

Davies, K.F., Melbourne, B.A. and Margules, C.R. 2001b. Effects of within- and between-patch processes on community dynamics in a fragmentation experiment. *Ecology,* **82**, 1830–1846.

Davis, L.S., Johnson, K.N., Bettinger, P.S. and Howard, T.E. 2001. *Forest Management to Sustain Ecological, Economic, and Social Values.* 4th ed. McGraw-Hill, New York.

Davis-Born, R. and Wolff, J.O. 2000. Age- and sex-specific responses of the gray-tailed vole *Microtus canicaudus,* to connected and unconnected habitat patches. *Canadian Journal of Zoology,* **78**, 864–870.

De Maynadier, P. and Hunter, M. 1998. Effects of silvicultural edges on the distribution and abundance of amphibians in Maine. *Conservation Biology,* **12**, 340–352.

De Solla, S.R., Bonduriansky, R. and Brooks, R.J. 1999. Eliminating autocorrelation reduces biological relevance of home range estimates. *Journal of Animal Ecology,* **68**, 221–234.

Deacon, J.N., and Mac Nally, R. 1998. Local extinction and nestedness of small-mammal fauna in fragmented forest of central Victoria, Australia. *Pacific Conservation Biology,* **4**, 122–131.

Dean, W.R.J., Milton, S.J. and Jeltsch, F. 1999. Large trees, fertile islands, and birds in arid savanna. *Journal of Arid Environments,* **41**, 61–78.

Debinski, D.M. and Holt, R.D. 2000. A survey and overview of habitat fragmentation experiments. *Conservation Biology,* **14**, 342–355.

Debinski, D.M., Ray, C. and Saveraid, E.H. 2001. Species diversity and the scale of the landscape mosaic: do scales of movement and patch size affect diversity? *Biological Conservation,* **98**, 179–190.

DeGraaf, R. and Miller, I., eds. 1996. *Conservation of Faunal Diversity in Forested Landscapes.* Chapman and Hall, London.

Department of Conservation and Land Management 2000. *Salinity Action Plan. Biological Survey of the Agricultural Zone: Status Report, June 2000.* Department of Conservation and Land Management, Perth.

Department of Environment and Heritage 2005. *Threatened Species and Threatened Ecological Communities.* Department of Environment and Heritage, Canberra. http://www.deh.gov.au/biodiversity/threatened/index.html.

Desrochers, A. and Hannon, S.J. 1997. Gap crossing decisions by forest songbirds during the post-fledging period. *Conservation Biology,* **11**, 1204–1210.

Dial, R.J. 1995. Species–area curves and Koopowitz et al.'s simulation of stochastic extinctions. *Conservation Biology,* **9**, 960–961.

Diamond, J. 1986. Overview: laboratory experiments, field experiments and natural experiments. Pp. 3–22 in *Community Ecology.* J. Diamond and T.J. Case, eds. Harper and Row, New York.

Diamond, J. 1989. Overview of recent extinctions. Pp. 37–41 in *Conservation for the Twenty-first Century.* D. Western and M. Pearl, eds. Oxford University Press, New York.

Diamond, J.M. 1973. Distributional ecology of New Guinea birds. *Science,* **179**, 759–769.

Diamond, J.M. 1975a. The island dilemma: lessons of modern biogeographic studies for the design of natural preserves. *Biological Conservation,* **7**, 129–146.

Diamond, J.M. 1975b. Assembly of species communities. Pp. 342–444 in *Ecology and Evolution of Communities.* M.L. Cody and J.M. Diamond, eds. Harvard University Press, Cambridge Massachusetts.

Diamond, J.M., Bishop, K.D. and van Balen, S. 1987. Bird survival in an isolated Javan woodlot: island or mirror. *Conservation Biology,* **2**, 132–142.

Díaz, J.A. 1997. Ecological correlates of the thermal quality of an ectotherm's habitat: a comparison between two temperate lizard populations. *Functional Ecology,* **11**, 79–89.

Dickman, C.R., Pressey, R.L., Lim, L. and Parnaby, H.E. 1993. Mammals of particular conservation concern in the Western Division of New South Wales. *Biological Conservation,* **65**, 219–248.

Didham, R.K., Ghazoul, J., Stork, N.E. and Davis, A.J. 1996. Insects in fragmented forests: a functional approach. *Trends in Ecology and Evolution,* **11**, 255–260.

Dieckmann, U., O'Hara, B. and Weisser, W. 1999. The evolutionary ecology of dispersal. *Trends in Ecology and Evolution,* **14**, 88–90.

Doak, D. and Mills, L.S. 1994. A useful role for theory in conservation. *Ecology,* **75**, 615–626.

Dobkin, D.S. and Wilcox, B.A. 1986. Analysis of natural forest fragments: riparian birds in Toiyabe Mountains, Nevada. Pp. 455–470 in *Wildlife 2000: Modelling Habitat Relationships of Terrestrial Vertebrates.* J. Verner, M.L. Morrison and C.J. Ralph, eds. University of Wisconsin Press, Madison, Wisconsin.

Dobson, A.P., Bradshaw, A.D. and Baker, A.J.M. 1997. Hopes for the future: restoration ecology and conservation biology. *Science,* **277**, 515–522.

Doeg, T.J. and Koehn, J.D. 1990. *A Review of Australian Studies on the Effects of Forestry Practices on Aquatic Values.* Silvicultural Systems Project Technical Report No. 5, Fisheries Division, Department of Conservation and Environment, Melbourne.

Doherty, P.F. and Grubb, T.C. 2003. Nest usurpation is an "edge effect" for Carolina chickadees *(Poecile carolinensis). Journal of Avian Ecology,* **33**, 77–82.

Donnelly, R. and Marzluff, J.M. 2004. Importance of reserve size and landscape context to urban bird conservation. *Conservation Biology,* **18**, 733–745.

Dorrough, J. and Ash, J.E. 1999. Using past and present habitat to predict the current distribution and abundance of a rare cryptic lizard, *Delma impar* (Pygopodidae). *Australian Journal of Ecology,* **24**, 614–624.

Dorrough, J. and Moxham, C. 2005. Eucalypt establishment in agricultural landscapes and implications for landscape-scale restoration. *Biological Conservation,* **123**, 55–66.

Dovers, S. and Wild River, S., eds. 2003. *Managing Australia's Environment.* Federation Press, Sydney.

Downes, S.J., Handasyde, K.A. and Elgar, M.A. 1997a. The use of corridors by mammals in fragmented Australian eucalypt forests. *Conservation Biology,* **11**, 718–726.

Downes, S.J., Handasyde, K.A. and Elgar, M.A. 1997b. Variation in the use of corridors by introduced and native rodents in southeastern Australia. *Biological Conservation,* **82**, 379–383.

Dramstad, W.E., Olson, J.D. and Forman, R.T. 1996. *Landscape Ecology Principles in Landscape Architecture and Land-Use Planning.* Harvard University Graduate School of Design, Island Press and American Society of Landscape Architects, Washington, D.C.

Drinnan, I.N. 2005. The search for fragmentation thresholds in a southern Sydney suburb. *Biological Conservation,* **124**, 339–349.

Driscoll, D. 2004. Extinction and outbreaks accompany fragmentation of a reptile community. *Ecological Applications,* **14**, 220–240.

Driscoll, D., Milkovits, G. and Freudenberger, D. 2000. *Impact and Use of Firewood in Australia.* CSIRO Sustainable Ecosystems Report, Canberra, November 2000.

Driscoll, D.A. and Weir, T. 2005. Beetle responses to habitat fragmentation depend on ecological traits, habitat condition, and remnant size. *Conservation Biology,* **19**, 182–194.

Duncan, R.S. and Chapman, C.A. 1999. Seed dispersal and potential forest succession in abandoned agriculture in tropical Africa. *Ecological Applications,* **9**, 998–1008.

Dunn, R.R. 2004. Managing the tropical landscape: a comparison of the effects of logging and forest conversion to agriculture on ants, birds, and lepidoptera. *Forest Ecology and Management,* **191**, 215–224.

Dunning, J.B., Borgella, R., Clements, K. and Meffe, G.K. 1995. Patch isolation, corridor effects, and colonization by a resident sparrow in a managed pine woodland. *Conservation Biology,* **9**, 542–550.

Dunstan, C.E. and Fox, B.J. 1996. The effects of fragmentation and disturbance of rainforest on ground-dwelling small mammals on the Robertson Plateau, New South Wales, Australia. *Journal of Biogeography,* **23**, 187–201.

Durant, S.M. 1998. Competition refuges and coexistence: an example from Serengeti carnivores. *Journal of Animal Ecology,* **67**, 370–386.

Dyrness, C.T., Viereck, L.A. and Van Cleve, K. 1986. Fire in taiga communities of interior Alaska. Pp. 74–86 in *Forest Ecosystems in the Alaskan Taiga: A Synthesis of Structure and Function.* K. Van Cleve,

F.S. Chapin, P.W. Flanagan, L.A. Viereck and C.T. Dyrness, eds. Springer-Verlag, New York.

Edenius, L. and Sjöberg, K. 1997. Distribution of birds in natural landscape mosaics of old-growth forests in northern Sweden: relations to habitat area and landscape context. *Ecography,* **20**, 425–431.

Eens, M. 1994. Bird-song as an indicator of habitat suitability. *Trends in Evolution and Ecology,* **9**, 63–64.

Ehmann, H. and Cogger, H. 1985. Australia's endangered herpetofauna: a review of criteria and policies. Pp. 435–447 in *Biology of Australasian Frogs and Reptiles.* G. Grigg, R. Shine and H. Ehmann, eds. Surrey Beatty and Sons, Sydney.

Ehrenfeld, D.W. 1976. The conservation of non-resources. *American Scientist,* **64**, 660–668.

Ehrlich, P.R. and Ehrlich, A.H. 1981. *Extinction: The Causes and Consequences of the Disappearance of Species.* Random House, New York.

Eldredge, L.G. and Miller, S.E. 1995. How many species are there in Hawaii? *Bishop Museum Occasional Paper,* **41**, 1–18.

Eldridge, M.D., King, J., Loupis, A.K., Spencer, P.B., Taylor, A.C., Pope, L. and Hall, G. 1999. Unprecedented low levels of genetic variation and inbreeding depression in an island population of the black-footed rock-wallaby. *Conservation Biology,* **13**, 531–541.

Elith, J. 2000. Quantitative methods for modeling species habitat: comparative performance and an application to Australian plants. Pp. 39–58 in *Quantitative Methods in Conservation Biology.* S. Ferson and M. Burgman, eds. Springer, Berlin.

Elliott, P.F. 1999. Killing of host nestlings by the brown-headed cowbird. *Journal of Field Ornithology,* **70**, 55–57.

Ellner, S.P., Fieberg, J., Ludwig, D. and Wilcox, C. 2002. Precision of population viability analysis. *Conservation Biology,* **16**, 258–261.

Elmes, G.W. and Thomas, J.A. 1992. Complexity of species conservation in managed habitats: interaction between *Maculina* butterflies and their ant hosts. *Biodiversity and Conservation,* **1**, 155–169.

Elmqvist, T., Wall, M., Berggren, A.L., Blix, L., Fritioff, A. and Rinman, U. 2002. Tropical forest reorganization after cyclone and fire disturbance in Samoa: remnant trees as biological legacies. Conservation *Ecology,* **5**.

Elmqvist, T., Folke, C., Nystrom, M., Peterson, G., Bengtsson, J., Walker, B. and Norberg, J. 2003. Response diversity, ecosystem change, and resilience. *Frontiers in Ecology and the Environment,* **1**, 488–494.

Elton, C.S. 1927. *Animal Ecology.* Methuen, London.

Engeman, R.M. and Linnell, M.A. 1998. Trapping strategies for deterring the spread of brown tree snakes from Guam. *Pacific Conservation Biology,* **4**, 348–353.

Engeman, R.M., Linnell, M.A., Aguon, P., Manibusan, A., Sayama, S. and Techaira, A 1999. Implications of brown tree snake captures from fences. *Wildlife Research,* **26**, 111–116.

Enoksson, B., Angelstam, P. and Larsson, K. 1995. Deciduous forest and resident birds: the problem of fragmentation within a coniferous forest landscape. *Landscape Ecology,* **10**, 267–275.

Epps, C.W., Paslbøll, P.J., Wehausen, J.D., Roderick, G.K., Ramey, R.R. and McCullough, D.R. 2005. Highways block gene flow and cause a rapid decline in genetic diversity in desert bighorn sheep. *Ecology Letters,* **8**, 1029–1038.

Ericsson, T.S., Berglund, H. and Östlund, L. 2005. History and forest biodiversity of woodland key habitats in south boreal Sweden. *Biological Conservation,* **122**, 289–303.

Esseen, P. 1994. Tree mortality patterns after experimental fragmentation of an old-growth conifer forest. *Biological Conservation,* **68**, 19–28.

Essen, P., Ehnstrom, B., Ericson, L. and Sjoberg, K. 1997. Boreal forests. Ecological Bulletins **46**: 16–47.

Esseen, P. and Renhorn, K.E. 1998. Edge effects on an epiphytic lichen in fragmented forests. *Conservation Biology,* **12**, 1307–1317.

Estades, C.F. and Temple, S.A. 1999. Deciduous-forest bird communities in a fragmented landscape dominated by exotic pine plantations. *Ecological Applications,* **9**, 573–585.

EU Habitat Directive. 2005. http://www.fern.org.pubs/archive/biohab.html.

Euskirchen, E.S., Chen, J.Q. and Bi, R.C. 2001. Effects of edges on plant communities in a managed landscape in northern Wisconsin. *Forest Ecology and Management,* **148**, 93–108.

Everett, R.A. 2000. Patterns and pathways of biological invasions. *Trends in Ecology and Evolution,* **15**, 177–178.

Ewers, R.M. 2005. Are conservation and development compatible? *Trends in Ecology and Evolution,* **20**, 159.

Fa, J.F., Peres, C.A. and Meeuwig, J. 2002. Bushmeat exploitation in tropical forests: an intercontinental comparison. *Conservation Biology*, **16**, 232–237.

Fagan, W.F., Meir, E., Prendergast, J., Folarin, A. and Karieva, P. 2001. Characterizing population vulnerability for 758 species. *Ecology Letters*, **4**, 132–138.

Fagan, W.F., Unmack, P.J., Burgess, C. and Minckley, W.L. 2002. Rarity, fragmentation and extinction risk in desert fishes. *Ecology*, **83**, 3250–3256.

Fahrig, L. 1992. Relative importance of spatial and temporal scales in a patchy environment. *Theoretical Population Biology*, **41**, 300–314.

Fahrig, L. 1997. Relative effects of habitat fragmentation and habitat loss on population extinction. *Journal of Wildlife Management*, **61**, 603–610.

Fahrig, L. 1999. Forest loss and fragmentation: which has the greater effect on persistence of forest-dwelling animals? Pp. 87–95 in *Forest Fragmentation: Wildlife Management Implications*. J.A. Rochelle, L.A. Lehmann and J. Wisniewski, eds. Brill, Leiden, Germany.

Fahrig, L. 2003. Effects of habitat fragmentation on biodiversity. *Annual Review of Ecology, Evolution and Systematics*, **34**, 487–515.

Fahrig, L. and Merriam, G. 1985. Habitat patch connectivity and population survival. *Ecology*, **66**, 1762–1768.

Fahrig, L. and Merriam, G. 1994. Conservation of fragmented populations. *Conservation Biology*, **8**, 50–59.

Falk, D.A. 1990. Endangered forest resources in the U.S.: Integrated strategies for conservation of rare species and genetic diversity. *Forest Ecology and Management*, **35**, 91–107.

FAO 1997. *State of the World's Forests*. Food and Agricultural Organisation, Rome, Italy.

FAO 2001. *State of the World's Forests*. Food and Agricultural Organisation, Rome, Italy.

Farina, A. 1995. Distribution and dynamics of birds in a rural sub-Mediterranean landscape. *Landscape and Urban Planning*, **31**, 269–280.

Fazey, I. 2005. *Understanding the Role and Value of Experience for Environmental Conservation*. PhD thesis, Australian National University, Canberra, Australia.

Fazey, I. and McQuie, A. 2005. Applying conservation theory in natural areas management. *Ecological Management and Restoration*, **6**, 147–149.

Fazey, I., Fischer, J. and Lindenmayer, D.B. 2005a. Who does all the research in conservation? *Biodiversity and Conservation*, **14**, 917–934.

Fazey, I., Salisbury, J., Lindenmayer, D.B., Douglas, R. and Maindonald, J. 2005b. Can methods applied in medicine be used to summarize and disseminate conservation research? *Environmental Conservation*, **31**, 190–198.

Fazey, I., Fazey, J.A., Salisbury, J.G. and Lindenmayer, D.B. 2006. The role and value of experience in conservation. *Environmental Conservation*.

Felton, A.M., Engstrom, L.M., Felton, A. and Knott, C.D. 2003. Orangutan population density, forest structure and fruit availability in hand-logged and unlogged peat swamp forests in West Kalimantan, Indonesia. *Biological Conservation*, **114**, 91–101.

Ferraz, S., Vettorazzi, Theobald, D.M. and Ballester. 2005. Landscape dynamics of Amazonian deforestation between 1984 and 2002 in central Rond_nia, Brazil: assessment and future scenarios. *Forest Ecology and Management*, **204**, 67–83.

Ferreras, P. 2001. Landscape structure and asymmetrical inter-patch connectivity in a metapopulation of the endangered Iberian Lynx. *Biological Conservation*, **100**, 125–136.

Fieberg, J. and Ellner, S.P. 2001. Stochastic matrix models for conservation and management: a comparative review of methods. *Ecology Letters*, **4**, 244–266.

Figgis, P. 1999. *Conservation on Private Lands: The Australian Experience*. IUCN, Gland, Switzerland.

Fischer, J. and Lindenmayer, D.B. 2002a. The conservation value of paddock trees for birds in a variegated landscape in southern New South Wales. I. Species composition and site occupancy patterns. *Biodiversity and Conservation*, **11**, 807–832.

Fischer, J. and Lindenmayer, D.B. 2002b. The conservation value of paddock trees for birds in a variegated landscape in southern New South Wales. II. Paddock trees as stepping stones. *Biodiversity and Conservation*, **11**, 832–849.

Fischer, J. and Lindenmayer, D.B. 2002c. The conservation value of small habitat patches: two case studies on birds from southeastern Australia. *Biological Conservation*, **106**, 129–136.

Fischer, J. and Lindenmayer, D.B. 2002d. Treating the nestedness temperature calculator as a black box can lead to false conclusions. *Oikos*, **99**, 193–199.

Fischer, J. and Lindenmayer, D.B. 2005a. Perfectly nested or significantly nested: an important difference for conservation management. *Oikos,* **109**, 485–494.

Fischer, J. and Lindenmayer, D.B. 2005b. Nestedness in fragmented landscapes: a case study on birds, arboreal marsupials and lizards. *Journal of Biogeography,* **32**, 1737–1750.

Fischer, J. and Lindenmayer, D.B. 2005c. The sensitivity of lizards to elevation: a case study from southeastern Australia. *Diversity and Distributions,* **11**, 225–233.

Fischer, J., Lindenmayer, D.B. and Fazey, I. 2004a. Appreciating ecological complexity: habitat contours as a conceptual model. *Conservation Biology,* **18**, 1245–1253.

Fischer, J., Lindenmayer, D.B. and Cowling, A. 2004b. The challenge of managing multiple species at multiple scales: reptiles in an Australian grazing landscape. *Journal of Applied Ecology,* **41**, 32–44.

Fischer, J., Lindenmayer, D.B., Barry, S. and Flowers, E. 2005a. Lizard distribution patterns in the Tumut fragmentation "Natural Experiment" in southeastern Australia. *Biological Conservation,* **123**, 301–315.

Fischer, J., Fazey, I., Briese, R. and Lindenmayer, D.B. 2005b. Making the matrix matter: challenges in Australian grazing landscapes. *Biodiversity and Conservation,* **14**, 561–578.

Fischer, W.C. and McClelland, B.R. 1983. *A Cavity-Nesting Bird Bibliography Including Related Titles on Forest Snags, Fire, Insects, Diseases and Decay.* General Technical Report INT-140, March 1983, Intermountain Forest and Range Experiment Station, Ogden, Utah.

Fisher, A.M. and Goldney, D.C. 1997. Use by birds of riparian vegetation in an extensively fragmented landscape. *Pacific Conservation Biology,* **3**, 275–288.

Flather, C.H. and Sauer, J.H. 1996. Using landscape ecology to test hypotheses about large-scale abundance patterns in migratory birds. *Ecology,* **77**, 28–35.

Fleishman, E., Betrus, C.J., Blair, R.B., MacNally, R. and Murphy, D.D. 2002. Nestedness analysis and conservation planning: the importance of place, environment, and life history across taxonomic groups. *Oecologia,* **133**, 78–89.

Fletcher, R.J. 2005. Multiple edge effects and their implications in fragmented landscapes. *Journal of Animal Ecology,* **74**, 342–352.

Foley, J.A., DeFries, R., Asner, G.P., Barford, C., Bonan, G., Carpenter, S.R., Chapin, F.S., Coe, M.T., Daily, G.C., Gibbs, H.K., Helkowski, J.H., Holloway, T., Howard, E.A., Kucharik, C.J., Monfreda, C., Patz, J.A., Prentice, I.C., Ramankutty, N. and Snyder, P.K. 2005. Global consequences of land use. *Science,* **309**, 570–574.

Folke, C., Carpenter, S., Walker, B., Scheffer, M., Elmqvist, T., Gunderson, L. and Holling, C.S. 2004. Regime shifts, resilience, and biodiversity in ecosystem management. *Annual Review of Ecology Evolution and Systematics,* **35**, 557–581.

Foran, B. and Poldy, F. 2002. *Future Dilemmas: Options to 2050 for Australia's Population, Technology, Resources and Environment.* Working paper 02/01. CSIRO Sustainable Ecosystems, Canberra.

Ford, H.A., and Barrett, G. 1995. The role of birds and their conservation in agricultural systems. Pp. 128–134 in *People and Nature Conservation.* A. Bennett, G. Backhouse and T. Clark, eds. Surrey Beatty and Sons, Chipping Norton, NSW.

Forest Ecosystem Management Assessment Team. 1993. *Forest Ecosystem Management: An Ecological, Economic, and Social Assessment.* USDA Forest Service, Portland, Oregon.

Forman, R.T. 1964. Growth under controlled conditions to explain the hierarchical distribution of a moss, Tetrapis pellucida. *Ecological Monographs,* **34**, 1–25.

Forman, R.T. 1995. *Land Mosaics: The Ecology of Landscapes and Regions.* Cambridge University Press, New York.

Forman, R.T. 1998. Roads and their major ecological effects. *Annual Review of Ecology and Systematics,* **29**, 207–231.

Forman, R.T. and Deblinger, R.D. 2000. The ecological road-effect zone of a Massachusetts (U.S.A.) Suburban Highway. *Conservation Biology,* **14**, 36–46.

Forman, R.T. and Godron, M. 1986. *Landscape Ecology.* Wiley and Sons, New York.

Forman, R.T., Sperling, D., Bissonette, J.A., Clevenger, A.P., Cutshall, C.D., Dale, V.H., Fahrig, L., France, R., Goldman, C.R., Heanue, K., Jones, J.A., Swanson, F.J., Turrentine, T. and Winter, T.C., eds. 2002. *Road Ecology. Science and Solutions.* Island Press, Washington, D.C.

Forney, K.A. and Gilpin, M.E. 1989. Spatial structure and population extinction: a study with *Drosophila* flies. *Conservation Biology,* **3**, 45–51.

Forshaw, J.M. 2002. *Australian Parrots.* 3rd ed. Avi-Trader Publishing, Brisbane.

Forsman, E.D., Meslow, E.C. and Wight, H. 1984. Distribution and ecology of the spotted owl in Oregon. *Wildlife Monographs,* **87**, 1–64.

Fortin, D. and Arnold, G.W. 1997. The influence of road verges on the use of nearby small shrubland remnants by birds in the central wheatbelt of western Australia. *Wildlife Research,* **24**, 679–689.

Fox, A.D. 2004. Has Danish agriculture maintained farmland bird populations? *Journal of Applied Ecology,* **41**, 427–439.

Fox, B.J. 1983. Mammal species diversity in Australian heathlands: the importance of pyric succession and habitat diversity. Pp. 473–489 in *Mediterranean-Type Ecosystems: The Role of Nutrients.* F.J. Kruger, D.T. Mitchell and J.U.M. Jarvis, eds. Springer-Verlag, Berlin.

Fox, B.J. 1987. Species assembly and the evolution of community structure. *Evolutionary Ecology,* **1**, 201–213.

Fox, B.J. and Brown, J.H. 1993. Assembly rules for functional groups in North American desert rodent communities. *Oikos,* **67**, 358–370.

Fox, M.D. and Fox, B.J. 1986. The susceptibility of natural communities to invasion. Pp. 57–66 in *Ecology of Biological Invasions: An Australian Perspective.* R.H. Groves and J.J. Burdon, eds. Australian Academy of Science, Canberra.

Frankham, R. 1996. Relationship of genetic variation to population size in wildlife. *Conservation Biology,* **10**, 1500–1508.

Frankham, R. 1998. Inbreeding and extinction: island populations. *Conservation Biology,* **12**, 665–675.

Franklin, J.F. 1993. Preserving biodiversity: species, ecosystems, or landscapes? *Ecological Applications,* **3**, 202–205.

Franklin, J.F. and Forman, R.T. 1987. Creating landscape patterns by forest cutting: ecological consequences and principles. *Landscape Ecology,* **1**, 5–18.

Franklin, J.F. and van Pelt, R., 2004. Spatial aspects of structural complexity in old-growth forests. *Journal of Forestry,* **102**, 22–27.

Franklin, J.F., Cromack, K., Denison, W., McKee, A., Maser, C., Sedell, J., Swanson, F. and Juday, G. 1981. *Ecological Attributes of Old-Growth Douglas-Fir Forests.* USDA Forest Service General Technical Report PNW-118, Pacific Northwest Forest and Range Experimental Station, Portland, Oregon.

Franklin, J.F., MacMahon, J.A., Swanson, F.J. and Sedell, J.R. 1985. Ecosystem responses to the eruption of Mount St. Helens. *National Geographic Research,* **Spring 1985**, 198–216.

Franklin, J.F., Berg, D.E., Thornburgh, D.A. and Tappeiner, J.C. 1997. Alternative silvicultural approaches to timber harvest: variable retention harvest systems. Pp. 111–139 in *Creating a Forestry for the 21st Century.* K.A. Kohm and J.F. Franklin, eds. Island Press, Covelo, California.

Franklin, J.F., Spies, T.A., van Pelt, R., Carey, A., Thornburgh, D., Berg, D.R., Lindenmayer, D.B., Harmon, M., Keeton, W. and Shaw, D.C. 2002. Disturbances and the structural development of natural forest ecosystems with some implications for silviculture. *Forest Ecology and Management,* **155**, 399–423.

Fraser, F. and Whitehead, P.J. 2005. Predation of artificial ground nests in Australian tropical savannas: inverse edge effects. *Wildlife Research,* **32**, 313–319.

Fraser, G.S. and Stutchbury, B.J.M. 2004. Area-sensitive forest birds move extensively among forest patches. *Biological Conservation,* **118**, 399–423.

Freemark, K. and Collins, B. 1992. Landscape ecology of birds breeding in temperate forest fragments. Pp. 443–454 in Conservation of Neotropical Migrants. J. Hagan and D.J. Johnston, eds. Smithsonian Institution, Washington, D.C.

Frelich, L.E., Calcote, R.R., Davis, M.B. and Pastor, J. 1993. Patch formation and maintenance in an old-growth hemlock hardwood forest. *Ecology,* **74**, 513–527.

Freudenberger, D. 1995. Separating the sheep from the 'roos: do we know the difference? Pp. 208–214 in *Conservation through Sustainable Use of Wildlife.* G.C. Grigg, P.T. Hale and D. Lunney, eds. University of Queensland Press, Brisbane.

Freudenberger, D. 2001. *Bush for the Birds: Biodiversity Enhancement Guidelines for the Saltshaker Project, Boorowa, NSW.* A report commissioned by Greening Australia ACT and SE NSW, CSIRO Sustainable Ecosystems, Canberra.

Friederici, P., ed. 2003. *Ecological Restoration of Southwestern Ponderosa Pine Forests.* Island Press, Washington, D.C.

Fries, C., Johansson, O., Petterson, B. and Simonsson, P. 1997. Silvicultural models to maintain and restore natural stand structures in Swedish boreal forests. *Forest Ecology and Management*, **94**, 89–103.

Fritz, R. 1979. Consequences of insular population structure: distribution and extinction of spruce grouse populations. *Oecologia*, **42**, 57–65.

Frumhoff, P.C. 1995. Conserving wildlife in tropical forests managed for timber. *BioScience*, **45**, 456–464.

Fuller, D.O., Jessup, T.C. and Salim, A. 2004. Loss of forest cover in Kalimantan, Indonesia, since the 1997–1998 El Niño. *Conservation Biology*, **18**, 249–254.

Functowicz, S.O. and Ravetz, J.R. 1991. A new scientific methodology for global environmental issues. Pp. 137–152 in *Ecological Economics: The Science and Management of Sustainability*. R. Costanza, ed. Columbia University Press, New York.

Galindo-Gonzalez, J. and Sosa, V.J. 2003. Frugivorous bats in isolated trees and riparian vegetation associated with human-made pastures in a fragmented tropical landscape. *Southwestern Naturalist*, **48**, 579–589.

Galindo-Gonzalez, J., Guevara, S. and Sosa, V.J. 2000. Bat- and bird-generated seed rains at isolated trees in pastures in a tropical rainforest. *Conservation Biology*, **14**, 1693–1703.

Gallent, N. and Kim, K.S. 2001. Land zoning and local discretion in the Korean planning system. *Land Use Policy*, **18**, 233–243.

Gambold, N. and Woinarski, J.C.Z. 1993. Distributional patterns of herpetofauna in monsoon rainforests of the Northern Territory, Australia. *Australian Journal of Ecology*, **18**, 431–449.

Ganzhorn, J.U. 1997. Tests of Fox's assembly rule for functional groups in lemur communities in Madagascar. *Journal of Zoology, London*, **241**, 533–542.

Gardener, C.J., McIvor, J.G. and Williams, J. 1990. Dry tropical rangelands: solving one problem and creating another. Pp. 279–286 in *Australian Ecosystems: 200 Years of Utilisation, Degradation and Reconstruction*. Proceedings of the Ecological Society of Australia, 16. D.A. Saunders, A.J.M. Hopkins and R.A. How, eds. Surrey Beatty, Chipping-Norton, Sydney, Australia.

Gardner, J.L. 2004. Winter flocking behaviour of speckled warblers and the Allee effect. *Biological Conservation*, **118**, 195–204.

Gardner, R.H., Milne, B.T., Turner, M.G. and O'Neill R.V. 1987. Neutral models for analysis of broad-scale landscape patterns. *Landscape Ecology*, **1**, 19–28.

Garnett, S.T. and Crowley, G.M. 2000. *The Action Plan for Australian Birds*. Natural Heritage Trust, Canberra.

Garrett, M.G. and Franklin, W.L. 1988. Behavioral ecology of dispersal in the black-tailed prairie dog. *Journal of Mammalogy*, **69**, 236–250.

Gascon, C. 1993. Breeding-habitat use by five Amazonian frogs at forest edge. *Biodiversity and Conservation*, **2**, 438–444.

Gascon, C. and Lovejoy, T.E. 1998. Ecological impacts of forest fragmentation in central Amazonia. *Ecology*, **101**, 273–280.

Gascon, C., Lovejoy, T.E., Bierregaard, R.O.J., Malcolm, J.R., Stouffer, P.C., Vasconcelos, H.L., Laurance, W.F., Zimmerman, B., Tocher, M. and Borges, S. 1999. Matrix habitat and species richness in tropical forest remnants. *Biological Conservation*, **91**, 223–229.

Gaston, K.J. 1994. *Rarity*. Chapman and Hall, London.

Gaston, K.J. 1996. Species richness: measure and measurement. Pp. 77–113 in *Biodiversity*. K.J. Gaston, ed. Blackwell, Oxford.

Gaston, K.J. and Blackburn, T. 2003. Macroecology and conservation biology. Pp. 345–367 in *Macroecology: Concepts and Consequences*. T.M. Blackburn and K.J. Gaston, eds. Blackwell Publishing, Oxford.

Gaston, K.J. and Spicer, J.I. 1998. *Biodiversity: An Introduction*. Blackwell Publishing, Oxford.

Gaston, K.J. and Spicer, J.I. 2004. *Biodiversity: An Introduction*. 2nd ed. Blackwell Publishing, Oxford.

Gates, J.E. and Gysel, L.W. 1978. Avian nest dispersion and fledging success in field-forest ecotones. *Ecology*, **59**, 871–883.

Gayer, K. 1886. *Der gemischte Wald: Seine Begründung und Pflege, insbesondere durch Horst- und Gruppenwirtschaft*. Verlag Paul Parey, Berlin.

Gentry, A.H. 1989. Speciation in tropical forests. Pp. 113–134 in *Tropical Forests, Botanical Dynamics,Sspeciation and Diversity*. L.B. Holm-Nielson and H. Balshev, eds. Academic Press, San Diego, California.

Ghassemi, F., Jakeman, A.J. and Nix, H.A. 1995. *Salinisation of Land and Water Resources*. University of New South Wales Press, Sydney.

Gibbons, J.W., Scott, D.E., Ryan, T.J., Buhlmann, K.A., Tuberville, T.D., Metts, B.S., Greene, J.L.,

Mills, T., Leiden, Y., Poppy, S. and Winne, C.T. 2000. The global decline of reptiles, deja vu amphibians. *BioScience*, **50**, 653–666.

Gibbons, P. and Boak, M. 2002. The value of paddock trees for regional conservation in an agricultural landscape. *Ecological Management and Restoration*, **3**, 205–210.

Gibbons, P. and Lindenmayer, D.B. 2002. *Tree Hollows and Wildlife Conservation in Australia*. CSIRO Publishing, Melbourne.

Gibbs, J.P. 1998. Amphibians movements in response to forest edges, roads, and streambeds in southern New England. *Journal of Wildlife Management*, **62**, 584–589.

Gibbs, J.P. and Faaborg, J. 1990. Estimating the viability of ovenbird and Kentucky warbler populations in forest fragments. *Conservation Biology*, **4**, 193–196.

Gibbs, J.P. and Shriver, W.G. 2002. Estimating the effects of road mortality on turtle populations. *Conservation Biology*, **16**, 1647–1652.

Gilbert, F., Gonzalez, A. and Evens-Freke, I. 1998. Corridors maintain species richness in the fragmented landscapes of a microsystem. *Proceedings of the Royal Society of London Series B*, **265**, 577–582.

Gilbert, L.E. 1980. The equilibrium theory of island biogeography: fact or fiction? *Journal of Biogeography*, **7**, 209–235.

Gill, A.M. 1975. Fire and the Australian flora: a review. *Australian Forestry*, **38**, 4–25.

Gill, A.M., Woinarski, J.C.Z. and York, A. 1999. *Australia's Biodiversity: Responses to Fire*. Biodiversity Technical Paper No. 1. Environment Australia, Canberra.

Gillespie, G.R. 1995. Tadpoles and trout don't mix. *Victorian Frog Group Newsletter*, **1**, 14–15.

Gilmore, A.M. 1985. The influence of vegetation structure on the density of insectivorous birds. Pp. 21–31 in *Birds of Eucalypt Forests and Woodlands: Ecology, Conservation, Management*. A. Keast, H.F. Recher, H. Ford and D. Saunders, eds. Surrey Beatty, Chipping Norton.

Gilmore, A.M. 1990. Plantation forestry: conservation impacts on terrestrial vertebrate fauna. Pp. 377–388 in *Prospects for Australian Plantations*. J. Dargavel and N. Semple, eds. Centre for Resource and Environmental Studies, Australian National University, Canberra.

Gilpin, M.E. 1987. Spatial structure and population vulnerability. Pp. 126–139 in *Viable Populations for Conservation*. M.E. Soulé, ed. Cambridge University Press, New York.

Gilpin, M.E. and Diamond, J.M. 1980. Subdivision of nature reserves and the maintenance of species diversity. *Nature*, **285**, 567–568.

Gilpin, M.E. and Soulé, M.E. 1986. Minimum viable populations: processes of species extinctions. Pp. 19–34 in *Conservation Biology: The Science of Scarcity and Diversity*. M.E. Soulé, ed. Sinauer Associates, Sunderland, Massachusetts.

Gleason, H.A. 1939. The individualistic concept of plant association. *American Midland Naturalist*, **21**, 92–110.

Godwin, H. 1975. *The History of the British Flora: A Factual Basis for Phytogeography*. 2nd ed. Cambridge University Press, Cambridge.

Golden, D.M. and Crist, T.O. 1999. Experimental effects of habitat fragmentation on old-field canopy insects: community, guild and species responses. *Oecologia*, **118**, 371–380.

Goldingay, R.L. and Quin, D.G. 2004. Components of the habitat of the yellow-bellied glider in north Queensland. Pp. 369–375 in R.L. Goldingay and S.M. Jackson, eds. *The Biology of Australian Possums and Gliders*. Surrey Beatty and Sons, Chipping Norton, Sydney.

Goldsborough, C.L., Hochuli, D.F. and Shine, R. 2003. Invertebrate biodiversity under hot rocks: habitat use by the fauna of sandstone outcrops in the Sydney region. *Biological Conservation*, **109**, 85–93.

Gonzalez, A. and Chaneton, E.J. 2002. Heterotroph species extinction, abundance and biomass dynamics in an experimentally fragmented microecosystem. *Journal of Animal Ecology*, **71**, 594–602.

Gonzalez, A., Lawton, J.H., Gilbert, F.S., Blackburn, T.M. and Evans-Freke, I. 1998. Metapopulation dynamics, abundance, and distribution in a microecosystem. *Science*, **281**, 2045–2047.

Goosem, M. 2000. Effects of tropical rainforest roads on small mammals: edge changes in community composition. *Wildlife Research*, **27**, 151–163.

Gopher Tortoise Council. 2005. www.gophertortoisecouncil.org/tortoise.html.

Gotelli, N.J. and Colwell, R.K.. 2001. Quantifying biodiversity: procedures and pitfalls in the measurement and comparison of species richness. *Ecology Letters*, **4**, 379–391.

Gotelli, N.J. and Graves, G.R. 1996. *Null Models in Ecology*. Smithsonian Institution Press, Washington and London.

Gotelli, N.J., Buckley, N.J. and Wiens, J.A. 1997. Co-occurrence of Australian land birds: Diamond's assembly rules revisited. *Oikos*, **80**, 311–324

Government of Brazil. 2005. Deforestation in the Amazon. http://www.mongabay.com/brazil.html.

Graham, C.H. 2001. Factors influencing movement patterns of keel-billed toucans in a fragmented tropical landscape in southern Mexico. *Conservation Biology*, **15**, 1789–1798.

Graves, G.R. and Gotelli, N.J. 1993. Assembly of avian mixed-species flocks in Amazonia. *Proceedings of the National Academy of Science USA*, **90**, 1388–1391.

Gray, M.J., Smith, L.M. and Leyva, R.I. 2004. Influence of agricultural landscape structure on a Southern High Plains, USA, amphibian assemblage. *Landscape Ecology*, **19**, 719–729.

Gray, P.A., Cameron, D. and Kirkham, I. 1996. Wildlife habitat evaluation in forested ecosystems: some examples from Canada and the United States. Pp. 407–533 in *Conservation of Faunal Diversity in Forested Landscapes*. R.M. Degraaf and R.I. Miller, eds. Chapman and Hall, New York.

Graynoth, E. 1989. Effects of logging on stream environments and faunas in Nelson. *New Zealand Journal of Marine and Freshwater Research*, **13**, 79–109.

Greenberg, C.H., Neary, D.G. and Harris, L.D. 1994. Effect of high-intensity wildfire and silvicultural treatments on reptile communities in sand-pine scrub. *Conservation Biology*, **8**, 1047–1057.

Greenwood, J.J. 1996. Basic techniques. Pp. 11–110 in *Ecological Census Techniques*. W.J. Sutherland, ed. Cambridge University Press, Cambridge.

Gregory, S.V. 1997. Riparian management in the 21st century. Pp. 69-85 in *Creating a Forestry for the 21st Century*. Kohm, K.A. and Franklin, J.F., eds. Island Press, Washington, D.C.

Gretton, A., Komdeur, J., and Kondeur, M. 1991. Saving the mapgie-robin. *World Birdwatch*, **13**, 10–11.

Grey, M.J., Clarke, M.F. and Loyn, R.H. 1997. Initial changes in the avian communities of remnant eucalypt woodlands following a reduction in the abundance of noisy miners, *Manorina melanocephala*. *Wildlife Research*, **24**, 631–648.

Grodinski, C. and Stüwe, M. 1987. With lots of help alpine ibex return to their mountains. *Smithsonian*, **18**, 68–77.

Groom, M.J. 1998. Allee effects limit population viability of an annual plant. *American Naturalist*, **151**, 487–496.

Groom, M.J., Meffe, G.K., and Carroll, R.C. ,eds. 2005. *Principles of Conservation Biology*. 3rd ed. Sinauer Associates, Sunderland Massachusetts.

Groombridge, B. and Jenkins, M.D. 2002. *World Atlas of Biodiversity: Earth's Living Resources in the 21st Century*. UNEP-WCMC. University of California Press, Berkeley.

Grover, J.P. 1994. Assembly rules for communities of nutrient-limited plants and specialist herbivores. *American Naturalist*, **143**, 258–282.

Guerry, A.D. and Hunter, M.L. 2002. Amphibian distributions in a landscape of forests and agriculture: an examination of landscape composition and configuration. *Conservation Biology*, **16**, 745–754.

Guevara, S. and Laborde, J. 1993. Monitoring seed dispersal at isolated standing trees in tropical pastures: consequences for local species availability. *Vegetatio*, **108**, 319–338.

Guevara, S., Laborde, J. and Sánchez, G. 1998. Are isolated remnant trees in pastures a fragmented canopy? *Selbyana*, **19**, 34–43.

Guisan, A. and Hofer, U. 2003. Predicting reptile distributions at the mesoscale: relation to climate and topography. *Journal of Biogeography*, **30**, 1233–1243.

Guisan, A. and Zimmermann, N.E. 2000. Predictive habitat distribution models in ecology. *Ecological Modelling*, **135**, 147–186.

Gurd, D.B., Nudds, T.D. and Rivard, D.H. 2001. Conservation of mammals in eastern North American reserves: how small is too small? *Conservation Biology*, **15**, 1355–1363.

Gustafsson, L. 2000. Red-listed species and indicators: vascular plants in woodland key habitats and surrounding production forests in Sweden. *Biological Conservation*, **92**, 35–43.

Gustaffson, L., de Jong, J. and Noren, M. 1999. Evaluation of Swedish woodland key habitats using red-listed bryophytes and lichens. *Biodiversity and Conservation*, **8**, 1101–1114.

Gustafson, E.J. 1998. Quantifying landscape spatial pattern: what is the state of the art? *Ecosystems*, **1**, 143–156.

Gustafson, E.J., and Gardner, R.H. 1996. The effect of landscape heterogeneity on the probability of patch colonization. *Ecology*, **77**, 94–107.

Gustafson, E.J. and Parker, G.R. 1992. Relationships between landcover proportion and indices of landscape spatial pattern. *Landscape Ecology*, **7**, 101–110.

Gutierrez, D., Fernandez, P., Seymour, A.S. and Jordano, D. 2005. Habitat distribution models: are mutualist distributions good predictors of their associates? *Ecological Applications*, **15**, 3–18.

Gutzwiller, K.J. and Anderson, S.H. 1987. Multiscale associations between cavity-nesting birds and features of Wyoming streamside woodlands. *Condor*, **89**, 534–548.

Haddad, N.M. 1999a. Corridor and distance effects on interpatch movements: a landscape experiment with butterflies. *Ecological Applications*, **9**, 612–622.

Haddad, N.M. 1999b. Corridor use predicted from behaviours at habitat boundaries. *American Naturalist*, **153**, 215–227.

Haddad, N.M. and Baum, K.A. 1999. An experimental test of corridor effects on butterfly densities. *Ecological Applications*, **9**, 623–633.

Haddad, N.M. and Tewksbury, J.J. 2005. Low-quality habitat corridors as movement conduits for two butterfly species. *Ecological Applications*, **15**, 250–257.

Haddad, N.M., Bowne, D.R., Cunningham, A., Danielson, B.J., Levey, D.J., Sargent, S. and Spira, T. 2003. Corridor use by diverse taxa. *Ecology*, **84**, 609–615.

Hadfield, M.G. 1986. Extinction in Hawaiian achatinelline snails. *Malacologia*, **27**, 67–81.

Hadfield, M.G., Miller, S.E. and Carwile, A.H. 1980. The decimation of endemic Hawaiian tree snails by alien predators. *American Zoologist*, **33**, 610–622.

Hagar, J.C. 1999. Influence of riparian buffer width on bird assemblages in western Oregon. *Journal of Wildlife Management*, **63**, 484–496.

Haila, Y. 1983. Land birds on northern islands: a sampling metaphor for insular colonization. *Oikos*, **41**, 334–351.

Haila, Y. 1986. North European land birds in forest fragments: evidence for area effects? Pp. 315–319 in *Wildlife 2000: Modeling Habitat Relationships of Terrestrial Vertebrates*. J. Verner, M.L. Morrison and C.J. Ralph, eds. University of Wisconsin Press, Madison, Wisconsin.

Haila, Y. 2002. A conceptual genealogy of fragmentation research from island biogeography to landscape ecology. *Ecological Applications*, **12**, 321–334.

Haila, Y., Järvinen, S. and Kuusela, S. 1983. Colonization of islands by land birds: prevalence functions in a Finnish archipelago. *Journal of Biogeography*, **10**, 499–531.

Haila, Y., Hanski, I.K. and Raivio, S. 1993. Turnover of breeding birds in small forest fragments: the sampling colonization hypothesis corroborated. *Ecology*, **74**, 714–725.

Haila, Y., Nicholls, A.O., Hanski, I.K. and Raivio, S. 1996. Stochasticity in bird habitat selection: year-to-year changes in territory locations in a boreal forest bird assemblage. *Oikos*, **76**, 536–552.

Haines-Young, R. and Chopping, M. 1996. Quantifying landscape structure: a review of landscape indices and their application to forested environments. *Progress in Physical Geography*, **20**, 418–445.

Hall, B., Motzkin, G., Foster, D.R., Syfert, M. and Burk, J. 2002. Three hundred years of forest and land-use change in Massachusetts, USA. *Journal of Biogeography*, **29**, 1319–1335.

Hall, C.M. 1988. The "worthless lands hypothesis" and Australia's national parks and reserves. Pp. 441–459 in *Australia's Ever Changing Forests*. K.J. Frawley and N.M. Semple, eds. Australian Defence Force Academy, Canberra.

Hall, R.J. and Henry, P.F.P. 1992. Assessing effects of pesticides on amphibians and reptiles: status and needs. *Herpetological Journal*, **2**, 65–71.

Halley, J.M., Thomas, C.F. and Jepson, P.C. 1996. A model for the spatial dynamics of linyphiid spiders in farmland. *Journal of Applied Ecology*, **33**, 471–492.

Halme, E. and Niemelä, J. 1993. Carabid beetles in fragments of coniferous forest. *Annales Zoologica Fennici*, **30**, 17–30.

Halpern, C.B. and Franklin, J.F. 1990. Physiognomic development of *Pseudotsuga* forests in relation to initial structure and disturbance intensity. *Journal of Vegetation Science*, **1**, 475–482.

Halpern, C.B. and Spies, T.A. 1995. Plant species diversity in natural and managed forests of the Pacific Northwest. *Ecological Applications,* **5**, 913–934.

Hamel, G., and Prahalad, C.K. 1989. Strategic intent. *Harvard Business Review,* **89**, 63–76.

Hamer, A.J., Lane, S.J. and Mahony, M.J. 2002. The role of introduced mosquitofish *(Gambusia holbrooki)* in excluding the native green and gold bell frog *(Litoria aurea)* from original habitats in southeastern Australia. *Oecologia,* **132**, 445–452.

Hampe, A. and Petit, R.J. 2005. Conserving biodiversity under climate change: the rear edge matters. *Ecology Letters,* **8**, 461–467.

Hannon, S.J. and Cotterill, S.E. 1998. Nest predation in aspen woodlots in an agricultural area in Alberta: the enemy from within. *Auk,* **115**, 16–25.

Hannon, S.J. and Schmiegelow, F. 2002. Corridors may not improve the conservation value of small reserves for most boreal birds. *Ecological Applications,* **12**, 1457–1468.

Hanski, I. 1994a. Patch occupancy dynamics in fragmented landscapes. *Trends in Evolution and Ecology,* **9**, 131–134.

Hanski, I. 1994b. A practical model of metapopulation dynamics. *Journal of Animal Ecology,* **63**, 151–162.

Hanski, I. 1997. Metapopulation dynamics: from concepts and observations to predictive models. Pp. 69–91 in *Metapopulation Biology: Ecology, Genetics and Evolution.* I. Hanksi and M.E. Gilpin, eds. Academic Press, San Diego.

Hanski, I. 1999a. *Metapopulation Ecology.* Oxford University Press, Oxford.

Hanski, I. 1999b. Habitat connectivity, habitat continuity, and metapopulations in dynamic landscapes. *Oikos,* **87**, 209–219.

Hanski, I. and Cambefort, Y. 1991. *Dung Beetle Ecology.* Princeton University Press, Princeton, NJ.

Hanski, I. and Gilpin, M. 1991. Metapopulation dynamics: brief history and conceptual domain. *Biological Journal of the Linnean Society,* **42**, 3–16.

Hanski, I. and Gyllenberg, M. 1993. Two general metapopulation models and the core-satellite hypothesis. *American Naturalist,* **142**, 17–41.

Hanski, I. and Simberloff, D. 1997. The metapopulation approach, its history, conceptual domain, and application to conservation. Pp. 5–26 in *Metapopulation Biology: Ecology, Genetic and Evolution.* I. Hanski and M.E. Gilpin, eds. Academic Press, San Diego.

Hanski, I. and Thomas, C.D. 1994. Metapopulation dynamics and conservation: a spatially explicit model applied to butterflies. *Biological Conservation,* **68**, 167–180.

Hanski, I.K., Fenske, T.J. and Niemi, G.J. 1996. Lack of edge effect in nesting success of breeding birds in managed forest landscapes. *Auk,* **113**, 578–585.

Hansson, L. 1983. Bird numbers across edges between mature conifer and clearcuts in central Sweden. *Ornis Scandinavia,* **14**, 97–103.

Hansson, L. 1998. Vertebrate distributions relative to clear-cut edges in a boreal forest landscape. *Landscape Ecology,* **9**, 105–115.

Hansson, L. 2002. Mammal movements and foraging at remnant woodlands inside coniferous forest landscapes. *Forest Ecology and Management,* **160**, 109–114.

Harcourt, A.H. and Parks, S.A. 2003. Threatened primates experience high human densities: adding an index of threat to the IUCN Red List criteria. *Biological Conservation,* **109**, 137–149.

Harcourt, A.H., Parks, S.A. and Woodroffe, R. 2001. Human density as an influence on species/area relationships: double jeopardy for small African reserves? *Biodiversity and Conservation,* **10**, 1011–1026.

Harcourt, C. 1992. Tropical moist forests. Pp. 256–275 in *Global diversity: Status of the Earth's Living Resources.* World Conservation Monitoring Centre, ed. Chapman and Hall, London.

Harley, K.L.S. and Forno, I.W. 1992. *Biological Control of Weeds: A Handbook for Practitioners and Students.* Inkata Press, Melbourne.

Harmon, M., Franklin, J.F., Swanson, F., Sollins, P., Gregory, S.V., Lattin, J.D., Anderson, N.H., Cline, S.P., Aumen, N.G., Sedell, J.R., Lienkaemper, G.W., Cromack, K. and Cummins, K. 1986. Ecology of coarse woody debris in temperate ecosystems. *Advances in Ecological Research,* **15**, 133–302.

Harner, R.F. and Harper, K.T. 1976. The role of area, heterogeneity, and favorability in plant species diversity of pinyon–juniper ecosystems. *Ecology,* **57**, 1254–1263.

Harper, J.L. 1982. After description. Pp. 11–25 in *The Plant Community as a Working Community.* E.I. Newman, ed. Blackwell Scientific Publications, Oxford.

Harper, K.A., Macdonald, S.E., Burton, P.J., Chen, J., Brosofske, K.D., Saunders, S.C., Euskirchen, E.S., Roberts, D., Jaiteh, M.S. and Essen, P-E. 2005. Edge influence on forest structure and composition in fragmented landscapes. *Conservation Biology*, **19**, 768–782.

Harris, L.D. 1984. *The Fragmented Forest: Island Biogeography Theory and the Preservation of Biotic Diversity.* University of Chicago Press, Chicago.

Harris, L.D., Maser, C. and McKee, A. 1982. Transactions of the North American Wildlife Natural Resources Couverence, **47**, 374–392.

Harris, L.D. and Scheck, J. 1991. From implications to applications: the dispersal corridor principle applied to the conservation of biology diversity. Pp. 189–220 in *Nature Conservation 2: The Role of Corridors.* D.A. Saunders and R.J. Hobbs, eds. Surrey Beatty and Sons, Chipping Norton.

Harris, S., Cresswell, W.J., Forde, P.G., Trewhella, W.J., Woollard, T. and Wray, S. 1990. Home-range analysis using radio-tracking data: a review of problems and techniques particularly as applied to the study of mammals. *Mammal Review*, **20**, 97–123.

Harrison, S. 1991. Local extinction in a metapopulation context: an empirical evaluation. *Biological Journal of the Linnean Society*, **42**, 73–88.

Harrison, S. and Bruna, E. 1999. Habitat fragmentation and large-scale conservation: what do we know for sure? *Ecography*, **22**, 225–232.

Harrison, S. and Taylor, A.D. 1997. Empirical evidence for metapopulation dynamics. Pp. 27–42 in *Metapopulation Biology: Ecology, Genetics and Evolution.* I. Hanski and M.E. Gilpin, eds. Academic Press, San Diego.

Harrison, S., Murphy, D.D. and Ehrlich, P. 1988. Distribution of the Bay checkerspot butterfly, *Euphydras editha bayensis*: evidence for a metapopulation model. *American Naturalist*, **132**, 360–382.

Harrison, S., Maron, J. and Huxel, G. 2000. Regional turnover and fluctuation of five plants confined to serpentine seeps. *Conservation Biology*, **14**, 769–779.

Harvey, C.A., and Haber, W.A. 1999. Remnant trees and the conservation of biodiversity in Costa Rican pastures. *Agroforestry Systems*, **44**, 37–68.

Haskell, D.G. 2000. Effects of forest roads on macroinvertebrate soil fauna of the southern Appalachian Mountains. *Conservation Biology*, **14**, 57–63.

Hastie, T.J. and Tibshirani, R.J. 1990. *Generalised Additive Models.* Chapman and Hall, London.

Hastings, A. 1993. Complex interactions between dispersal and dynamics: lessons from coupled logistic equations. *Ecology*, **74**, 1362–1372.

Hastings, A. and Harrison, S. 1994. Metapopulation dynamics and genetics. *Annual Review of Ecology and Systematics*, **25**, 167–188.

Haughland, D.L. and Larsen, K.W. 2004. Exploration correlates with settlement: red squirrel dispersal in contrasting habitats. *Journal of Animal Ecology*, **73**, 1024–1034.

Hayward, M.W., Tores de, P.J., Dillon, M.J. and Fox, B.J. 2003. Local population structure of a naturally occurring metapopulation of the quokka (*Setonix brachyurus* Macropodidae: Marsupialia). *Biological Conservation*, **110**, 343–355.

Hazell, D., Cunnningham, R., Lindenmayer, D., Mackey, B. and Osborne, W. 2001. Use of farm dams as frog habitat in an Australian agricultural landscape: factors affecting species richness and distribution. *Biological Conservation*, **102**, 155–169.

Hazell, D., Osborne, W. and Lindenmayer, D.B. 2003. Impact of post-European stream change on frog habitat: southeastern Australia. *Biodiversity and Conservation*, **12**, 301–320.

Hazell, D., Hero, J., Lindenmayer, D.B. and Cunningham, R.B. 2004. A comparison of constructed and natural habitat for frog conservation in an Australian agricultural landscape *Biological Conservation*, **119**, 61–71.

Hedrick, P.W. and Kalinowski, S.T. 2000. Inbreeding depression in conservation biology. *Annual Review of Ecology and Systematics*, **31**, 139–162.

Hels, T. and Buchwald, E. 2001. The effect of road kills on amphibian populations. *Biological Conservation*, **99**, 331–340.

Herkert, J.R. 1994. The effects of habitat fragmentation on midwestern grassland bird communities. *Ecological Applications*, **4**, 461–471.

Hewittson, H. 1997. *The Genetic Consequences of Habitat Fragmentation on the Bush Rat* (Rattus fuscipes) *in a Pine Plantation near Tumut, NSW.* Honours thesis, Division of Botany and Zoology, Australian National University, Canberra, Australia.

Higgs, E. 2005. The two-culture problem: ecological restoration and the integration of knowledge. *Restoration Ecology*, **13**, 159–164.

Hill, C.J. 1995. Linear strips of rain forest vegetation as potential dispersal corridors for rain forest insects. *Conservation Biology,* **9**, 1559–1566.

Hill, J.K., Thomas, C.D. and Lewis, O.T. 1996. Effects of habitat patch size and isolation on dispersal by *Hesperia comma* butterflies: implications for metapopulation structure. *Journal of Animal Ecology,* **65**, 725–735.

Hill, J.K., Thomas, C.D. and Lewis, O.T. 1999. Flight morphology in fragmented populations of a rare British butterfly, *Hesperia comma. Biological Conservation,* **87**, 277–283.

Hilty, J. and Merenlender, A.M. 2003. Studying biodiversity on private lands. *Conservation Biology,* **17**, 132–137.

Hilty, J. and Merenlender, A.M. 2004. Use of riparian corridors and vineyards by mammalian predators in northern California. *Conservation Biology,* **18**, 126–135.

Hinsley, S.A., Bellamy, P.E., Newton, I. and Sparks, T.H. 1995. Habitat and landscape factors influencing the presence of individual breeding bird species in woodland fragments. *Journal of Avian Biology,* **26**, 94–104.

Hinsley, S.A., Rothery, P. and Bellamy, P.E. 1999. Influence of woodland area on breeding success in great tits *Parus major* and blue tits *Parus caeruleus. Journal of Avian Biology,* **30**, 271–281.

Hobbs, R. 1997. Future landscapes and the future of landscape ecology. *Landscape and Urban Planning,* **37**, 1–9.

Hobbs, R.J. 1992. The role of corridors in conservation: solution or bandwagon? *Trends in Ecology and Evolution,* **7**, 389–392.

Hobbs, R.J. 2000. Fire regimes and their effects in Australian temperate woodlands. Pp. 305–326 in *Flammable Australia. The fire Regimes and Biodiversity of a Continent.* R.A. Bradstock, J.E. Williams, and A.M. Gill, eds. Cambridge University Press, Melbourne.

Hobbs, R.J. 2001. Synergisms among habitat fragmentation, livestock grazing, and biotic invasions in southwestern Australia. *Conservation Biology,* **15**, 1522–1528.

Hobbs, R.J. 2003. Ecological management and restoration: assessment, setting goals, and measuring success. *Ecological Management and Restoration,* **4(Supplement)**, S2–S3.

Hobbs, R.J. and Hopkins, A.J.M. 1990. From frontier to fragments: European impact on Australia's vegetation. Pp. 93–114 in *Australian Ecosystems: 200 Years of Utilisation, Degradation and Reconstruction.* Proceedings of the Ecological Society of Australia, 16, D.A. Saunders, A.J.M. Hopkins and R.A. How, eds. Surrey Beatty, Chipping-Norton.

Hobbs, R.J. and Yates, C.J., eds. 2000. *Temperate Eucalypt Woodlands in Australia: Biology, Conservation, Management and Restoration.* Surrey Beatty and Sons, Chipping Norton.

Hobbs, R.J. and Yates, C.J. 2003. Impacts of ecosystem fragmentation on plant populations: generalising the idiosyncratic. *Australian Journal of Botany,* **51**, 471–488.

Hobbs, R.J., Saunders, D.A. and Arnold, G.W. 1993. Integrated landscape ecology: a western-Australian perspective. *Biological Conservation,* **64**, 231–233.

Hobson, K.A. and Bayne, E. 2000. The effects of stand age on avian communities in aspen-dominated forests of central Saskatchewan, Canada. *Forest Ecology and Management,* **136**, 121–134.

Hodges, M.F. and Krementz, D.G. 1996. Neotropical migratory breeding bird communities in riparian forests of different widths along the Altamaha River, Georgia. *Wilson Bulletin,* **108**, 498–506.

Hoekstra, J.M., Boucher, T.M., Ricketts, T.H. and Roberts, C. 2005. Confronting a biome crisis: global disparities of habitat loss and protection. *Ecology Letters,* **8**, 23–29.

Hokit, D.G., Stith, B.M. and Branch, L.C. 1999. Effects of landscape structure in Florida scrub: a population perspective. *Ecological Applications,* **9**, 124–134.

Holdaway, R.N. and Jacomb, C. 2000. Rapid extinction of the moas (Aves: Dinornithiformers): model, test and implications. *Science,* **287**, 2250–2254.

Hole, D.G., Perkins, A.J., Wilson, J.D., Alexander, I.H., Grice, F. and Evans, A.D. 2005. Does organic farming benefit biodiversity? *Biological Conservation,* **122**, 113–130.

Holekamp, K.E. 1984. Natal dispersal in Belding's ground squirrel *(Spermophilus beldingi). Behavioural Ecology and Sociobiology,* **16**, 21–30.

Holling, C.S. 1973. Resilience and stability of ecological systems. *Annual Review of Ecology and Systematics,* **4**, 1–23.

Holt, R.D. 1977. Predation, apparent competition and the structure of prey communities. *Theoretical Population Biology,* **12**, 197–229.

Holt, R.D., Grover, J. and Tilman, D. 1994. Simple rules for interspecific dominance in systems with exploitative and apparent competition. *American Naturalist,* **144**, 741–771.

Holt, R.D., Debinski, D.M., Diffendorfer, J., Gaines, M., Matinko, E., Robinson, G., and Ward, G. 1995. Perspectives from an experimental study of habitat fragmentation in an agroecosystem. Pp. 147–175 in *Ecology and Integrated Farming Systems.* D.M. Gren, M.P. Greaves and H.M. Anderson, eds. Wiley, New York.

Holyoak, M. 2000. Habitat subdivision causes changes in food web structure. *Ecology Letters,* **3**, 509–515.

Homan, R.N., Windmiller, B.S. and Reed, J.M. 2004. Critical thresholds associated with habitat loss for two vernal pool-breeding amphibians. *Ecological Applications,* **14**, 1547–1553.

Honnay, O., Hermy, M. and Coppin, P. 1999. Effects of area, age and diversity of forest patches in Belgium on plant species richness, and implications for conservation and reafforestation. *Biological Conservation,* **87**, 73–84.

Honnay, O., Verheyen, K. and Hermy, M. 2002. Permeability of ancient forest edges for weedy plant species invasion. *Forest Ecology and Management,* **161**, 109–122.

Hopper, S.D. 2000. Floristics of Australian granitoid inselberg vegetation. Pp. 391–407 in *Inselbergs.* Ecological Studies, vol. 146. S. Porembski and W. Barthlott, eds. Springer-Verlag, Berlin.

Horner-Devine, M.C., Daily, G.C., Ehrlich, P.R. and Boggs, C.L. 2003. Countryside biogeography of tropical butterflies. *Conservation Biology,* **17**, 168–177.

Horskins, K. 2004. *The Effectiveness of Wildlife Corridors in Facilitating Connectivity: Assessment of a Model System from the Australian Wet Tropics.* PhD thesis, Queensland University of Technology, Brisbane, Australia.

How, R.A. Barnett, J.L. Bradley, A.J. Humphreys, W.F. and Martin, R. 1984. The population biology of *Pseudocheirus peregrinus* in a *Leptospermum laevigatum* thicket. Pp. 261–268 in *Possums and Gliders.* A.P. Smith and I.D. Hume, eds. Surrey Beatty and Sons, Sydney, Australia.

Howe, R.W. 1984. Local dynamics of bird assemblages in small forest habitat islands in Australia and North America. *Ecology,* **65**, 1585–1601.

Hoyle, M. 2005. Experimentally fragmented communities are more aggregated. *Journal of Animal Ecology,* **74**, 430–442.

Huang, C.M., Wei, F.W., Li, M., Quan, G.Q. and Li, H.H. 2002. Current status and conservation of white-headed langur *(Trachypithecus leucocephalus)* in China. *Biological Conservation,* **104**, 221–225.

Huggard, O.J. 1993. Prey selectivity of wolves in Banff National Park. II. Age, sex and condition of elk. *Canadian Journal of Zoology,* **71**, 140–147.

Huggett, A.J. 2005. The concept and utility of "ecological thresholds" in biodiversity conservation. *Biological Conservation,* **124**, 301–310.

Huggett, R. and Cheesman, J. 2002. *Topography and the Environment.* Prentice Hall, London.

Hughes, J.B., Daily, G.C. and Ehrlich, P.R. 2002. Conservation of tropical forest birds in countryside habitats. *Ecology Letters,* **5**, 121–129.

Hughes, L. 2000. Biological consequences of global warming: is the signal already apparent? *Trends in Ecology and Evolution,* **15**, 56–61.

Hughes, L. 2003. Climate change and Australia: trends, projections and impacts. *Austral Ecology,* **28**, 423–443.

Huhta, E., Jokimäki, J. and Rahko, P. 1999. Breeding success of pied flycatchers in artificial forest edges: the effect of a suboptimally shaped foraging area. *Auk,* **116**, 528–535.

Huhta, E., Aho, T., Jantti, A., Suorsa, P., Kuitunen, M., Nikula, A. and Hakkarainnen, H. 2004. Forest fragmentation increases nest predation in the Eurasian treecreeper. *Conservation Biology,* **18**, 148–155.

Huijser, M.P. and Bergers, P.J. 2000. The effect of roads and traffic on hedgehog *(Erinaceus europaeus)* populations. *Biological Conservation,* **95**, 111–116.

Hunter, M.L. 1990. *Wildlife, Forests and Forestry: Principles for Managing Forests for Biological Diversity.* Prentice Hall, Englewood Cliffs.

Hunter, M.L. 1993. Natural fire regimes as spatial models for managing boreal forests. *Biological Conservation,* **65**, 115–120.

Hunter, M.L. 1994. *Fundamentals of Conservation Biology.* Blackwell, Cambridge, Melbourne.

Hunter, M.L. 1997. The biological landscape. Pp. 57–67 in *Creating a Forestry for the 21st Century: The Science of Ecosystem Management.* K.A. Kohm and J.F. Franklin, eds. Island Press, Washington, D.C.

Hunter, M.L., ed. 1999. *Managing Biodiversity in Forest Ecosystems.* Cambridge University Press, London.

Hunter, M.L. 2002. *Fundamentals of Conservation Biology.* 2nd ed. Blackwell Science, Melbourne.

Hunter, M.L. 2005. A mesofilter conservation strategy to complement fine and coarse filters. *Conservation Biology,* **19**, 1025–1029.

Hurlbert, S.H. 1984. Pseduoreplcation and the design of ecological field experiments. *Ecological Monographs,* **54**, 187–211.

Huston, M.A. 1997. Hidden treatments in ecological experiments: reevaluating the ecosystem function of biodiversity. *Oecologia,* **110**, 449–460.

Hutchinson, G.E. 1958. Concluding remarks. *Cold Spring Harbor Symposia on Quantitative Biology,* **22**, 415–427.

Hylander, K. 2005. Aspect modifies the magnitude of edge effects on bryophyte growth in boreal forests. *Journal of Applied Ecology,* **42**, 518–525.

Hylander, K., Jonsson, B.G. and Nilsson, C. 2002. Evaluating buffer strips along boreal streams using bryophytes as indicators. *Ecological Applications,* **12**, 797–806.

Hylander, K., Nilsson, C. and Göthner, T. 2004. Effects of buffer-strip retention and clearcutting on land snails in boreal riparian forests. *Conservation Biology,* **18**, 1052–1062.

Ims, R.A., Rolstad, J. and Wegge, P. 1993. Predicting space use responses to habitat fragmentation: can voles *Microtus oeconomus* serve as an experimental model system (EMS) for capercaillie grouse *Tetrao urogallus* in boreal forest? *Biological Conservation,* **63**, 261–268.

Incoll, R.D., Loyn, R.H., Ward, S.J., Cunningham, R.B. and Donnelly, C.F. 2000. The occurrence of gliding possums in old-growth patches of mountain ash *(Eucalyptus regnans)* in the Central Highlands of Victoria. *Biological Conservation,* **98**, 77–88.

Ingham, D.S. and Samways, M.J. 1996. Application of fragmentation and variegation models to epigaeic invertebrates in South Africa. *Conservation Biology,* **10**, 1353–1358.

Isaac, N.J.B., Mallet, J. and Mace, G.M. 2004. Taxonomic inflation: its influence on macroecology and conservation. *Trends in Ecology and Evolution,* **19**, 464–469.

IUCN (1998). *Guidelines for Reintroductions.* IUCN/SSC-Reintroduction Specialist Group, Gland, Switzerland.

Jablonski, D. and Valentine, J.W. 1990. From regional to total geographic ranges: testing the relationship in recent bivalves. *Paleobiology,* **16**, 126–142.

Jackson, D.R. and Milstrey, E.G. 1989. The fauna of gopher tortoise burrows. Pp. 86–98 in *Gopher Tortoise Relocation Symposium Proceedings.* Nongame Wildlife Technical Report No. 5. J.E. Diemer, D.R. Jackson, J.L. Landers, J.N. Layne and D.A. Woods, eds. Florida Game and Fresh Water Fish Commission, Tallahassee, Florida.

Jackson, J.A. 1978. Red-cockaded woodpeckers and pine red heart disease. *Auk,* **94**, 160–163.

Jackson, J.A. 1994. The red-cockaded woodpecker recovery program: professional obstacles to cooperation. Pp. 157–181 in *Endangered Species Recovery: Finding the Lessons, Improving the Process.* T.W. Clark, R.P. Reading and A.I. Clarke, eds. Island Press, Washington, D.C.

Jackson, S.J. 2000. Home-range and den use of the mahogany glider, *Petaurus gracilis. Wildlife Research,* **27**, 49–60.

Jäggi, C. and Baur, B. 1999. Overgrowing forest as a possible cause for the local extinction of Vipera aspis in the northern Swiss Jura mountains. *Amphibia-Reptilia,* **20**, 25–34.

James, C.D. 1991. Population dynamics, demography, and life history of sympatric scincid lizards (Ctenotus) in Central Australia. *Herpetologica,* **47**, 194–210.

James, C., Landsberg, J. and Morton, S. 1999. Provision of watering points in Australian arid zone: a review of effects on the biota. *Journal of Arid Environments,* **41**, 87–121.

James, S.E. and M'Closkey, R.T.M. 2003. Lizard microhabitat and fire fuel management. *Biological Conservation,* **114**, 293–297.

Jansen, A. and Robertson, A.I. 2001. Relationships between livestock management and the ecological condition of riparian habitats along an Australian floodplain river. *Journal of Applied Ecology,* **38**, 63–75.

Janzen, D.H. 1983. No park is an island: increase in

interference from outside as park size decreases. *Oikos*, **41**, 402–410.

Jaquet, N. 1996. How spatial and temporal scales influence understanding of sperm whale distribution: a review. *Mammal Review*, **26**, 51–65.

Jeffries, M.J. and Lawton, J.H. 1984. Enemy free space and the structure of ecological communities. *Biological Journal of the Linnaean Society*, **23**, 269–286.

Jeffries, M.J. and Lawton, J.H. 1985. Predator–prey rates in communities of freshwater invertebrates: the role of enemy free space. *Freshwater Biology*, **15**, 105–112.

Jellinek, S., Driscoll, D.A. and Kirkpatrick, J.B. 2004. Environmental and vegetation variables have a greater influence than habitat fragmentation in structuring lizard communities in remnant urban bushland. *Austral Ecology*, **29**, 294–304.

Jeltsch, F., Milton, S.J., Dean, W.R.J. and VanRooyen, N. 1996. Tree spacing and coexistence in semiarid savannas. *Journal of Ecology*, **84**, 583–595.

Jha, D.K., Sharma, G.D. and Mishra, R.R. 1992. Ecology of soil microflora and mycorrhizal symbionts in degraded forests at two altitudes. *Biology and Fertility of Soils*, **12**, 272–278.

Johns, A.D. and Skorupa, J.P. 1987. Responses of rain-forest primates to habitat disturbance: a review. *International Journal of Primatology*, **8**, 157–191.

Johnson, A.S., Hale, P.E., Ford, W.M., Wentworth, J.M., French, J.R., Anderson, O.F. and Pullen, G.B. 1995. White-tailed deer in relation to successional stage, overstorey type and management of southern Appalachian forests. *American Midland Naturalist*, **133**, 18–35.

Johnson, C.J., Boyce, M.S., Mulders, R., Gunn, A., Gau, R.J., Cluff, H.D. and Case, R.L. 2004. Quantifying patch distribution at multiple spatial scales: applications to wildlife-habitat models. *Landscape Ecology*, **19**, 869–882.

Johnson, D. 1980. The comparison of useage and availability measurements for evaluating resource preference. *Ecology*, **6**, 65–71.

Johnson, K.N. and Herring, J. 1999. Understanding bioregional assessments. Pp. 341–376 in *Bioregional Assessments: Science at the Crossroads of Management and Policy*. K.N. Johnson, F. Swanson, J. Herring and S. Greene, eds. Island Press, Washington, D.C.

Johnson, M.L. and. Gaines, M.S. 1985. Selective basis for emigration of the Prairie Vole, *Microtus ochrogaster*: an open field experiment. *Journal of Animal Ecology*, **54**, 399–410.

Johnson, M.L. and. Gaines, M.S. 1990. Evolution of dispersal: theoretical models and empirical tests using birds and mammals. *Annual Review of Ecology and Systematics*, **21**, 449–480.

Joly, P., Miaud, C., Lehmann, A. and Grolet, O. 2001. Habitat matrix effects on pond occupancy in newts. *Conservation Biology*, **15**, 239–248.

Jones, C.G., and Duffy, K. 1993. Conservation management of the echo parakeet. *Dodo*, **29**, 126–148.

Jones, J.A. and Grant, G.E.. 1996. Peak flow responses to clear-cutting and roads in small and large basins, Western Cascades, Oregon. *Water Resources Research*, **32**, 959–974.

Jones, M.E. 2000. Road upgrade, road mortality and remedial measures: impacts on a population of eastern quolls and Tasmanian devils. *Wildlife Research*, **27**, 289–296.

Jonsson, B.G. 2001. A null model for randomization tests of nestedness in species assemblages. *Oecologia*, **127**, 309–313.

Jonsson, B.G. and Jonsell, M. 1999. Exploring potential biodiversity indicators in boreal forests. *Biodiversity and Conservation*, **8**, 1417–1433.

Joyal, L., McCollough, A.M. and Hunter, M.L. 2001. Landscape ecology approaches to wetland species conservation: a case study of two turtle species in southern Maine. *Conservation Biology*, **15**, 1755–1762.

Kadmon, R. 1995. Nested species subsets and geographic isolation: a case study. *Ecology*, **76**, 458–465.

Kapos, V. 1989. Effects of isolation on the water status of forest patches in Brazilian Amazonia. *Journal of Tropical Ecology*, **5**, 173–185.

Kareiva, P. 1987. Habitat fragmentation and the stability of predator–prey interactions. *Nature*, **26**, 388–390.

Kareiva, P., Skelly, D. and Ruckleshaus, M. 1997. Reevaluation of the use of models to predict the consequences of habitat loss and fragmentation. Pp. 156–166 in *The Ecological Basis for Conservation*. S.T.A. Pickett, R.S. Ostfeld, M. Shachak and G. Likens, eds. Chapman and Hall, New York.

Kattan, G.H., Alvarezlopez, H. and Giraldo, M. 1994. Forest fragmentation and bird extinctions: San-

Antonio 80 years later. *Conservation Biology*, **8**, 138–146.

Katti, M. and Warren, P.S. 2004. Tits, noise and urban bioacoustics. *Trends in Ecology and Evolution*, **19**, 109–110.

Kauffman, J.B. and C. Uhl. 1991. Interactions of anthropogenic activities, fire, and rainforests in the Amazon Basin. Pp. 117–134 in *Fire in the Tropical Biota*. J.G. Goldammer, ed. Springer-Verlag, New York.

Kearns, C.A. and Inouye, D.W. 1997. Pollinators, flowering plants, conservation biology. *BioScience*, **47**, 297–307.

Kearns, C.A., Inouye, D.W. and Waser, N.M. 1998. Endangered mutualisms: the conservation of plant-pollinator interactions. *Annual Review of Ecology and Systematics*, **29**, 83–112.

Keast, A. 1968. Seasonal movements in the Australian honeyeaters (Meliphagidae) and their ecological significance. *Emu*, **67**, 159–210.

Keast, A., ed. 1981. *Ecological Biogeography of Australia*. Junk, The Hague.

Keenan, R., Lamb, D., Woldring, O., Irvine, T. and Jensen, R., 1997. Restoration of plant biodiversity beneath tropical tree plantations in Northern Australia. *Forest Ecology and Management*, **99**, 117–131.

Keith, D., McCaw, W.L. and Whelan, R.J. 2002a. Fire regimes in Australian heathlands and their effects on plants. Pp. 199-237 in *Flammable Australia: The Fire Regimes and Biodiversity of a Continent*. Bradstock, R., Williams, J. and Gill, A.M., eds. Cambridge University Press, Cambridge.

Keith, D., Williams, J. and Woinarski, J. 2002b. Fire management and biodiversity conservation: key approaches and principles. Pp. 401–425 in *Flammable Australia: The Fire Regimes and Biodiversity of a Continent*. R. Bradstock, J. Williams and A.M. Gill, eds. Cambridge University Press, Cambridge.

Kellas, J.D. and Hateley, A.J.M. 1991. Management of dry sclerophyll forests in Victoria. I. The low elevation mixed forests. Pp. 142–162 in *Forest Management in Australia*. F.H. McKinnell and J.E.D. Fox, eds. Surrey Beatty, Chipping Norton.

Keller, C.M., Robbins, C.S. and Hatfield, J.S. 1993. Avian communities in riparian forests of different widths in Maryland and Delaware. *Wetlands*, **13**, 137–144.

Keller, L.F. 1998. Inbreeding and its fitness effects in an insular population of song sparrows *(Melospiza melodia). Evolution*, **52**, 240–250.

Keller, L.F. and Waller, D.M. 2002. Inbreeding effects in wild populations. *Trends in Ecology and Evolution*, **17**, 230–241.

Kelly, P.A. and Rotenberry, J.T. 1993. Buffer zones and ecological reserves in California: replacing guesswork with science. Pp. 85–92 in *Interface between Ecology and Land Development in California*. J.E. Keeley, ed. Southern California Academy of Sciences, Los Angeles.

Kerr, J.T. and Deguise, I. 2004. Habitat loss and the limits to endangered species recovery. *Ecology Letters*, **7**, 1163–1169.

Kilgo, J.C., Sargent, R.A., Chapmen, B.R. and Miller, K.V. 1998. Effect of stand width and adjacent habitat on breeding bird communities in bottomland hardwoods. *Journal of Wildlife Management*, **62**, 72–83.

Kindvall, O. 1999. Dispersal in a metapopulation of the bush cricket *Metrioptera bicolour* (Orthoptera: Tettigonidae). *Journal of Animal Ecology*, **68**, 172–185.

Kindvall, O. and Ahlen, I. 1992. Geometrical factors and metapopulation dynamics of the bush cricket, *Metrioptera* bicolor Philippi (Orthoptera: Tettigonidae). *Conservation Biology*, **6**, 520–529.

Kinley, T. and Newhouse, N.J. 1997. Relationship of riparian reserve zone width to bird density and diversity in southeastern British Columbia. *Northwest Science*, **71**, 75–86.

Kinnear, J.E., Sumner, N.R. and Onus, M.L. 2002. The red fox in Australia: an exotic predator turned biocontrol agent. *Biological Conservation*, **108**, 335–359.

Kinnunen, H., Tiainen, J. and Tukia, H. 2001. Farmland carabid beetle communities at multiple levels of spatial scale. *Ecography*, **24**, 189–197.

Kirchner, F., Ferdy, J-B., Andalo, C., Colas, B. and Moret, J. 2003. Role of corridors in plant dispersal: an example with the endangered *Ranunculus nodiflorus*. *Conservation Biology*, **17**, 401–410.

Kirkpatrick, J. 1994. *A Continent Transformed: Human Impact on the Natural Vegetation of Australia*. Oxford University Press, Melbourne.

Kirkpatrick, J.B. 1998. Nature conservation and the Regional Forest Agreement process. *Australian Journal of Environmental Management*, **5**, 31–37.

Kirkpatrick, J. and Gilfedder, L. 1995. Maintaining integrity compared with maintaining rare and threatened taxa in remnant bushland in subhumid Tasmania. *Biological Conservation,* **74**, 1–8.

Kitahara, M. and Watanabe, M. 2003. Diversity and rarity hotspots and conservation of butterfly communities in and around the Aokigahara woodland of Mount Fuji, central Japan. *Ecological Research,* **18**, 503–522.

Kitchener, D.J. and How, R.A. 1982. Lizard species in small mainland habitat isolates and islands off southwestern Western Australia. *Australian Wildlife Research,* **9**, 357–363.

Kitchener, D.J., Chapman, A., Dell, J., Muir, B.G. and Palmer, M. 1980a. Lizard assemblage and reserve size and structure in the western Australian wheatbelt: some implications for conservation. *Biological Conservation,* **17**, 25–62.

Kitchener, D.J., Chapman, A., Muir, B.G. and Palmer, M. 1980b. The conservation value for mammals of reserves in the Western Australian wheatbelt. *Biological Conservation,* **18**, 179–207.

Kitchener, D.J., Dell, J., Muir, B.G. and Palmer, M. 1982. Birds in western Australian wheatbelt reserves: implications for conservation. *Biological Conservation,* **22**, 127–163.

Kjoss, V.A. and Litvaitis, J.A. 2001. Community structure of snakes in a human-dominated landscape. *Biological Conservation,* **98**, 285–292.

Kleijn, D. and Sutherland, W.J. 2003. How effective are European agri-environment schemes in conserving and promoting biodiversity? *Journal of Applied Ecology,* **40**, 947–969.

Kleijn, D., Berendse, F., Smit, R., Gilissen, N., Smit, J., Brak, B. and Groeneveld, R. 2004. Ecological effectiveness of agri-environment schemes in different agricultural landscapes in the Netherlands. *Conservation Biology,* **18**, 775–786.

Klein, B.C. 1989. Effects of forest fragmentation on dung and carrion beetle communities in central Amazonia. *Ecology,* **70**, 1715–1725.

Klenner, W. and Kroeker, D.W. 1990. Denning behaviour of black bears, *Ursus americanus,* in western Manitoba. *Canadian Field-Naturalist,* **104**, 540–544.

Klink, C.A. and Machado, R.B. 2005. Conservation of the Brazilian Cerrado. *Conservation Biology,* **19**, 707–713.

Knaapen, J.P., Bottom, M., and Harms, B. 1992. Estimating habitat isolation in landscape planning. *Landscape and Urban Planning,* **23**, 1–16.

Knapp, A.K., Blair, J.M., Briggs, J.M., Collins, S.L., Hartnett, D.C., Johnson, L.C. and Towne, E.G. 1999. The keystone role of bison in north American tallgrass prairie: bison increase habitat heterogeneity and alter a broad array of plant, community, and ecosystem processes. *BioScience,* **49**, 39–50.

Knight, E.H. and Fox, B.J. 2000. Does habitat structure mediate the effects of forest fragmentation and human-induced disturbance on the abundance of *Antechinus stuartii? Australian Journal of Zoology* **48**, 577–595.

Knight, R.L. 1999. Private lands: the neglected geography. *Conservation Biology,* **13**, 223–224.

Knopf, F.L. 1992. Faunal mixing, faunal integrity, and the biopolitical template for diversity conservation. *Transactions of the 57th North American Wildlife and Natural Resource Conference,* 330–342.

Koenig, W.D. 1998. Spatial autocorrelation in California land birds. *Conservation Biology,* **12**, 612–620.

Koh, L.P., Sodhi, N.S. and Brook, B.W. 2004. Ecological correlates of extinction proneness in tropical butterflies. *Conservation Biology,* **18**, 1571–1578.

Kolbe, J.J. and Janzen, F.J. 2002. Spatial and temporal dynamics of turtle nest predation: edge effects. *Oikos,* **99**, 538–544.

Komdeur, J. 1996. Breeding of the Seychelles magpie-robin *Copsychus sechellarum* and implications for conservation. *Ibis,* **138**, 485–498.

Koopowitz, H., Thornhill, A.D. and Anderson, M. 1994. A general stochastic model for the prediction of biodiversity losses based on habitat conservation. *Conservation Biology,* **8**, 425–438.

Korpilahti, E. and T. Kuuluvainen, eds. 2002. Disturbance dynamics in boreal forests: defining the ecological basis of restoration and management of biodiversity. *Silva Fennica,* **36**, 1–447.

Kotiaho, J.S., Kaitala, V., Kolmonen, A. and Päivinen, J. 2005. Predicting the risk of extinction from shared ecological characteristics. *Proceedings of the National Academy of Sciences of the United States of America,* **102**, 1963–1967.

Kotliar, N.B. and Wiens, J.A. 1990. Multiple scales of patchiness and patch structure: a hierarchical framework for the study of heterogeneity. *Oikos,* **59**, 253–260.

Kozakiewicz, M. and Kanopka, J. 1991. Effect of habitat isolation on genetic divergence of bank vole populations. *Acta Theriologica,* **36**, 363–367.

Krebs, C.J. 1978. *Ecology: The Experimental Analysis of Distribution and Abundance.* 2nd ed. Harper International, New York.

Krebs, C.J. 1985. *Ecology: The Experimental Analysis of Distribution and Abundance.* 3rd ed. Harper and Row, New York.

Krebs, C. 1999. Current paradigms of rodent population dynamics: what are we missing? Pp. 33–48 in *Ecologically Based Management of Rodent Pests.* G.R. Singleton, L.A. Hinds, H. Leirs and Z. Zhang, eds. Australian Centre for International Agricultural Research, Canberra.

Krebs, C.J. 1992. The role of dispersal in cyclic rodent populations. Pp. 160–175 in *Animal Dispersal.* N. Stenseth and W. Lidicker, eds. Chapman and Hall, London.

Krebs, C.J., Keller, B.L. and Tamarin, R.H. 1969. *Microtus* population biology: demographic changes in fluctuating populations of *M. ochrogaster* and *M. pennsylvanicanicus* in southern Indiana. *Ecology,* **50**, 587–607.

Kremsater, L. and Bunnell, F.L. 1999. Edge effects: theory, evidence and implications to management of western North American forests. Pp. 117–153 in *Forest Wildlife and Fragmentation: Management Implications.* J. Rochelle, L.A. Lehmann and J. Wisniewski, eds. Brill, Leiden, Germany.

Kreuzer, M.P. and Huntly, N.J. 2003. Habitat-specific demography: evidence for source-sink population structure in a mammal, the pika. *Oecologia,* **134**, 343–349.

Krohne, D.T. 1997. Dynamics of metapopulations of small mammals. *Journal of Mammalogy,* **78**, 1014–1026.

Kruuk, L.E.B., Sheldon, B.C. and Merila, J. 2002. Severe inbreeding depression in collared flycatchers *(Ficedula albicollis). Proceedings of the Royal Society of London Series B-Biological Sciences,* **269**, 1581–1589.

Kuchling, G., Dejose, J.P., Burbidge, A.A. and Bradshaw, S.D. 1992. Beyond captive breeding: the western swamp tortoise *Pseudemydura umbrina* recovery programme. *International Zoo YearBook,* **31**, 37–41.

Kurki, S., Nikula, A., Helle, P. and Linden, H. 2000. Landscape fragmentation and forest composition effects on grouse breeding success in boreal forests. *Ecology,* **81**, 1985–1997.

La Polla, V.N. and Barrett, G.W. 1993. Effects of corridor width and presence on the population dynamics of the meadow vole *(Microtus pennsylvanicus). Landscape Ecology,* **8**, 25–37.

Lacy, R.C. 1987. Loss of genetic diversity from managed populations: interacting effects of drift, mutation, immigration, selection and population subdivision. *Conservation Biology,* **1**, 143–158.

Lacy, R.C. 1993. Impacts of inbreeding in natural and captive populations of vertebrates: implications for conservation. *Perspectives in Biology and Medicine,* **36**, 480–496.

Lacy, R.C. and Clark, T.W. 1990. Population viability assessment of eastern barred bandicoot. Pp. 131–146 in *The Management and Conservation of Small Populations.* T.W. Clark and J.H. Seebeck, eds. Proceedings of a Conference, Melbourne, September 26–27, 1989.

Lacy, R.C., Flesness, N.R. and Seal, U.S. 1989. *Puerto Rican Parrot Population Viability Analysis.* Report to the U.S. Fish and Wildlife Service. Captive Breeding Specialist Group, Species Survival Commission, I.U.C.N., Apple Valley, Minnesota.

Lahti, D.C. 2001. The "edge effect on nest predation" hypothesis after twenty years. *Biological Conservation,* **99**, 365–374.

Lair, H. 1987. Estimating the location of the focal center in red squirrel home ranges. *Ecology,* **68**, 1092–1101.

Lajeunesse, M.J. and Forbes, M.R. 2003. Variable reporting and quantitative reviews: a comparison of three meta-analytical techniques. *Ecology Letters,* **6**, 448–454.

Lambeck, R.J. 1997. Focal species: a multi-species umbrella for nature conservation. *Conservation Biology,* **11**, 849–856.

Lamberson, R.H., Noon, B.R., Voss, C. and McKelvey, R. 1994. Reserve design for territorial species: the effects of patch size and spacing on the viability of the northern spotted owl. *Conservation Biology,* **8**, 185–195.

Landres, P.B., Verner, J. and Thomas, J.W. 1988. Ecological uses of vertebrate indicator species: a critique. *Conservation Biology,* **2**, 316–328.

Landsberg, J. 1999. Status of temperate eucalypt woodlands in the Australian Capital Territory region. Pp. 32–44 in *Temperate Eucalypt Woodlands in Australia: Biology, Conservation, Management and Restoration*. R.J. Hobbs and C.J. Yates, eds. Surrey Beatty and Sons, Chipping Norton.

Landsberg, J., James, C.D., Morton, S.R., Hobbs, T.J., Stohl, J., Drew, A. and Tongway, H. 1997. *The Effects of Artificial Sources of Water on Rangeland Biodiversity*. Biodiversity Unit of Environment Australia and CSIRO Division of Wildlife and Ecology, Canberra.

Laurance, S.G. 2004. Responses of understorey rain forest birds to road edges in Central Amazonia. *Ecological Applications*, **14**, 1344–1357.

Laurance, W.F. 1990. Comparative responses of five arboreal marsupials to tropical forest fragmentation. *Journal of Mammalogy*, **71**, 641–653.

Laurance, W.F. 1991. Ecological correlates of extinction proneness in Australian tropical rain forest mammals. *Conservation Biology*, **5**, 79–89.

Laurance, W.F. 1997. Responses of mammals to rainforest fragmentation in tropical Queensland: a review and synthesis. *Wildlife Research*, **24**, 603–612.

Laurance, W.F. 1999. Introduction and synthesis. *Biological Conservation*, **91**, 101–107.

Laurance, W.F. 2000. Do edge effects occur over large spatial scales? *Trends in Ecology and Evolution*, **15**, 134–135.

Laurance, W.F. and Bierregaard, R.O., eds. 1997. *Tropical Forest Remnants: Ecology, Management and Conservation of Fragmented Communities*. University of Chicago Press, Chicago.

Laurance, W.F. and Yensen, E. 1991. Predicting the impacts of edge effects in fragmented habitats. *Biological Conservation*, **55**, 77–92.

Laurance, W.F., Bierregaard, R.O., Gascon, C., Didham, R.K., Smith, A.P., Lynam, A.J., Viana, V. M., Lovejoy, T.E., Sieving, K.E., Sites, J.W., Andersen, M., Tocher, M.D., Kramer, E.A., Restrepo, C. and Moritz, C. 1997. Tropical forest fragmentation: synthesis of a diverse and dynamic discipline. Pp. 502–525 in *Tropical Forest Remnants. Ecology, Management and Conservation of Fragmented Communities*. W.F. Laurance and R.O. Bierregaard, eds. University of Chicago Press, Chicago.

Laurance, W.F., Pérez-Slicrup, D., Delamônica, P., Fearnside, P.M., D'Angelo, S., Jerozolinski, A., Pohl, L. and Lovejoy, T.E. 2001. Rain forest fragmentation and the structure of Amazonian liana communities. *Ecology*, **82**, 105–116.

Laurance, W.F., Lovejoy, T.E., Vasconcelos, H.L., Bruna, E.M., Didham, R.K., Stouffer, P.C., Gascon, C., Bierregaard, R.O., Laurance, S.G. and Sampaio, E. 2002. Ecosystem decay of Amazonian forest fragments: a 22-year investigation. *Conservation Biology*, **16**, 605–618.

Laurance, W.F., Oliveira, A.A., Laurance, S.G., Condit, R., Nascimento, H.E.M., Sanchez-Thorin, A.C., Lovejoy, T.E., Andrade, A., D'Angelo, S., Ribeiro, J.E. and Dick, C.W. 2004. Pervasive alteration of tree communities in undisturbed Amazonian forests. *Nature*, **428**, 171–175.

Law, B.S., Anderson, J. and Chidel, M. 1999. Bat communities in a fragmented forest landscape on the southwest slopes of New South Wales, Australia. *Biological Conservation*, **88**, 333–345.

Leathwick, J.R. and Austin, M.P. 2001. Competitive interactions between tree species in New Zealand's old-growth indigenous forests. *Ecology*, **82**, 2560–2573.

Lebreton, J.D., Burnham, K.P., Clobert, J. and Anderson, D.R. (1992). Modelling survival and testing biological hypotheses using marked animals: a unique approach with case studies. *Ecological Monographs*, **62**, 67–118.

Lee, A.K. and Martin, R. 1988. *The Koala*. Australian Natural History Series. University of New South Wales Press, Sydney.

Lehtinen, R.M., Ramanamanjato, J-B. and Raveloarison, J.G. 2003. Edge effects and extinction proneness in a herpetofauna from Madagascar. *Biodiversity and Conservation*, **12**, 1357–1370.

Lennon, J.J., Koleff, P., Greenwood, J.J.D. and Gaston, K.J. 2004. Contribution of rarity and commonness to patterns of species richness. *Ecology Letters*, **7**, 81–87.

Lenoir, L., Bengtsson, J. and Persson, T. 2003. Effects of formica ants on soil fauna-results from a short-term exclusion and a long-term natural experiment. *Oecologia*, **134**, 423–430.

Letnic, M.I. and Fox, B.J. 1997. The impact of industrial fluoride fallout on faunal succession following sand mining of dry sclerophyll forest at Tomago, NSW. I. Lizard recolonisation. *Biological Conservation*, **80**, 63–81.

Leung, K.P., Dickman, C.R. and Moore, L.A. 1993. Genetic variation in fragmented populations of an Australian rainforest rodent, *Melomys cervinipes*. *Pacific Conservation Biology*, **1**, 58–65.

Lever, C. 2001. *The Cane Toad: The History and Ecology of a Successful Colonist.* Westbury Academic and Scientific Publishing, London.

Levey, D.J., Bolker, B.M., Tewksbury, J.J., Sargent, S. and Haddad, N.M. 2005. Effects of landscape corridors on seed dispersal by birds. *Science,* **309**, 146–148.

Levins, R.A. 1970. Extinction. *Lecture Notes in Mathematics and Life Sciences,* **2**, 75–107.

Lidicker, W.Z. 1999. Responses of mammals to habitat edges: an overview. *Landscape Ecology,* **14**, 333–343.

Lienert, J. and Fischer, M. 2003. Habitat fragmentation effects the common wetland specialist *Primula farinosa* in northeast Switzerland. *Journal of Ecology,* **91**, 587–599.

Lindenmayer, D.B. 1994. Wildlife corridors and the mitigation of logging impacts on forest fauna in southeastern Australia. *Wildlife Research,* **21**, 323–340.

Lindenmayer, D.B. 1997. Differences in the biology and ecology of arboreal marsupials in forests of southeastern Australia. *Journal of Mammalogy,* **78**, 1117–1127.

Lindenmayer, D.B. 1998. The design of wildlife corridors in wood production forests. N.S.W. National Parks and Wildlife Service, Occasional Paper Series. *Forest Issues Paper,* **4**, 1–41.

Lindenmayer, D.B. 2000. Factors at multiple scales affecting distribution patterns and its implications for animal conservation: Leadbeater's possum as a case study. *Biodiversity and Conservation,* **9**, 15–35.

Lindenmayer, D.B. 2002. The greater glider as a model to examine key issues in Australian forest ecology and management. Pp. 46–58 in *Perspectives on Wildlife Research. Celebrating 50 years of CSIRO Wildlife and Ecology.* D.A. Saunders, D. Spratt and M. van Wensveen, eds. Surrey Beatty and Sons, Chipping Norton, Sydney, Australia.

Lindenmayer, D.B. and Burgman, M.A. 2005. *Practical Conservation Biology.* CSIRO Publishing, Melbourne, Australia.

Lindenmayer, D. B. and Fischer, J. 2003. Sound science or social hook: a response to Brooker's application of the focal species approach. *Landscape and Urban Planning,* **62**, 149–158.

Lindenmayer, D.B. and Franklin, J.F. 2002. *Conserving Forest Biodiversity: A Comprehensive Multiscaled Approach.* Island Press, Washington, D.C.

Lindenmayer, D.B. and Franklin, J.F., eds. 2003. *Towards Forest Sustainability.* Island Press, Washington, D.C.

Lindenmayer, D.B. and Hobbs, R.J. 2004. Biodiversity conservation in plantation forests: a review with special reference to Australia. *Biological Conservation,* **119**, 151–168.

Lindenmayer, D.B. and Luck, G. 2005. Ecological thresholds: a synthesis. *Biological Conservation,* **124**, 351–354.

Lindenmayer, D.B. and McCarthy, M.A. 2001. The spatial distribution of non-native plant invaders in a pine–eucalypt landscape mosaic in southeastern Australia. *Biological Conservation,* **102**, 77–87.

Lindenmayer, D.B. and McCarthy, M.A. 2002. Congruence between natural and human forest disturbance: an Australian perspective. *Forest Ecology and Management,* **155**, 319–335.

Lindenmayer, D.B. and Nix, H.A. 1993. Ecological principles for the design of wildlife corridors. *Conservation Biology,* **7**, 627–630.

Lindenmayer, D.B. and Peakall, R.H. 2000. The Tumut experiment: integrating demographic and genetic studies to unravel fragmentation effects. Pp. 173–201 in *Genetics, Demography and Viability of Fragmented Populations.* A. Young and G. Clarke, eds. Cambridge University Press, Cambridge.

Lindenmayer, D.B. and Possingham, H.P. 1995. The conservation of arboreal marsupials in the montane ash forests of the Central Highlands of Victoria, south-east Australia. VII. The contribution of changed wood production practices to metapopulation persistence. *Biological Conservation,* **73**, 239–257.

Lindenmayer, D.B. and Possingham, H.P. 1996. Modeling the relationships between habitat connectivity, corridor design and wildlife conservation within intensively logged wood production forests of southeastern Australia. *Landscape Ecology,* **11**, 79–105.

Lindenmayer, D.B., Cunningham, R.B., Tanton, M.T., Nix, H.A. and Smith, A.P. 1991a. The conservation of arboreal marsupials in the montane ash forests of the Central Highlands of Victoria, south-east Australia. III. The habitat requirements of Leadbeaters possum, Gymnobelideus leadbeateri McCoy and models of the diversity and abundance of arboreal marsupials. *Biological Conservation,* **56**, 295–315.

Lindenmayer, D.B., Cunningham, R.B., Nix, H.A., Tanton, M.T. and Smith, A.P. 1991b. Predicting the abundance of hollow-bearing trees in montane ash forests of southeastern Australia. *Australian Journal of Ecology,* **16**, 91–98.

Lindenmayer, D.B., Cunningham, R.B. and Donnelly, C.F. 1993. The conservation of arboreal marsupials in the montane ash forests of the Central Highlands of Victoria, southeast Australia. IV. The distribution and abundance of arboreal marsupials in retained linear strips (wildlife corridors) in timber production forests. *Biological Conservation,* **66**, 207–221.

Lindenmayer, D.B., Ritman, K., Cunningham, R.B., Smith, J.D.B. and Horvath, D. 1995a. A method for predicting the spatial distribution of arboreal marsupials. *Wildlife Research,* **22**, 445–456.

Lindenmayer, D.B., Burgman, M.A., Akçakaya, H.R., Lacy, R.C. and Possingham, H.P. 1995b. A review of the generic computer programs ALEX, RAMAS/space and VORTEX for modelling the viability of metapopulations. *Ecological Modelling,* **82**, 161–174.

Lindenmayer, D.B., Welsh, A., Donnelly, C.F. and Meggs, R. 1996. The use of nest trees by the mountain brushtail possum *(Trichosurus caninus)* (Phalangeridae: Marsupialia). I. Number of occupied trees and frequency of tree use. *Wildlife Research,* **23**, 343–361.

Lindenmayer, D.B., Cunningham, R.B. and Donnelly, C.F. 1997. Tree decline and collapse in Australian forests: implications for arboreal marsupials. *Ecological Applications,* **7**, 625–641.

Lindenmayer, D.B., Cunningham, R.B. and McCarthy, M.A. 1999a. The conservation of arboreal marsupials in the montane ash forests of the Central Highlands of Victoria, southeastern Australia. VIII. Landscape analysis of the occurrence of arboreal marsupials in the montane ash forests. *Biological Conservation,* **89**, 83–92.

Lindenmayer, D.B., Cunningham, R.B., Pope, M. and Donnelly, C.F. 1999b. The response of arboreal marsupials to landscape context: a large-scale fragmentation study. *Ecological Applications,* **9**, 594–611.

Lindenmayer, D.B., McCarthy, M.A. and Pope, M.L. 1999c. A test of Hanski's simple model for metapopulation model. *Oikos,* **84**, 99–109.

Lindenmayer, D.B., Pope, M.L. and Cunningham, R.B. 1999e. Roads and nest predation: an experimental study in a modified forest system. *Emu,* **99**, 148–152.

Lindenmayer, D.B., Margules, C.R. and Botkin, D. 2000. Indicators of forest sustainability biodiversity: the selection of forest indicator species. *Conservation Biology,* **14**, 941–950.

Lindenmayer, D.B., Cunningham, R.B., MacGregor, C., Tribolet, C.R. and Donnelly. 2001. A prospective longitudinal study of landscape matrix effects on woodland remnants: experimental design and baseline data for mammals, reptiles and nocturnal birds. *Biological Conservation,* **101**, 157–169.

Lindenmayer, D.B., Cunningham, R.B., Donnelly, C.F., Nix, H.A. and Lindenmayer, B.D 2002a. The distribution of birds in a novel landscape context. *Ecological Monographs,* **72**, 1–18.

Lindenmayer, D.B., Cunningham, R.B., Donnelly, C.F. and Lesslie, R. 2002b. On the use of landscape indices as ecological indicators in fragmented forests. *Forest Ecology and Management,* **159**, 203–216.

Lindenmayer, D.B., Manning, A., Smith, P.L, McCarthy, M., Possingham, H.P., Fischer, J. and Oliver, I. 2002c. The focal species approach and landscape restoration: a critique. *Conservation Biology,* **16**, 338–345.

Lindenmayer, D.B., McIntyre, S. and Fischer, J. 2003a. Birds in eucalypt and pine forests: landscape alteration and its implications for research models of faunal habitat use. *Biological Conservation,* **110**, 45–53.

Lindenmayer, D.B., Claridge, A.W., Hazell, D., Michael, D.R., Crane, M., MacGregor, C.I. and Cunningham, R.B. 2003b. *Wildlife on Farms: How to Conserve Native Wildlife.* CSIRO Publishing, Melbourne.

Lindenmayer, D.B., Possingham, H.P., Lacy, R.C., McCarthy, M.A. and Pope, M.L. 2003c. How accurate are population models? Lessons from landscape-scale population tests in a fragmented system. *Ecology Letters,* **6**, 41–47.

Lindenmayer, D.B., Pope, M.L., and Cunningham, R.B. 2004a. Patch use by the greater glider in a fragmented forest ecosystem. II. Den tree use. *Wildlife Research,* **31**, 569–577.

Lindenmayer, D.B., Cunningham, R.B. and Lindenmayer, B.D. 2004b. Sound recording of bird vocalisations in forests. II. Longitudinal profiles in vocal activity. *Wildlife Research,* **31**, 209–217.

Lindenmayer, D.B., Cunningham, R.B. and Peakall, R. 2005a. On the recovery of populations of small mammals in forest fragments following major population reduction. *Journal of Applied Ecology,* **42**, 649–658.

Lindenmayer, D.B., Beaton, E., Crane, M., Michael, D., McGregor, C. and Cunningham, R. 2005b. *Woodlands: A Disappearing Landscape.* CSIRO Publishing, Melbourne.

Lindenmayer, D.B., Cunningham, R.B. and Fischer, J. 2005c. Vegetation cover thresholds and species responses. *Biological Conservation,* **124**, 311–316.

Linder, P. and Östlund, L. 1998. Structural changes in three midboreal Swedish forest landscapes, 1885–1996. *Biological Conservation,* **85**, 9–19.

Liu, J.G., Ouyang, Z., Taylor, W.W., Groop, R., Tan, K.C. and Zhang, H.M. 1999. A framework for evaluating the effects of human factors on wildlife habitat: the case of giant pandas. *Conservation Biology,* **13**, 1360–1370.

Liu, J.G., Linderman, M., Ouyang, Z.Y., An, L., Yang, J. and Zhang, H.M. 2001. Ecological degradation in protected areas: the case of Wolong Nature Reserve for giant pandas. *Science,* **292**, 98–101.

Liu, J.G., Daily, G.C., Ehrlich, P.R. and Luck, G.W. 2003. Effects of household dynamics on resource consumption and biodiversity. *Nature,* **421**, 530–533.

Lomolino, M.V. 1996. Investigating causality of nestedness of insular communities: selective immigrations or extinctions? *Journal of Biogeography,* **23**, 699–703.

Lonsdale, W.M. 1999. Global patterns of plant invasions and the concept of invasibility. *Ecology,* **80**, 1522–1536.

Lovejoy, T.E. 1980. *A Projection of Species Extinctions,* Vol. 2. U.S. Government Printing Office, Washington D.C.

Lovejoy, T.E. and Oren, D.C. 1981. The minimum critical size of ecosystems. Pp. 7–12 in *Forest Island Dynamics in Man-Dominated Landscapes.* R.L. Burgess and D.M. Sharpe, eds. Springer-Verlag, New York.

Lovejoy, T.E., Rankin, J.M., Bierregaard, R.O., Brown, K.S., Emmons, L.H., and van der Voort, M.E. 1984. Ecosystem decay of Amazon remnants. Pp. 295–325 in *Extinctions.* M.H. Niteki, ed. University of Chicago Press, Chicago.

Lovejoy, T.E., Bierregaard, R.O.J., Rylands, A.B., Malcolm, J.R., Quintela, C.E., Harper, L.H., Brown, K.S.J., Powell, A.H., Powell, G.V.N., Schubart, H.O.R. and Hays, M.B. 1986. Edge and other effects of isolation on Amazon forest fragments. Pp. 258–285 in *Conservation Biology: The Science of Scarcity and Diversity.* M.E. Soulé, ed. Sinauer, Sunderland, Massachusetts.

Low, T. 1999. *Feral Future.* Viking Penguin Books, Ringwood, Australia.

Loyn, R.H. 1987. Effects of patch area and habitat on bird abundances, species numbers and tree health in fragmented Victorian forests. Pp. 65–77 in *Nature Conservation: The Role of Remnants of Native Vegetation.* D.A. Saunders, G. Arnold, A. Burbidge and A. Hopkins, eds. Surrey Beatty, Chipping Norton, NSW, Australia.

Loyn, R.H., Chesterfield, E.A. and Macfarlane, M.A. 1980. *Forest Utilization and the Flora and Fauna in Boola Boola State Forest in Southeastern Victoria.* Forest Commission of Victoria Bulletin, 28. Forest Commission of Victoria, Melbourne, Victoria.

Lubchenco, J., Olson, A.M., Brubaker, L.B., Carpenter, S.R., Holland, M.M., Hubbell, S.P., Levin, S.A., MacMahon, J.A., Matson, P.A., Melillo, J.M., Mooney, H.A., Peterson, C.H., Pulliam, H.R., Real, L.A., Regal, P.J. and Risser, P.G. 1991. The sustainable biosphere initiative: An ecological research agenda: a report from The Ecological Society of America. *Ecology,* **72**, 371–412.

Luck, G. 2002a. The habitat requirements of the rufous treecreeper *(Climacteris rufa)*. I. Preferential habitat use demonstrated at multiple spatial scales. *Biological Conservation,* **105**, 383–394.

Luck, G. 2002b. The habitat requirements of the rufous treecreeper *(Climacteris rufa)*. I. Validating predictive habitat models. *Biological Conservation,* **105**, 395–403.

Luck, G.W. and Daily, G.C. 2003. Tropical countryside bird assemblages: richness, composition, and foraging differ by landscape context. *Ecological Applications,* **13**, 235–247.

Luck, G.W., Possingham, H.P. and Paton, D.C. 1999a. Bird responses at inherent and induced edges in the Murray Mallee, South Australia. I. Differences in abundance and diversity. *Emu,* **99**, 157–169.

Luck, G.W., Possingham, H.P. and Paton, D.C. 1999b. Bird responses at inherent and induced edges in the Murray Mallee, South Australia. II. Nest predation as an edge effect. *Emu,* **99**, 170–175.

Luck, G., Ricketts, T.H., Daily, G. and Imhoff, M. 2004. Alleviating spatial conflict between people and biodiversity. *Proceedings of the National Academy of Sciences,* **101**, 182–186.

Ludwig, D. 1999. Is it meaningful to estimate a probability of extinction? *Ecology,* **80**, 298–310.

Ludwig, D., Hilborn, R. and Walters, C. 1993. Uncertainty, resource exploitation and conservation: lessons from history. *Science,* **260**, 17–20.

Ludwig, D., Mangel, M. and Haddad, B. 2001. Ecology, conservation and public policy. *Annual Reviews of Ecology and Systematics,* **32**, 481–517.

Lugo, A.E. and Helmer, E. 2004. Emerging forests on abandoned land: Puerto Rico's new forests. *Forest Ecology and Management,* **190**, 145–161.

Lumsden, L.F. and Bennett, A.F. 2005. Scattered trees in rural landscapes: foraging habitat for insectivorous bats in southeastern Australia. *Biological Conservation,* **122**, 205–222.

Lumsden, L., Bennett, A. and Silins, J. 2002. Location of roosts of the lesser long-eared bat *Nyctophilus geoffroyii* and Gould's wattled bat *Chalinolobus gouldii* in a fragmented landscape in southeastern Australia. *Biological Conservation,* **106**, 237–249.

Lumsden, L.F., Bennett, A.F., Silins, J. and Krasna, S. 1994. *Fauna in a Remnant Vegetation–Farmland Mosaic, Movement, Roosts and Foraging Ecology of Bats.* A report to the Australian Nature Conservation Agency "Save the Bush Program." Flora and Fauna Branch, Department of Conservation and Natural Resources, Melbourne.

Lyles, A. 2002. Zoos and zoological parks. Pp. 901-912 in *Encyclopedia of Biodiversity,* vol. 5. S.A. Levin, ed. Academic Press, San Diego.

Lynam, A.J. and Billick, I. 1999. Differential responses of small mammals to fragmentation in a Thailand tropical forest. *Biological Conservation,* **91**, 191–200.

Lynch, J.F. and Whigham, D.F. 1984. Effects of forest fragmentation on breeding bird communities in Maryland, USA. *Biological Conservation,* **28**, 287–324.

Lyons, K.G., Brigham, C.A., Traut, B.H. and Schwartz, M.W. 2005. Rare species and ecosystem functioning. *Conservation Biology,* **19**, 1019–1024.

M'Closkey, R.T. 1978. Niche separation and assembly in four species of Sonoran desert rodents. *American Naturalist,* **112**, 683–694.

Maass, J.M., Balvanera, P., Castillo, A., Daily, G. C., Mooney, H.A., Ehrlich, P., Quesada, M., Miranda, A., Jaramillo V.J., Garcia-Oliva, F., Martinez-Yrizar, A., Cotler, H., Lopez-Blanco, J., Perez-Jimenez, A., Burquez, A., Tinoco, C., Ceballos, G., Barraza, L., Ayala, R. and Sarukhan, J. 2005. Ecosystem services of tropical dry forests: insights from long-term ecological and social research on the Pacific Coast of Mexico. *Ecology and Society,* **10**, 17. http://www.ecologyandsociety.org/vol10/iss1/art17/.

Mabry, K. and Barrett, G.W. 2002. Effects of corridors on home range sizes and interpatch movements of three small mammal species. *Landscape Ecology,* **17**, 629–636.

Mac Nally, R. and Bennett A.F. 1997. Species-specific predictions of the impact of habitat fragmentation: local extinction of birds in the box–ironbark forests of central Victoria, Australia. *Biological Conservation,* **82**, 147–155.

Mac Nally, R., Bennett, A.F. and Horrocks, G. 2000. Forecasting the impacts of habitat fragmentation: evaluation of species-specific predictions of the impact of habitat fragmentation on birds in the box–ironbark forests of central Victoria, Australia. *Biological Conservation,* **95**, 7–29.

Mac Nally, R., Bennett, A.F., Brown, G.W., Lumsden, L.F., Yen, A., Hinkley, S., Lillywhite, P. and Ward. D.A. 2002. How well do ecosystem-based planning units represent different components of biodiversity? *Ecological Applications,* **12**, 900–912.

MacArthur, R.H. 1964. Environmental factors affecting bird species diversity. *American Naturalist,* **98**, 387–397.

MacArthur, R.H. 1972. *Geographical Ecology: Patterns in the Distribution of Species.* Harper and Row, New York.

MacArthur, R.H. and MacArthur, J.W. 1961. On bird species diversity. *Ecology,* **42**, 594–598.

MacArthur, R.H. and Wilson, E.O. 1963. An equilibrium theory of insular zoogeography. *Evolution,* **17**, 373–387.

MacArthur, R.H. and Wilson, E.O. 1967. *The Theory of Island Biogeography.* Princeton University Press, Princeton.

MacCleery, D.W. 1996. *American Forests: A History of Resiliency and Recovery.* Forest History Society Issues Series, Durham, North Carolina.

MacDonald, D., ed. 2001. *The New Encyclopaedia of Mammals.* Oxford University Press, Oxford.

Machtans, C.S., Villard, M.A. and Hannon, S.J. 1996. Use of riparian buffer strips as movement corridors for birds. *Conservation Biology,* **10**, 1366–1379.

Mackey, B.G. and Lindenmayer, D.B. 2001. Towards a hierarchical framework for modelling the spatial distribution of animals. *Journal of Biogeography,* **28**, 1147–1166.

Mackey, B.G., Lindenmayer, D.B., Gill, A.M., McCarthy, M.A. and Lindesay, J. 2002. *Wildlife, Fire and Future Climate: A Forest Ecosystem Analysis.* CSIRO Publishing, Melbourne.

MacNally, R., Parkinson, A. Horrocks, G., Conole, L. and Tzaros, C. 2001. Relationships between terrestrial vertebrate diversity, abundance, and availability of coarse woody debris on south-eastern Australian floodplains. *Biological Conservation,* 99: 191–205.

MacMahon, J.A. and Holl, K.D. 2001. Ecological restoration: a key to conservation biology's future. Pp. 245–264 in *Conservation Biology for the Next Decade.* M.E. Soulé and G.H. Orians, eds. Island Press, Washington, D.C.

Maddox, J. 2000. Positioning the goal posts. *Nature,* **403**, 139.

Mader, H.J. 1984. Animal habitat isolation by roads and agricultural fields. *Biological Conservation,* **29**, 81–96.

Madsen, T., Shine, R., Olsson, M. and Wittzell, H. 1999. Restoration of an inbred adder population. *Nature,* **402**, 34–35.

Magara, T. 2002. Carabids and forest edge: spatial pattern and edge effect. *Forest Ecology and Management,* **157**, 23–37.

Majer, J.D., Recher, H.F. and Postle, A.C. 1994. Comparison of arthropod species richness in eastern and western Australian canopies: a contribution to the species number debate. *Memoirs of the Queensland Museum,* **36**, 121–131.

Major, R.E. and Kendal, C.E. 1996. The contribution of artificial nest experiments to understanding avian reproductive success: a review of methods and conclusions. *Ibis,* **138**, 298–307.

Malcolm, S.B., and M.P. Zalucki. 1993. *Biology and Conservation of the Monarch Butterfly.* Natural History Muesum of Los Angeles County. Los Angeles, California.

Mallet, J. 1995. A species definition for the modern synthesis. *Trends in Ecology and Evolution,* **10**, 294–299.

Manel, S., Schwartz, M., Luikart, G. and Taberlet, P. 2003. Landscape genetics: combining landscape ecology and population genetics. *Trends in Ecology and Evolution,* **18**, 18–197.

Manning, A.D. 2004. *A Multiscale Study of the Superb Parrot* (Polytelis Swainsonii): *Implications for Landscape-Scale Ecological Restoration.* PhD thesis, Australian National University, Canberra.

Manning, A.D., Lindenmayer, D.B. and Nix, H.A. 2004a. Continua and Umwelt: alternative ways of viewing landscapes. *Oikos,* **104**, 621–628.

Manning, A., Lindenmayer, D.B. and Barry, S. 2004b. The conservation implications of bird reproduction in the agricultural "matrix": a case study of the vulnerable superb parrot *(Polytelis swainsonii)* of southeastern Australia. *Biological Conservation,* **120**, 367–378.

Mansergh, I. and Scotts, D. 1989. Habitat continuity and social organisations of the mountain pygmy possum restored by tunnel. *Journal of Wildlife Management,* **53**, 701–707.

Marcot, B.G. 1997. Biodiversity of old forests of the west: a lesson from our elders. Pp. 87–105 in *Creating a Forestry for the 21st Century.* K.A. Kohm and J.F. Franklin, eds. Island Press, Washington, D.C.

Margules, C.R. and Austin, M.P., eds. 1991. *Nature Conservation: Cost Effective Biological Surveys and Data Analysis.* CSIRO, Canberra, Australia.

Margules, C.R. and Pressey, R.L. 2000. Systematic conservation planning. *Nature,* **405**, 243–253.

Margules, C.R., Milkovits, G.A. and Smith, G.T. 1994. Contrasting effects of habitat fragmentation on the scorpion *Cercophonius squama* and an amphipod. *Ecology,* **75**, 2033–2042.

Margules, C.R., Davies, K.F., Meyers, J.A. and Milkovits, G.A. 1995. The responses of some selected arthropods and the frog *Crinia signifera* to habitat fragmentation. Pp. 94–103 in *Conserving Biodiversity: Threats and Solutions.* R.A. Bradstock, T.A. Auld, D.A. Keith, R.T. Kingsford, D. Lunney and D.P. Sivertsen, eds. Surrey Beatty, Chipping Norton, Sydney.

Marine Stewardship Council 2005. *MSC Online.* Marine Stewardship Council, London. http://www.msc.org.

Martín, J. and Salvador, A. 1997. Microhabitat selection by the Iberian rock lizard Lacerta monticola: effects on density and spatial distribution of individuals. *Biological Conservation,* **79**, 303–307.

Martin, T.E. and Karr, J.R. 1986. Patch utilisation by migrating birds: resource orientated? *Ornis Scandinavia,* **17**, 165–174.

Martin, T.G., Kuhnert, P.M., Mengersen, K. and Possingham, H.P. 2005. The power of expert opinion in ecological models using Bayesian methods: impact of grazing on birds. *Ecological Applications,* **15**, 266–280

Martin, W.K., Eyears-Chaddock, M., Wilson, B.R. and Lemon, J. 2004. The value of habitat reconstruction to birds at Gunnedah, New South Wales. *Emu,* **104**, 177–189.

Marzluff, J.M. and Restani, M. 1999. The effect of forest fragmentation on avian nest predation. Pp. 155–169 in *Forest Wildlife and Fragmentation: Management Implications.* J. Rochelle, L.A. Lehmann and J. Wisniewski, eds. Brill, Leiden, Germany.

Maser, C., Trappe, J.M. and Nussbaum, R.A. 1978. Fungi–small mammal interrelationships with emphasis on Oregon coniferous forests. *Ecology,* **59**, 799–809.

Mather, A.S. 1990. *Global Forest Resources.* Belhaven Press, London.

Matlack, G.R. 1993. Microenvironment variation within and among forest edge sites in the eastern United States. *Biological Conservation,* **66**, 185–194.

Matlack, G.R. and Litvaitis, J.A. 1999. Forest edges. Pp. 210–233 in *Managing Biodiversity in Forest Ecosystems.* M. Hunter III, ed. Cambridge University Press, Cambridge.

Matthysen, E., Adriaensen, F. and Dhondt, A.A. 1995. Dispersal distances of nuthatches, *Sitta europaea,* in a highly fragmented habitat. *Oikos,* **72**, 375–381.

May, R.M. 1973. *Stability and Complexity in Model Ecosystems.* Princeton University Press, Princeton, New Jersey.

May, R.M. 1978. Evolution of ecological systems. *Scientific American,* **239**, 160–168.

May, R.M., Lawton, J.H. and Stork, N.E. 1995. Assessing extinction rates. Pp. 1–24 in *Extinction Rates.* J.H. Lawton and R.M. May, eds. Oxford University Press, Oxford.

May, S. and Norton, T.W. 1996. Influence of fragmentation and disturbance on the potential impact of feral predators on native fauna in Australian forest ecosystems. *Wildlife Research,* **23**, 387–400.

May, S.A. 2001. *Aspects of the Ecology of the Cat, Dog and Fox in the Southeast Forests of N.S.W.* PhD thesis, Australian National University, Canberra.

Mayfield, M.M. and Daily, G.C. 2005. Countryside biogeography of neotropical herbaceous and shrubby plants. *Ecological Applications,* **15**, 423–439.

Mayr, E. 1942. *Systematics and the Origin of Species.* Columbia University Press, New York.

McAlpine, C.A., Mott, J.J., Grigg, G.C. and Sharma, P. 1999. The influence of landscape structure on kangaroo abundance in disturbed semiarid woodland. *Rangeland Journal,* **21**, 104–134.

McAlpine, C.M., Lindenmayer, D.B., Eyre, T. and Phinn, S. 2002a. Landscape surrogates for conserving Australia's forest fauna: synthesis of Montreal Process case studies. *Pacific Conservation Biology,* **8**, 108–120.

McAlpine, C.M., Fensham, R.J. and Temple-Smith, D.E. 2002b. Biodiversity conservation and vegetation clearing in Queensland: principles and thresholds. *Rangelands Journal,* **24**, 36–55.

McArthur, C. 2000. Balancing browsing damage management and fauna conservation in plantation forestry. *Tasforests,* **12**, 167–169.

McCallum, H. 2000. *Population Parameters. Estimation for Ecological Models.* Blackwell Science, Oxford, United Kingdom.

McCarthy, M., Lindenmayer, D.B. and Dreschler, M. 1997. Extinction debts and the risks faced by abundant species. *Conservation Biology,* **11**, 221–226.

McCarthy, M.A. 1997. The Allee effect: finding mates and theoretical models. *Ecological Modelling,* **103**, 99–102.

McCarthy, M.A. and Lindenmayer, D.B. 1999. Incorporating metapopulation dynamics of greater gliders into reserve design in disturbed landscapes. *Ecology,* **80**, 651–667.

McCarthy, M.A. and Parris, K.M. 2004. Clarifying the effect of toe clipping on frogs with Bayesian statistics. *Journal of Applied Ecology,* **41**, 780–786.

McCarthy, M.A., Franklin, D.C. and Burgman, M.A. 1994. The importance of demographic uncertainty: an example from the helmeted honeyeater. *Biological Conservation,* **67**, 135–142.

McCarthy, M.A., Lindenmayer, D.B. and Possingham, H.P. 2000. Australian treecrepers and landscape fragmentation: a test of a spatially explicit PVA model. *Ecological Applications*, **10**, 1722–1731.

McCarthy, M.A., Possingham, H.P., Day, J.R. and Tyre, A.J. 2001. Testing the accuracy of population viability analysis. *Conservation Biology*, **73**, 143–150.

McCarthy, M.A., Andelman, S.J. and Possingham, H.P. 2003. Reliability of relative predictions in population viability analysis. *Conservation Biology*, **17**, 982–989.

McCarthy, M.A., Menkhorst, P.W., Quin, B.R., Smales, I.J. and Burgman, M.A. 2004. Assessing options for establishing a new population of the helmeted honeyeater *Lichenostomus melanops cassidix*. Pp. 191–200 in *Species Conservation and Management*. H.R. Akçakaya, M.A. Burgman, O. Kindvall, C.C. Wood, P. Sjorgen-Gulve, J. Hatfield and M.A. McCarthy, eds. Oxford University Press, New York.

McCarty, J.P. 2001. Ecological consequences of recent climate change. *Conservation Biology*, **15**, 320–331.

McCollin, D. 1998. Forest edges and habitat selection in birds: a functional approach. *Ecography*, **21**, 247–260.

McComb, W.C., McGarigal, K., Fraser, J.D. and Davis, W.H. 1991. Planning for basin-level cumulative effects in the Appalachian coal fields. Pp. 138–151 in *Wildlife Habitats in Managed Landscapes*. J.E. Rodiek and E.C. Bolen, eds. Island Press, Covelo, California.

McCoy, E.D. and Mushinsky, H.R. 1999. Habitat fragmentation and the abundances of vertebrates in the Florida scrub. *Ecology*, **80**, 2526–2538.

McCullagh, P. and Nelder, J.A. 1988. *Generalized Linear Models*. 2nd ed. Chapman and Hall, New York.

McGarigal, K. and Cushman, S.A. 2002. Comparative evaluation of experimental approaches to the study of fragmentation studies. *Ecological Applications*, **12**, 335–345.

McGarigal, K. and McComb, W.C. 1992. Streamside versus upslope breeding bird communities in the central Oregon coast range. *Journal of Wildlife Management*, **56**, 10–23.

McGarigal, K. and McComb, W.C. 1995. Relationships between landscape structure and breeding birds in the Oregon Coast Range. *Ecological Monographs*, **65**, 235–260.

McGarigal, K. and McComb, W.C. 1999. Forest fragmentation effects on breeding birds in the Oregon Coast Range. Pp. 223–246 in *Forest Fragmentation: Wildlife and Management Implications*. J.A. Rochelle, L.A. Lehman and J. Wisniewski, eds. Koninklijke Brill NV, Leiden, the Netherlands.

McGrady-Steed, J., Harris, P.M. and Morin, P.J. 1997. Biodiversity regulates ecosystem predictability. *Nature*, **390**, 162–165.

McGuinness, K.A. 1984. Equations and explanations in the study of species–area curves. *Biological Reviews*, **59**, 423–440.

McIntyre, S. 1994. Integrating agriculture and land use and management for conservation of a native grassland flora in a variegated landscape. *Pacific Conservation Biology*, **1**, 236–244.

McIntyre, S. and Barrett, G.W. 1992. Habitat variegation, an alternative to fragmentation. *Conservation Biology*, **6**, 146–147.

McIntyre, S. and Hobbs, R. 1999. A framework for conceptualizing human effects on landscapes and its relevance to management and research models. *Conservation Biology*, **13**, 1282–1292.

McIntyre, S., and Lavorel, S. 1994. Predicting richness of native, rare, and exotic plants in response to habitat and disturbance variables across a variegated landscape. *Conservation Biology*, **8**, 521–531.

McIntyre, S., Barrett, G.W. and Ford, H.A. 1996. Communities and ecosystems. Pp. 154–170 in *Conservation Biology*. I.F. Spellerberg, ed. Longman, Harlow, England.

McIntyre, S., McIvor, J.G. and MacLeod, N.D. 2000. Principles for sustainable grazing in eucalypt woodlands: landscape-scale indicators and the search for thresholds. Pp. 92–100 in *Management for Sustainable Ecosystems*. P. Hale, D. Moloney and P. Sattler, eds. Centre for *Conservation Biology*, University of Queensland, Brisbane, Australia.

McIntyre, S., McIvor, J.G. and Heard, K.M., eds. 2002. *Managing and Conserving Grassy Woodlands*. CSIRO Publishing, Melbourne.

McKenzie, N.L., Burbidge, A. and Rolfe, J.K. 2003. Effect of salinity on small, ground-dwelling animals in the Western Australian wheatbelt. *Australian Journal of Botany*, **51**, 725–740.

McKinney, M.L. 1997. Extinction vulnerability and selectivity: combining ecological and paleantological views. *Annual Review of Ecology and Systematics*, **28**, 495–516.

McNay, R.S., Morgan, J.A. and Bunnell, F.L. 1994. Characterizing independence of observations in movements of Colombian black-tailed deer. *Journal of Wildlife Management,* **58**, 422–429.

McNeely, J.A., Miller, K.R., Reid, W.V., Mittermeier, R.A. and Werner, T.B. 1990. *Conserving the World's Biological Diversity.* International Union for the Conservation of Nature and Natural Resources, Gland, Switzerland.

McNeely, J.A., Neville, L.E. and Rejmanek, M. 2003. When is eradication a sound investment? *Conservation in Practice,* **4**, 30–41.

McShea, W.J. and Rappole, J.H. 2000. Managing the abundance and diversity of breeding bird populations through manipulation of deer populations. *Conservation Biology,* **14**, 1161–1170.

Mech, L.D. 1970. *The Wolf.* Natural History Press, New York.

Mech, S.G. and Hallett, J.G. 2001. Evaluating the effectiveness of corridors: a genetic approach. *Conservation Biology,* **15**, 467–474.

Meggs, J.M. 1997. *Simsons Stag Beetle,* Hoplogonus simonsi, *in Northeast Tasmania: Distribution, Habitat Characteristics and Conservation Requirements.* A report to the Forest Practices Board of Tasmania. Hobart, Tasmania, November, 1997.

Meik, J.M., Jeo, R.M., Mendelson, J.R. and Jenks, K.E. 2002. Effects of bush encroachment on an assemblage of diurnal lizard species in central Namibia. *Biological Conservation,* **106**, 29–36.

Menges, E. 1990. Population viability analyses for an endangered plant. *Conservation Biology,* **4**, 52–62.

Mercer, D. 1995. *A Question of Balance: Natural Resource Conflict Issues in Australia.* 2nd ed. Federation Press, Sydney.

Merritt, B. and Wallis, R. 2004. Are wide revegetated riparian strips better for birds and frogs than narrow ones? *Victorian Naturalist,* **121**, 288–292.

Mesquita, R.C., Delamonica, P. and Laurance, W.F. 1999. Effect of surrounding vegetation on edge-related tree mortality in Amazonian forest fragments. *Biological Conservation,* **91**, 129–134.

Metzger, J.P. 1997. Relationships between landscape structure and tree species diversity in tropical forests of southeast Brazil. *Landscape and Urban Planning,* **37**, 29–35.

Meyer, C.L., Sisk, T.D. and Covington, W.W. 2001. Microclimatic changes induced by ecological restoration of ponderosa pine forests in northern Arizona. *Restoration Ecology,* **9**, 443–452.

Middleton, J. and Merriam, G. 1981. Woodland mice in a farmland mosaic. *Journal of Applied Ecology,* **18**, 703–710.

Mikkelson, G.M. 1993. How do food webs fall apart? A study of changes in trophic structure during relaxation on habitat fragments. *Oikos,* **67**, 539–547.

Milberg, P. and Tyrberg, T. 1993. Native birds and noble savages: a review of man-caused pre-historic extinctions of island birds. *Ecography,* **16**, 229–250.

Milledge, D.R., Palmer, C.L. and Nelson, J.L. 1991. "Barometers of change": the distribution of large owls and gliders in mountain ash forests of the Victorian Central Highlands and their potential as management indicators. Pp. 55–65 in *Conservation of Australia's Forest Fauna.* D. Lunney, ed. Royal Zoological Society of NSW, Sydney, Australia.

Miller, B. and Mullette, K.J. 1985. Rehabilitation of an endangered Australian bird: the Lord Howe Island woodhen *Tricholimnas sylivestris* (Sclater). *Biological Conservation,* **34**, 55–95.

Miller, S.G., Bratton, S.P. and Hadidian, J. 1992. Impacts of white-tailed deer on endangered and threatened vascular plants. *Natural Areas Journal,* **12**, 67–74.

Mills, L.S. and Allendorf, F.W. 1996. The one-migrant-per-generation rule in conservation and management. *Conservation Biology,* **10**, 1509–1518.

Milne, B.T., Turner, M.G., Wiens, J.A. and Johnson, A.R. 1992. Interactions between fractal geometry of landscapes and allometric herbivory. *Theoretical Population Biology,* **41**, 337–353.

Milton, S.J., Dean, W.R.J., Duplessis, M.A. and Siegfried, W.R. 1994. A conceptual model of arid rangeland degradation: the escalating cost of declining productivity. *BioScience,* **44**, 70–76.

Minta, S.C. and Clark, T.W. 1989. Habitat suitability analysis of potential translocation sites for black-footed ferrets in northcentral Montana. Pp. 29–45 in *The Prarie Dog Ecosystem.* T.W. Clark, D. Hinckley and T. Rich, eds. Technical Bulletin No. 2, MT5907. Montana Bureau Land Management Wildlife, Billings, Montana.

Mirande, C., Lacy, R.C. and Seal, U.S. 1991. *Whooping Crane Population Viability Analysis and Species Survival Plan*. Captive Breeding Specialist Group, Species Survival Commission, IUCN, Apple Valley, Minnesota.

Mistry, J. 2000. Savannas. *Progress in Physical Geography,* **24**, 601–608.

Mithen, S.J. and Lawton, J.H. 1986. Food-web models that generate constant predator–prey ratios. *Oecologia,* **69**, 542–550.

Mittelbach, G.G., Steiner, C.F., Scheiner, S.M., Gross, K.L., Reynolds, H.L., Waide, R.B., Willig, M.R., Dodson, S.I., and Gough, L. 2001. What is the observed relationship between species richness and productivity? *Ecology,* **82**, 2381–2396.

Mizutani, F. and Jewell, P.A. 1998. Home-range and movement of leopards *(Panthera pardus)* on a livestock ranch in Kenya. *Journal of Zoology, London,* **244**, 269–286.

Mladenoff, D.J., White, J., Crow, T.R. and Pastor, J. 1994. Applying principles of landscape design and management to integrate old-growth forest enhancement and commodity use. *Conservation Biology,* **8**, 752–762.

Moilanen, A. and Nieminen, M. 2002. Simple connectivity measures in spatial ecology. *Ecology,* **83**, 1131–1145.

Mönkkönen, M. and Mutanen, M. 2003. Occurrence of moths in boreal forest corridors. *Conservation Biology,* **17**, 468–475.

Mönkkönen, M. and Reunanen, P. 1999. On critical thresholds in landscape connectivity: a management perspective. *Oikos,* **84**, 302–305.

Mooney, H.A. and Hobbs, R.J., eds. 2000. *Invasive Species in a Changing World.* Island Press, Washington, D.C.

Moore, M.K. 1977. Factors contributing to blowdown in streamside leave strips on Vancouver Island. *Land Management Report,* **3**, 1–34. British Columbia Forest Service, Vancouver.

Moore, M.M., Huffman, D.W., Fulé, P.Z., Covington, W.W. and Crouse, J.E. 2004. Comparison of historical and contemporary forest structure and composition on permanent plots in southwestern ponderosa pine forests. *Forest Science,* **50**, 162–176.

Morgan, G. 2001. *Landscape Health in Australia. A Rapid Assessment of the Relative Condition of Australia's Bioregions and Subregions.* Environment Australia and National Land and Water Resources Audit, Canberra.

Moritz, C. 1994. Defining "Evolutionarily Significant Units" for conservation. *Trends in Ecology and Evolution,* **9**, 373–375.

Moritz, C. 1999. Conservation units and translocations: strategies for conserving evolutionary processes. *Hereditas,* **130**, 217–228.

Morris, D.W. and Knight, T.W. 1996. Can consumer-resource dynamics explain patterns of guild assembly? *American Naturalist,* **147**, 558–575.

Morrison, M.L., Marcot, B.G. and Mannan, R.W. 1992. *Wildlife Habitat Relationships: Concepts and Applications.* University of Wisconsin Press, Madison.

Morrison, M.L., Hall, L.S., Robinson, S.K., Rothstein, S.I., Hahn, D.C. and Rich, T.D., eds. 1999. Research and management of the brown-headed cowbird in western landscapes. *Studies in Avian Biology,* **18**, 1–312.

Morton, A.C. 1984. The effects of marking and handling on recapture frequencies of butterflies. Pp. 55–58 in *The Biology of Butterflies.* R. Vane-Wright and P.R. Vickery, eds. Academic Press, London.

Mueck, S.G., Ough, K. and Banks, J.C. 1996. How old are wet forest understories? *Australian Journal of Ecology,* **21**, 345–348.

Munro, H.L. and Rounds, R.C. 1985. Selection of artificial nest sites by five sympatric passerines. *Journal of Wildlife Management,* **49**, 264–276.

Murcia, C. 1995. Edge effects on fragmented forests: implications for conservation. *Trends in Ecology and Evolution,* **10**, 58–62.

Murphy, D.D. 1989. Conservation and confusion: wrong species, wrong scale, wrong conclusion. *Conservation Biology,* **3**, 82–84.

Murphy, D.D. and Noon, B.R. 1992. Integrating scientific methods with habitat conservation planning: reserve design for northern spotted owls. *Ecological Applications,* **2**, 3–17.

Murphy, D.M., Freas, K.E. and Weiss, S.T. 1990. An environment-metapopulation approach to population viability for a threatened invertebrate. *Conservation Biology,* **4**, 41–51.

Murtaugh, P.A. 2002. Journal quality, effect size and publication bias in meta-analysis. *Ecology,* **83**, 1162–1166.

Myers, J.H., Simberloff, D., Kuris, A.M. and Carey, J.R. 2000. Eradication revisited: dealing with exotic species. *Trends in Ecology and Evolution,* **15**, 196–199.

Naeem, S., Chain, F.S., Costanza, R., Ehrlich, P., Golley, F., Hoper, D.U., Kawton, J.H., O'Neill, R.V., Mooney, H.A., Sala, O.E., Symstad, A.J. and Tilman, D. 1999. Biodiversity and ecosystem functioning: maintaining natural life support systems. *Issues in Ecology,* **4**, Fall 1999.

Naiman, R.J., ed. 1992. *Watershed Management: Balancing Sustainability and Environmental Change.* Springer-Verlag, New York.

Naiman, R.J., Decamps, H. and Pollock, H. 1993. The role of riparian corridors in maintaining regional biodiversity. *Ecological Applications,* **3**, 209–212.

National Board of Forestry [Sweden]. 1995. *A Richer Forest.* National Board of Forestry, Jonkoping, Sweden.

National Board of Forestry [Sweden]. 1996. *The Swedish Forest: A Compilation of Facts on Forestry and the Forest Industries in Sweden.* National Board of Forestry, Jonkoping, Sweden.

Nautiyal, S., Rao, K.S., Maikhuri, R.K. and Saxena, K.G. 2003. Transhuman pastoralism in the Nanda Devi Biosphere Reserve, India: a case study in the buffer zone. *Mountain Research and Development,* **23**, 255–262.

Nelson, M.E. 1993. Natal dispersal and gene flow in white-tailed deer in northeastern Minnesota. *Journal of Mammalogy,* **74**, 316–322.

Ness, J.H. 2004. Forest edges and fire ants alter the seed shadow of an ant-dispersed plant. *Oecologia,* **138**, 448–454.

Nève, G., Barascud, B., Hughes, R., Aubert, J., Descimon, H., Lebrun, P. and Baguette, M. 1996. Dispersal, colonization power and metapopulation structure in the vulnerable butterfly *Proclossiana enomia* (Lepidoptera: Nymphalidae). *Journal of Applied Ecology,* **33**, 14–22.

New, T. 2000. *Conservation Biology: An Introduction for Southern Australia.* Oxford University Press, Melbourne.

Newdick, M.T. 1983. *The Behavioural Ecology of Urban Foxes,* Vulpes vulpes, *in Oxford.* PhD Thesis, University of Oxford.

Newman, D. and Tallmon, D.A. 2001. Experimental evidence for beneficial fitness effects of gene flow in recently isolated populations. *Conservation Biology,* **15**, 1054–1063.

Newmark, W.D. 1987. A land-bridge island perspective on mammalian extinctions in western North American parks. *Nature,* **325**, 430–432.

Newton, I. 1994. The role of nest sites in limiting the numbers of hole-nesting birds: a review. *Biological Conservation,* **70**, 265–276.

Newton, I. 1995. The contribution of some recent research on birds to ecological understanding. *Journal of Animal Ecology,* **64**, 675–696.

Nicholls, A.O. 1989. How to make biological surveys go further with generalised linear models. *Biological Conservation,* **50**, 51–75.

Nicholls, A.O. 1991. Examples of the use of generalised linear models in analysis of survey data for conservation evaluation. Pp. 54-63 in *Nature Conservation: Cost Effective Biological Surveys and Data Analysis.* Margules, C.R. and Austin, M.P., eds. CSIRO Australia, Melbourne.

Nicholls, A.O. 1994. Variation in mosaic diversity in the forests of coastal northern New South Wales. *Pacific Conservation Biology,* **1**, 177–182.

Niemelä, J., Langor, D. and Spence, J.R. 1993. Effects of clear-cut harvesting on boreal ground-beetle assemblages (Coleoptera: Carabidae) in western Canada. *Conservation Biology,* **7**, 551–561.

Niemi, G.J., Hanowski, J.M., Lima, A.R., Nicholls, T. and Weiland, N. 1997. A critical analysis on the use of indicator species in management. *Journal of Wildlife Management,* **61**, 1240–1252.

Noor, M.A.F. 2002. Is the biological species concept showing its age? *Trends in Ecology and Evolution,* **17**, 153–154.

Nores, M. 1995. Insular biogeography of birds on mountain-tops in northwestern Argentina. *Journal of Biogeography,* **22**, 61–70.

Norris, R.D. and Stutchbury, B.J. 2001. Extraterritorial movements of a forest songbird in a fragmented landscape. *Conservation Biology,* **15**, 729–736.

Norton, D.A., Hobbs, J.J. and Atkins, L. 1995. Fragmentation, disturbance, and plant distribution: mistletoes in woodland remnants in the western Australian wheatbelt. *Conservation Biology,* **9**, 426–438.

Noss, R.F. 1987 Corridors in real landscapes: a reply to Simberloff and Cox. *Conservation Biology,* **1**, 159–164.

Noss, R.F. 1990. Indicators for monitoring biodiversity: a hierarchical approach. *Conservation Biology,* **4**, 355–364.

Noss, R.F. 1991. Landscape connectivity: different functions at different scales. Pp. 27–39 in *Landscape Linkages and Biodiversity.* W.E. Hudson, ed. Island Press, Covelo, California.

Noss, R.F. and Beier, P. 2000. Arguing over little things: response to Haddad et al. *Conservation Biology,* **15**, 1546–1548.

Noss, R.F. and Cooperrider, A.Y. 1994. *Saving Nature's Legacy: Protecting and Restoring Biodiversity.* Island Press, Covelo, California.

Noss, R.F. and Harris, L.D. 1986. Nodes, networks, and MUMs: preserving diversity at all scales. *Environmental Management,* **10**, 299–309.

Noss, R.F., Csuti, B. and Groom, M.J. 2005. Habitat fragmentation. Pp. 213–251 in *Principles of Conservation Biology.* 3rd ed. M.J. Groom, G.K. Meffe, and R.C. Carroll, eds. Sinauer Associates, Sunderland, Massachusetts.

Noss, R.F., Beier, P., Covington, W., Grumbine, E., Lindenmayer, D.B., Prather, J., Schmiegelow, F., Sisk, T. and Vosick, D. 2006. Integrating ecological restoration and conservation biology: a case study for ponderosa pine ecosystems of the Southwest. *Restoration Ecology,* **14**, 4–10.

Nupp, T.E. and Swihart, R.K. 2000. Landscape-level correlates of small-mammal assemblages in forest fragments of farmland. *Journal of Mammalogy,* **81**, 512–526.

O'Grady, J.J., Reed, D.H., Brook, B.W. and Frankham, R. 2004. What are the best correlates of predicted extinction risk? *Biological Conservation,* **118**, 513–520.

O'Dowd, D.J., Green, P.T. and Lake, P.S. 2003. Invasional "meltdown" on an oceanic island. *Ecology Letters,* **6**, 812–818.

Oakleaf, B., Luce, B., Thorne, E.T. and Williams, B. 1993. *Black-Footed Ferret Reintroduction in Shirley Basin: 1992 Completion Report.* Wyoming Game and Fish Department, Cheyene, Wyoming.

Oaks, J.L., Gilbert, M., Virani, M.Z., Watson, R.T., Meteyer, C.U., Rideout, B.A., Shivaprasad, H.L., Ahmed, S., Chaudhry, M.J.I., Arshad, M., Mahmood, S., Ali, A. and Khan, A.A. 2004. Diclofenac residues as the cause of vulture population decline in Pakistan. *Nature,* **427**, 630–633.

Odum, E.P. 1971. *Fundamentals of Ecology.* W.B. Saunders, Philadelphia.

Økland, B. 1996. Unlogged forests: important sites for preserving the diversity of mycetophilids (Diptera: Sciaroidea). *Biological Conservation,* **76**, 297–310.

Oksanen, L. 2001. Logic of experiments in ecology: is pseudoreplication a pseudoissue? *Oikos,* **94**, 27–38.

Olesen, J.M. and Valido, A. 2003. Lizards as pollinators and seed dispersers: an island phenomenon. *Trends in Ecology and Evolution,* **18**, 177–181.

Oliver, C.D. and Larson, B.C. 1996. *Forest Stand Dynamics.* John Wiley and Sons, New York.

Oliver, C.D., Adams, D., Bonnicksen, T., Bowyer, J., Cubbage, F., Sampson, N., Schlarblum, S., Whaley, R. and Wiant, H. 1997. *Report on Forest Health of the United States.* Chartered by Charles Taylor, Member, United States Congress, 11th District, North Carolina.

Oliver, W.L. 1984. The distribution and status of the hispid hare *Caprolagus hispidis:* the summarized findings of the 1984 pigmy hog and hispid hare field survey in northern Bangaldesh, southern Nepal and northern India. *Jersey Wildlife Preservation Trust,* **21**, 6–32.

Oliver, W.L. 1987. The pigmy hog and the hispid hare: case histories of conservation problems and related considerations in northeastern India. Pp. 67–82 in *The Conservation of Indian Heritage.* B. Allchin, F.R. Allchin and B.K. Thepar, eds. Cosmo, New Dehli.

Olsen, P., Fuller, P. and Marples, T.G. 1993. Pesticide-related eggshell thinning in Australian raptors. *Emu,* **93**, 1–11.

Olsson, M., Gullberg, A. and Tegelstrom, H. 1996. Malformed offspring, sibling matings, and selection against inbreeding in the sand lizard *(Lacerta agilis). Journal of Evolutionary Biology,* **9**, 229–242.

O'Neill, R.V., Krummel, J.R., Gardener, R.H., Sugihara, G., DeAngelis, D.L., Milne, B.T., Turner, M.G., Zysmunt, B., Christensen, S.W., Dale, V.H. and Graham, R.L. 1988. Indices of landscape pattern. *Landscape Ecology,* **1**, 153–162.

Osenberg, C.W., Sarnelle, O., Cooper, S.D. and Holt, R.D. 1999. Resolving ecological questions through meta-analysis: goals, metrics, and models. *Ecology,* **80**, 1105–1117.

Ostfeld, R.S. and LoGuidice, K. 2003. Community disassembly, biodiversity loss, and the erosion of an ecosystem service. *Ecology,* **84**, 1421–1427.

Ostrowski, S., Bedin, E., Lenain, D.M. and Abuzinada, A.H. 1998. Ten years of Arabian oryx conservation breeding in Saudi Arabia: achievements and regional perspectives. *Oryx,* **32**, 209–222.

Ough, K. 2001. Regeneration of wet forest flora a decade after clearfelling or wildfire: is there a difference? *Australian Journal of Botany*, **49**, 645–664.

Pacala, S.W. and Tilman, D. 1994. Limiting similarity in mechanistic and spatial models of plant competition on heterogenous environments. *American Naturalist*, **143**, 222–257.

Paine, R.T. 1969. A note on trophic complexity and community stability. *American Naturalist*, **103**, 91–93.

Paine, R.T., Tegner, M.J. and Johnson, E.A. 1998. Compounded perturbations yield ecological surprises. *Ecosystems*, **1**, 535–545.

Palmer, M.W. and White, P.S. 1994. Scale dependence and the species–area relationship. *American Naturalist*, **144**, 717–740.

Palomares, F. 2001. Vegetation structure and prey abundance requirements of the Iberian lynx: implications for the design of reserves and corridors. *Journal of Applied Ecology*, **38**, 9–18.

Palomares, F., Delibes, M., Ferreras, P., Fedriani, J.M., Calzada, J. and Revilla, E. 2000. Iberian lynx in a fragmented landscape: predispersal, dispersal and postdispersal habitats. *Conservation Biology*, **14**, 809–818.

Pardini, R. 2004. Effects of forest fragmentation on small mammals in an Atlantic forest landscape. *Biodiversity and Conservation*, **13**, 2567–2586.

Parker, L. 1997. Restaging an evolutionary drama: thinking big on the Chequamegon and Nicolet National forests. Pp. 218–219 in *Creating a Forestry for the 21st Century*. K.A. Kohm and J.F. Franklin, eds. Island Press, Covelo, California.

Parker, M. and Mac Nally, R. 2002. Habitat loss and the habitat fragmentation threshold: an experimental evaluation of impacts on richness and total abundances using grassland invertebrates. *Biological Conservation*, **105**, 217–229.

Parker, T.H., Stansberry, B.M., Becker, C.D. and Gipson, P.S. 2005. Edge and area effects on the occurrence of migrant songbirds. *Conservation Biology*, **19**, 1157–1167.

Parks, S.A. and Harcourt, A.H. 2002. Reserve size, local human density, and mammalian extinction in U.S. protected areas. *Conservation Biology*, **16**, 800–808.

Parmesan, C. 1996. Climate and species' range. *Nature*, **383**, 765–766.

Parry, B.B. 1997. *Abiotic Edge Effects in Wet Sclerophyll Forest in the Central Highlands of Victoria*. MSc. thesis, School of Botany, University of Melbourne.

Paton, D.C. 2000. Disruption of bird–plant pollination systems in southern Australia. *Conservation Biology*, **14**, 1232–1234.

Paton, P.W. 1994. The effect of edge on avian nest success: how strong is the evidence? *Conservation Biology*, **8**, 17–26.

Patterson, B.D. 1987. The principle of nested subsets and its implications for biological conservation. *Conservation Biology*, **1**, 247–293.

Patterson, B.D. and Atmar, W. 1986. Nested subsets and the structure of insular mammalian faunas and archipelagos. *Biological Journal of the Linnean Society*, **28**, 65–82.

Patterson, B.D. and Brown, J.H. 1991. Regionally nested patterns of species composition in granivorous rodent assemblages. *Journal of Biogeography*, **18**, 395–402.

Patton, D.R. 1974. Cutting increases deer and edges of a pine forest in Arizona. *Journal of Forestry*, **December 1974**, 764–766.

Peacock, M.M. and Smith, A.T. 1997. The effect of habitat fragmentation on dispersal patterns, mating behaviour, and genetic variation in a pika *(Ochotona princeps)* metapopulation. *Oecologia*, **112**, 524–533.

Peakall, R., Ruibal, M. and Lindenmayer, D.B. 2003. Spatial autocorrelation analysis offers new insights into gene flow in the Australian bush rat *(Rattus fuscipes)* biology. *Evolution*, **57**, 1182–1195.

Pearce, J. and Ferrier, S. 2000. Evaluating the predictive performance of habitat models developed using logistic regression. *Ecological Modelling*, **133**, 225–245.

Pears, A. 2000. *Wood Heating: Its Impacts, Its Use and Its Future: A Burning Issue*. Bendigo, Victorian National Parks Association, Australia.

Pearson, S.M., Turner, M.G., Gardner, R.H. and O'Neill, R.V. 1996. An organism perspective of habitat fragmentation. Pp. 77–95 in *Biodiversity in Managed Landscapes: Theory and Practice*. R.C. Szaro and D.W. Johnston, eds. Oxford University Press, New York.

Peek, M.S., Leffler, A.J., Flint, S.D. and Ryel, R.J. 2003. How much variance can be explained by ecologists and evolutionary biologists? Additional perspectives. *Oecologia*, **137**, 161–170.

Perault, D.R. and Lomolino, M. 2000. Corridors and mammal community structure across a fragmented, old-growth forest landscape. *Ecological Monographs,* **70**, 401–422.

Perera, A.H., Buse, L.J. and Weber, M.G., eds. 2004. *Emulating Natural Forest Landscape Disturbances: Concepts and Applications.* Columbia University Press, New York.

Perry, D.A. 1994. *Forest Ecosystems.* Johns Hopkins University Press, Baltimore.

Perry, G., Rodda, G.H., Fritts, T.H. and Sharp, T.R. 1998. The lizard fauna of Guam's fringing islets: island biogeography, phylogenetic history, and conservation implications. *Global Ecology and Biogeography,* **7**, 353–365.

Peterken, G.F. 1996. *Natural Woodland: Ecology and Conservation in Northern Temperate Regions.* Cambridge University Press, Cambridge.

Peterken, G.F. and Francis, J.L. 1999. Open spaces as habitats for vascular ground flora species in the woods of central Lincolnshire, UK. *Biological Conservation,* **91**, 55–72.

Peterken, G.F. and Ratcliffe, P.R., 1995. The potential of biodiversity in British upland spruce forests. *Forest Ecology and Management,* **79**, 153–160.

Peters, R.H. 1991. *A Critique for Ecology.* Cambridge University Press, Cambridge.

Peters, R.L. and Lovejoy, T.E., eds. 1992. *Global Warming and Biological Diversity.* Yale University Press, New Haven.

Peterson, G., Allen, C.R. and Holling, C.S. 1998. Ecological resilience, biodiversity, and scale. *Ecosystems,* **1**, 6–18.

Peterson, G.D., Cumming, G.S. and Carpenter, S.R. 2003a. Scenario planning: a tool for conservation in an uncertain world. *Conservation Biology,* **17**, 358–366.

Peterson, G.D., Beard, T.D., Beisner, B.E., Bennet, E.M., Carpenter, S.R., Cumming, G.S., Dent, C.L. and Havlicek, T.D. 2003b. Assessing future ecosystem services a case study of the Northern Highlands Lake District, Wisconsin. *Conservation Ecology,* **7**, **1**. http://www.ecologyandsociety.org/vol7/iss3/art1/.

Petit, L.J. and Petit, D.R. 2003. Evaluating the importance of human-modified lands for neo-tropical bird conservation. *Conservation Biology,* **17**, 687–694.

Petit, S. and Burel, F. 1998. Effects of landscape dynamics on the metapopulation of a ground beetle (Coleoptera, Carabidae) in a hedgerow network. *Agriculture, Ecosystems and Environment,* **69**, 243–252.

Pharo, E., Lindenmayer, D.B. and Taws, N. 2004. The response of bryophytes to landscape context: a large-scale fragmentation study. *Journal of Applied Ecology,* **41**, 910–921.

Pianka, E.R. 1973. The structure of lizard communities. *Annual Review of Ecology and Systematics,* **4**, 53–74.

Pickett, S. Kolasa, T.A.J. and Jones, C.G. 1994. *Ecological Understanding.* Academic Press, San Diego.

Pickett, S.T. and Thompson, J.H. 1978. Patch dynamics and the design of nature reserves. *Biological Conservation,* **13**, 27–37.

Pimentel, D., Lach, L., Zuniga, R. and Morrison, D. 2000. Environmental and economic costs of non-indigenous species introductions in the United States. *BioScience,* **50**, 53–65.

Pimm, S.L. 1992. *Balance of Nature.* University of Chicago Press, Chicago.

Pimm, S.L. 1995. Dead reckoning: getting hard numbers on extinction rates in paradise. *Sciences-New York,* **35**, 15–17.

Pimm, S.L., Jones, H.L. and Diamond, J. 1988. On the risk of extinction. *American Naturalist,* **132**, 757–785.

Pimm, S.L., Russell, G.J., Gittleman, J.L. and Brooks, T.M. 1995. The future of biodiversity. *Science,* **269**, 347–350.

Piper, S.D. and Catterall, C.P. 2003. A particular case and a general pattern: hyperaggressive behaviour by one species mediates avifaunal decreases in fragmented Australian forests. *Oikos,* **101**, 602–614.

Piper, S., Catterall, C.P. and Olsen, M. 2002. Does adjacent land use affect predation of artificial shrub-nests near eucalypt forest edges? *Wildlife Research,* **29**, 127–133.

Pither, J. and Taylor, P.D. 1998. An experimental assessment of landscape connectivity. *Oikos,* **83**, 166–174.

Polhemus, D.A. 1993. Conservation of aquatic insects: worldwide crisis or localized threats? *American Zoologist,* **33**, 588–598.

Pope, M.L. 2003. *A Study of the Greater Glider* (Petauroides volans) *Persisting in Remnant Eucalypt*

Patches Surrounded by a Softwood Plantation. MSc thesis, Australian National University, Canberra, Australia.

Pope, M.L., Lindenmayer, D.B. and Cunningham, R.B. 2004 Patch use by the greater glider in a fragmented forest ecosystem. I. Home range size and movements. *Wildlife Research,* **31**, 559–568.

Pope, S.E., Fahrig, L. and Merriam, H.G. 2000. Landscape complementation and metapopulation effects on leopard frog populations. *Ecology,* **81**, 2498–2508.

Population Action International 2000. *Nature's Place: Human Population and the Future of Biological Diversity.* Population Action International, Washington, D.C.

Possingham, H.P. and Davies, I. 1995. ALEX: A model for the viability analysis of spatially structured populations. *Biological Conservation,* **73**, 143–150.

Possingham, H.P., Lindenmayer, D.B. and McCarthy, M.A. 2001. Population viability analysis. Pp. 831–843 in *Encyclopaedia of Biodiversity,* vol. 4. S. Levin, ed. Academic Press, San Diego.

Potter, M.A. 1990. Movement of North Island brown kiwi *(Apteryx australis mantelli)* between forest remnants. *New Zealand Journal of Ecology,* **14**, 17–24.

Powell, A.H. and Powell, G.V. 1987. Population dynamics of male euglossine bees in Amazonian forest fragments. *Biotropica,* **19**, 176–179.

Power, M.E., Tilman, D., Estes, J.A., Menge, B.A., Bond, W.J., Mills, L.S., Daily, G., Castilla, J.C., Lubchenco, J. and Paine, R.T. 1996. Challenges in the quest for keystones. *BioScience,* **46**, 609–620.

Prance, G.T. 1991. Rates of loss of biodiversity: a global view. Pp. 27–44 in *The Scientific Management of Temperature Forest Communities for Conservation.* I.F. Spellerberg, F.B. Goldsmith and M.G. Morris, eds. Blackwell Scientific, Oxford.

Pressey, R.L. 1995. Conservation reserves in N.S.W.: crown jewels or leftovers? *Search,* **26**, 47–51.

Preston, F.W. 1962. The canonical distribution of commonness and rarity. *Ecology,* **43**, 185–215.

Price, O.F., Woinarski, J.C.Z. and Robinson. D. 1999. Very large requirements for frugivorous birds in monsoon rainforests of the Northern Territory, Australia. *Biological Conservation,* **91**, 169–180.

Primack, R. 2001. Causes of extinction. Pp. 697–713. in *Encyclopaedia of Biodiversity,* vol. 2. S.A. Levin, ed. Academic Press, San Diego.

Primack, R. 2002. *Essentials of Conservation Biology.* 3rd ed. Sinauer Associates, Sunderland, Massachusetts.

Primack, R.B. 1993. *Essentials of Conservation Biology.* Sinauer Associates, Sunderland, Massachusetts.

Prober, S. and Thiele, K.R. 1995. Conservation of the grassy white box woodlands: relative contributions of size and disturbance to floristic composition and diversity of remnants. *Australian Journal of Botany,* **43**, 349–366.

Pulliam, H.R., Dunning, J.B. and Liu, J. 1992. Population dynamics in complex landscapes: a case study. *Ecological Applications,* **2**, 165–177.

Pullin, A.A. and Knight, T.M. 2001. Effectiveness in conservation practice: pointers from medicine and public health. *Conservation Biology,* **15**, 50–54.

Pullin, A.S. 2002. *Conservation Biology.* Cambridge University Press, Cambridge.

Putz, F.E., Redford, K.H., Robinson, J.G., Fimbel, R. and Bate, G.M. 2000. *Biodiversity Conservation in the Context of Tropical Forest Management.* Paper No. 75. Biodiversity Series—Impact studies. World Bank Environment Department Papers, The World Bank, Washington D.C., U.S.A. September, 2000.

Pyke, G., Saillard, R. and Smith, R. 1995. Abundance of eastern bristlebirds in relation to habitat and fire history. *Emu,* **95**, 106–110.

Quinn, G.P. and Keough, M.J. 2002. *Experimental Design and Data Analysis for Biologists.* Cambridge University Press, Cambridge.

Quinn, J.F. and Hastings, A. 1987. Extinction in subdivided habitats. *Conservation Biology,* **1**, 198–208.

Quinn, J.M., Boothroyd, I.K. and Smith, B.J. 2004. Riparian buffers mitigate effects of pine plantation logging on New Zealand streams. II. Invertebrate communities. *Forest Ecology and Management,* **191**, 129–146.

Rabinowitz, A. 1995. Helping a species go extinct: the Sumatran Rhino in Borneo. *Conservation Biology,* **9**, 482–488.

RAC (Resource Assessment Commission). 1991. *Forest and Timber Inquiry, Draft Report,* Vol. 1. Australian Government Publishing Service, Canberra.

Radford, J.Q. and Bennett, A.F. 2004. Thresholds in landscape parameters: occurrence of the white-browed treecreeper *Climacteris affinis* in Victoria, Australia. *Biological Conservation*, **117**, 375–391.

Radford, J.Q., Bennett, A.F. and Cheers, G.J. 2005. Landscape-level thresholds of habitat cover for woodland-dependent birds. *Biological Conservation*, **124**, 317–337.

Raff, M. 2003. *Private Property and Environmental Responsibility: A Comparative Study of German Real Property Law.* Aspen Law and Business.

Ralls, K., Harvey, P.H. and Lyles, A.M. 1986. Heterozygosity and fitness in natural populations of animals. Pp. 35–56 in *Conservation Biology: The Science of Scarcity and Diversity.* M.E. Soulé, ed. Sinauer, Sunderland, Massachusetts.

Ralph, C.J., Sauer, J.R. and Droege, S. 1997. *Monitoring Bird Populations by Point Counts.* USDA Forest Service. General Technical Report PSW-GTR-149. Pacific Southwest Research Station, Albany, California.

Ramirez-Smith, C. and Armesto, J.J. 2003. Foraging behavior of bird pollinators on *Embothrium coccineum* (Proteaceae) trees in forest fragments and pastures in southern Chile. *Austral Ecology*, **28**, 53–60.

Rand, G.M. and Newman, J.R. 1998. The applicability of habitat evaluation methodologies in ecological risk assessment. *Human and Ecological Risk Assessment*, **4**, 905–929.

Ratcliffe, D. 1979. The end of the large blue butterfly. *New Scientist*, **8**, 457–458.

Read, J.L. 1998. Are geckos useful bioindicators of air pollution? *Oecologia*, **114**, 180–187.

Reading, R.P., Clark, T.W., Seebeck, J.H. and Pearce, J. 1996. Habitat suitability index model for the eastern barred bandicoot, *Parameles gunnii*. *Wildlife Research*, **23**, 221–235.

Recher, H.F. 1988. Counting terrestrial birds: use and application of census procedures in Australia. *Australian Zoological Reviews*, **1**, 25–45.

Recher, H.F. 1999. The state of Australia's avifauna: a personal opinion and prediction for new millennium. *Australian Zoologist*, **31**, 11–29.

Recher, H.F., Shields, J., Kavanagh, R.P. and Webb, G. 1987. Retaining remnant mature forest for nature conservation at Eden, New South Wales: a review of theory and practice. Pp. 177–94 in *Nature Conservation: The Role of Remnants of Vegetation.* D.A. Saunders, G.W. Arnold, A.A. Burbidge and A.J. Hopkins, eds. Surrey Beatty and Sons, Chipping Norton.

Redford, K.H. 1992. The empty forest. *BioScience*, **42**, 412–422.

Redpath, S.M. 1995. Habitat fragmentation and the individual: tawny owls *(Strix aluco)* in woodland patches. *Journal of Animal Ecology*, **64**, 652–661.

Reed, D.H., O'Grady, J.J., Brook, B.W., Ballou, J.D. and Frankham, R. 2003. Estimates of minimum viable population sizes for vertebrates and factors influencing those estimates. *Biological Conservation*, **113**, 23–34.

Reed, J.M. and Dobson, A.P. 1993. Behavioural constraints and conservation biology: conspecific attraction and recruitment. *Trends in Evolution and Ecology*, **8**, 397–401.

Reed, J.M., Mills, L.S., Dunning, J.B., Menges, E.S., McKelvey, K.S., Frye, R., Beissenger, S.R., Antett, M. and Miller, P. 2002. Emerging issues in population viability analysis. *Conservation Biology*, **16**, 7–19.

Regan, H.M., Colyvan, M. and Burgman, M.A. 2002. A taxonomy and treatment of uncertainty for ecology and conservation biology. *Ecological Applications*, **12**, 618–628.

Reid, J.R. 1999. *Threatened and Declining Birds in the New South Wales Sheep–Wheat Belt. I. Diagnosis, Characteristics and Management.* Consulting Report to NSW National Parks and Wildlife Service, Canberra.

Renjifo, L.M. 2001. Effect of natural and anthropogenic landscape matrices on the abundance of sub-Andean bird species. *Ecological Applications*, **11**, 14–31.

Revilla, E., Palomares, F. and Delibes, M. 2001. Edge–core effects and the effectiveness of traditional reserves in conservation: Eurasian badgers in Donana National Park. *Conservation Biology*, **15**, 148–158.

Reynolds, J.D. 2003. Life histories and extinction risk. Pp. 195–217 in *Macroecology: Concepts and Consequences.* T.M. Blackburn and K.J. Gaston, eds. Blackwell, Oxford.

Ribon, R., Simon, J.E. and De Mattos, G.T. 2003. Bird extinctions in Atlantic forest fragments of the Vicosa region, Southeastern Brazil. *Conservation Biology*, **17**, 1827–1839.

Richardson, D.M., Williams, P.A. and Hobbs, R.J. 1994. Pine invasions in the southern hemisphere: determinants of spread and invadability. *Journal of Biogeography*, **21**, 511–527.

Ricketts, T.H. 2001. The matrix matters: effective isolation in fragmented landscapes. *American Naturalist,* **158**, 87–99.

Ricketts, T.H., Daily, G.C., Ehrlich, P.R. and Fay, J.P. 2001. Countryside biogeography of moths in a fragmented landscape: biodiversity in native and agricultural habitats. *Conservation Biology,* **15**, 378–388.

Ricketts, T.H., Daily, G.C., Ehrlich, P.R. and Michener, C.D. 2004. Economic value of tropical forest to coffee production. *Proceedings of the National Academy of Sciences of the United States of America,* **101**, 12579–12582.

Ries, L. and Debinski, D.M. 2001. Butterfly responses to habitat edges in the highly fragmented prairies of central Iowa. *Journal of Animal Ecology,* **70**, 840–852.

Ries, L., Fletcher, R.J., Battin, J. and Sisk, T.D. 2004. Ecological responses to habitat edges: mechanisms, models, and variability explained. *Annual Review of Ecology Evolution and Systematics,* **35**, 491–522.

Ripple, W.J., and Beschta, R.L. 2005. Linking wolves and plants: Aldo Leopold on trophic cascades. *BioScience,* **55**, 613–621.

Roach, J.L. 2001. Genetic structure of a metapopulation of black-tailed prarie dogs. *Journal of Mammalogy,* **82**, 946–959.

Robbins, C.S. 1978. Census techniques for forest birds. Pp. 142–163 in *Proceedings of the Workshop: Management of Southern Forests for Nongame Birds.* USDA, Forest Service General Technical Report SE-14.

Robbins, C.S. 1980. Effect of forest fragmentation on breeding bird populations in the Piedmont of the mid-Atlantic region. *Atlantic Naturalist,* **33**, 31–36.

Roberge, J-M. and Angelstam, P. 2004. Usefulness of the umbrella species concept as a conservation tool. *Conservation Biology,* **18**, 76–85.

Roberts, W.B. 1973. Air movements within a plantation and an open area and their effects on fire behaviour. *Australian Forest Research,* **4**, 41–47.

Robertshaw, J.D. and Harden, R.H. 1989. Predation on Macropodoidea: a review. Pp. 735–753 in *Kangaroos, Wallabies and Rat Kangaroos.* G. Grigg, P. Jarman and I. Hume, eds. Surrey Beatty and Sons, Chipping Norton, Sydney.

Robertson, A.W., Kelly, D., Ladley, J. and Sparrow, A.D. 1999. Effects of pollinator loss on endemic New Zealand mistletoes (Loranthaceae). *Conservation Biology,* **13**, 499–508.

Robertson, K.R., Anderson, R.C. and Schwartz, M.W. 1997. The tallgrass prairie mosaic. Pp. 55–87 in *Conservation in Highly Fragmented Landscapes.* M.W. Schwartz, ed. Chapmen and Hall, New York.

Robertson, M., Nichols, P., Horwitz, P., Bradby, K. and MacKintosh, D. 2000. Environmental narratives and the need for multiple perspectives to restore degraded landscapes in Australia. *Ecosystem Health,* **6**, 119–133.

Robinson, G.R., Holt, R.D., Gaines, M.S., Hamburg, S.P., Johnson, M.L., Fitch, H.S. and Martinko, E.A. 1992. Diverse and contrasting effects of habitat fragmentation. *Science,* **257**, 524–526.

Robinson, S.K., Thompson, E.R., Donovan, T.M., Whitehead, D.R. and Faaborg J. 1995. Regional forest fragmentation and the nesting success of migratory birds. *Science,* **267**, 1987–1990.

Robinson, W.D. 1999. Long-term changes in the avifauna of Barro Colorado Island, Panama, a tropical forest isolate. *Conservation Biology,* **13**, 85–97.

Rodrigues, A.S.L., Andelman, S.J., Bakarr, M.I., Biotani, L., Brooks, T.M., Cowling, R.M., Fishpool, L.D.C., da Fonseca, G.A.B., Gaston, K.J., Hoffmann, M., Long, J.S., Marquet, P.A., Pilgrim, J.D., Pressey, R.L., Schipper, J., Sechrest, W., Stuart, S.N., Underhill, L.G., Waller, R.W., Watts, M.E.J. and Yan, X. 2004. Effectiveness of the global protected area network in representing species diversity. *Nature,* **428**, 640–643.

Rodriguez, A., Crema, G. and Delibes, M. 1996. Use of non-wildlife passages across a high speed railway by terrestrial vertebrates. *Journal of Applied Ecology,* **33**, 1527–1540.

Rodríguez, A., Andrén, H. and Jansson, G. 2001. Habitat-mediated predation risk and decision-making of small birds at forest edges. *Oikos,* **95**, 383–396.

Roemer, G.W., Donlan, C.J. and Courchamp, F. 2002. Golden eagles, feral pigs, and insular carnivores: how exotic species turn native predators into prey. *Proceedings of the National Academy of Sciences of the United States of America,* **99**, 791–796.

Rolstad, J. and Wegge, P. 1987. Distribution and size of capercaillie leks in relation to old forest fragmentation. *Oecologia,* **72**, 389–394.

Rolstad, J., Gjerdde, I., Gundersen, V.S. and Saetersal, M. 2002. Use of indicator species to assess forest continuity: a critique. *Conservation Biology*, **16**, 253–257.

Rolston, H. 1985. Duties to endangered species. *BioScience*, **35**, 718–726.

Romme, W.H. and Despain, D.G. 1989. Historical perspectives on the Yellowstone fires of 1988. *BioScience*, **39**, 695–699.

Rooney, S.M., Wolfe, A. and Hayden, T.J. 1998. Autocorrelated data in telemetry studies: time to independence and the problem of behavioural effects. *Mammal Review*, **29**, 89–98.

Roos, S. 2002. Functional response, seasonal decline and landscape differences in nest predation risk. *Oecologia*, **133**, 606–615.

Roper, J.J. 1992. Nest predation experiments with quail eggs: too much to swallow? *Oikos*, **65**, 528–530.

Rosenberg, D.K., Noon, B.R. and Meslow, E.C. 1997. Biological corridors: form, function and efficacy. *BioScience*, **47**, 677–687.

Rosenberg, D.K., Noon, B.R., Megahan, J.W. and Meslow, E.C. 1998. Compensatory behaviour of *Ensatina eschscholtzii* in biological corridors: a field experiment. *Canadian Journal of Zoology*, **76**, 117–133.

Rosenberg, M.S., Adams, D.C. and Gurewich, J. 2000. *Metawin: Statistical Software for Meta-Analysis*. Version 2.0. Sinauer Associates, Sunderland, Massachusetts.

Rosenblatt, D.L., Heske, E.J., Nelson, S.L., Barber, D.H., Miller, M.A. and MacAllister, B. 1999. Forest fragments in east-central Illinois: islands or habitat patches for mammals? *American Midland Naturalist*, **141**, 115–123.

Rosenzweig, M.L. 1995. *Species Diversity in Space and Time*. Cambridge University Press, Cambridge.

Roughgarden, J. 1975. A simple model for population dynamics in stochastic environments. *American Naturalist*, **109**, 713–736.

Rowley, I. 1990. *The Galah: Behavioural Ecology of Galahs*. Surrey Beatty, Chipping Norton.

Rubinoff, D. and Powell, J.A. 2004. Conservation of fragmented small populations: endemic species persistence on California's smallest channel island. *Biodiversity and Conservation*, **13**, 2537–2550.

Rubio, J.L. and Carrascal, L.M. 1994. Habitat selection and conservation of an endemic Spanish lizard *Algyroides marchi* (Reptilia, Lacertidae). *Biological Conservation*, **70**, 245–250.

Rudnicky, T.C. and Hunter, M.L. 1993. Avian nest predation in clearcuts, forests, and edges in a forest-dominated landscape. *Journal of Wildlife Management*, **57**, 358–364.

Rudolph, D.C. and Connor, R.N. 1996. Red-cockaded woodpeckers and silvicultural practice: is uneven-aged silviculture preferable to even-aged? *Wildlife Society Bulletin*, **24**, 330–333.

Rülcker, C., Angelstam, P. and Rosenberg, P. 1994. Natural forest-fire dynamics can guide conservation and silviculture in boreal forests. *SkogForsk*, **2**, 1–4.

Rushton, S.P., Ormerod, S.J. and Kerby, G. 2004. New paradigms for modelling species distributions? *Journal of Applied Ecology*, **41**, 193–200.

Russell, R.W., Carpenter, F.L., Hixon, M.A. and Paton, D.C. 1994. The impact of variation in stopover habitat quality on migrant rufous hummingbirds. *Conservation Biology*, **8**, 483–490.

Saab, V. 1999. Importance of spatial scale to habitat use by breeding birds in riparian forests: a hierarchical approach. *Ecological Applications*, **9**, 135–151.

Sabo, J.L., Sponseller, R., Dixon, M., Gade, K., Harms, T., Heffernan, J., Jani, A., Katz, G., Soykan, C., Watts, J. and Welter, A. 2005. Riparian zones increase regional species richness by harboring different, not more, species. *Ecology*, **86**, 56–62.

Saccheri, I., Kuussaari, M., Kankare, M., Vikman, P., Fortelius, W. and Hanski, I. 1998. Inbreeding and extinction in a butterfly metapopulation. *Nature*, **392**, 491–494.

Sackett, D.L., Strauss, S.E., Richardson, W.S., Rosenberg, W. and Haynes, B. 2000. *Evidence-Based Medicine: How to Practice and Teach EB*. 2nd ed. Churchill Livingstone, Edinburgh, United Kingdom.

Saetersdal, M., Gjerde, I. and Blom, H.H. 2005. Indicator species and the problem of spatial inconsistency in nestedness patterns. *Biological Conservation*, **122**, 305–316.

Safford, R.J. 1997. Distribution studies of the forest-living native passerines of Mauritius. *Biological Conservation*, **80**, 189–198.

Saj, T., Sicotte, P. and Paterson, J.D. 2001. The conflict between vervet monkeys and farmers at the forest edge in Entebbe, Uganda. *African Journal of Ecology*, **39**, 195–199.

Sala, O.E., Chapin, F.S., Armesto, J.J., Berlow, E., Bloomfield, J., Dirzo, R., Huber-Sanwald,

E., Huenneke, L.F., Jackson, R.B., Kinzig, A., Leemans, R., Lodge, D.M., Mooney, H.A., Oesterheld, M., Poff, N.L., Sykes, M.T., Walker, B.H., Walker, M. and Wall, D.H. 2000. Biodiversity: global biodiversity scenarios for the year 2100. *Science*, **287**, 1770–1774.

Salafasky, N. 1993. Mammalian use of a buffer zone agroforestry system bordering Gunung Palung National Park, West Kalimantan, Indonesia. *Conservation Biology*, **7**, 928–933.

Salt, D., Lindenmayer, D.B. and Hobbs, R.J. 2004. *Trees and Biodiversity: A Guide for Australian Farm Forestry*. Rural Industries Research and Development Corporation, Canberra, Australia.

Samuelsson, J., Gustafsson, L. and Ingelög, T. 1994. *Dying and Dead Trees: A Review of Their Importance for Biodiversity*. Swedish Threatened Species Unit, Uppsala, Sweden.

Santelmann, M.V., White, D., Freemark, K., Nassauer, J.I., Eilers, J.M., Vache, K.B., Danielson, B.J., Corry, R.C., Clark, M.E., Polasky, S., Cruse, R.M., Sifneos, J., Rustigian, H., Coiner, C., Wu, J. and Debinski, D. 2004. Assessing alternative futures for agriculture in Iowa, USA. *Landscape Ecology*, **19**, 357–374.

Sargent, R.A., Kilgo, J.C., Chapman, B.R. and Miller, K.V. 1998. Predation of artificial nests in hardwood fragments enclosed by pine and agricultural habitats. *Journal of Wildlife Management*, **62**, 1438–1442.

Sarre, S. 1996. Habitat fragmentation promotes fluctuating asymmetry but not morphological divergence in two geckos. *Researches on Population Ecology*, **38**, 57–64.

Sarre, S. and Dearn, J.M. 1991. Morphological variation and fluctuating asymmetry among insular populations of the sleepy lizard, *Trachydosaurus rugosus* Gray (Squamata, Scincidae). *Australian Journal of Zoology*, **39**, 91–104.

Sarre, S., Dearn, J.M. and Georges, A. 1994. The application of fluctuating asymmetry in the monitoring of animal populations. *Pacific Conservation Biology*, **1**, 118–122.

Sarre, S., Smith, G.T. and Meyers, J.A. 1995. Persistence of two species of gecko (*Oedura reticulata* and *Gehyra variegata*) in remnant habitat. *Biological Conservation*, **71**, 25–33.

Saunders, A. 1994. Translocations in New Zealand: an overview. Pp. 43–46 in *Reintroduction Biology of Australian and New Zealand Fauna*. M. Serena, ed. Surrey Beatty and Sons, Chipping Norton, Sydney.

Saunders, D.A. 1977. The effect of agricultural clearing on the breeding success of the white-tailed black cockatoo. *Emu*, **77**, 180–184.

Saunders, D.A. 1980. The breeding behaviour and biology of the short-billed form of the white-tailed black cockatoo. *Ibis*, **124**, 422–455.

Saunders, D.A. 1989. Changes in the avifauna of a region, district and remnant as a result of fragmentation of native vegetation: the wheatbelt of western Australia: a case study. *Biological Conservation*, **50**, 99–135.

Saunders, D.A. 1996. Does our lack of vision threaten the viability of restoration of disturbed ecosystems? *Pacific Conservation Biology*, **2**, 321–326.

Saunders, D.A., and de Rebeira, C.P. 1991. Values of corridors to avian populations in a fragmented landscape. Pp. 221–240 in *Nature Conservation 2: The Role of Corridors*. D.A. Saunders, and R.J. Hobbs, eds. Surrey Beatty and Sons, Chipping Norton.

Saunders, D.A. and Hobbs, R.J., eds. 1991. *Nature Conservation 2: The Role of Corridors*. Surrey Beatty, Chipping Norton.

Saunders, D.A. and Ingram, J. 1987. Factors affecting survival of breeding populations of Carnaby's cockatoo *Calyptorhyncus funerus latirostris* in remnants of native vegetation. Pp. 249–258 in *Nature Conservation: The Role of Remnants of Native Vegetation*. D.A. Saunders, G.W. Arnold, A.A. Burbidge and A.J.M. Hopkins, eds. Surrey Beatty, Chipping Norton, NSW.

Saunders, D.A. and Ingram, J. 1995. *Birds of Southwestern Australia*. Surrey Beatty, Chipping Norton.

Saunders, D.A., Rowley, I. and Smith, G.T. 1985. The effects of clearing for agriculture on the distribution of cockatoos in the southwest of Western Australia. Pp. 309–321 in *Birds of Eucalypt Forests and Woodlands: Ecology, Conservation and Management*. A. Keast, H.F. Recher, H. Ford and D. Saunders, eds. Surrey Beatty, Chipping Norton.

Saunders, D.A., Arnold, G.W., Burbidge, A.A. and Hopkins, A.J., eds. 1987. *Nature Conservation: The Role of Remnants of Native Vegetation*. Surrey Beatty and Sons, Chipping Norton, Sydney, Australia.

Saunders, D.A., Hobbs, R.J. and Margules, C.R. 1991. Biological consequences of ecosystem fragmentation: a review. *Conservation Biology,* **5**, 18–32.

Saunders, D.A., Hobbs, R.J. and Ehrlich, P., eds. 1993a. *Nature Conservation 3: Reconstruction of Fragmented Ecosystems.* Surrey Beatty, Chipping Norton, Sydney.

Saunders, D.A., Hobbs, R.J. and Arnold, G.W. 1993b. The Kellerberrin project on fragmented landscapes: a review of current information. *Biological Conservation,* **64**, 185–192.

Saunders, D.A., Smith, G.T., Ingram, J.A. and Forrester, R.I. 2003. Changes in a remnant salmon gum *Eucalyptus salmonophloia* and York gum *E. loxophleba* woodland, 1978 to 1997: implications for woodland conservation in the wheat–sheep regions of Australia. *Biological Conservation,* **110**, 245–256.

Saurez, A.V., Bolger, D.T. and Case, T.J. 1998. Effects of fragmentation and invasion on native ant communities. *Ecology,* **79**, 2041–2056.

Savidge, J.A. 1987. Extinction of an island forest avifauna by an introduced snake. *Ecology,* **68**, 660–668.

Savill, P.S. 1983. Silviculture in windy climates. *Commonwealth Forestry Bureau,* **44**, 473–488.

Schall, R. 1991. Estimation of generalised linear models with random effects. *Biometrika,* **78**, 719–727.

Scheffer, M., Carpenter, S., Foley, J.A., Folke, C. and Walker, B. 2001. Catastrophic shifts in ecosystems. *Nature,* **413**, 591–596.

Scheiner, S.M. 2003. Six types of species–area curves. *Global Ecology and Biogeography,* **12**, 441–447.

Schilthuizen, M., Liew, T-S., Elahan, B.B., and Lackman-Ancrenaz, I. 2005. Effects of karst forest degradation on pulmonate and prosobranch land snail communities in Sabah, Malaysian Borneo. *Conservation Biology,* **19**, 949–954.

Schlaepfer, M.A. and Gavin, T.A. 2001. Edge effects on lizards and frogs in tropical forest fragments. *Conservation Biology,* **15**, 1079–1090.

Schmiegelow, F.K. and Hannon, S.J. 1993. Adaptive management, adaptive science and the effects of forest fragmentation on boreal birds in northern Alberta. *Transactions of the North American Wildlife and Natural Resources Conference,* **58**, 584–598.

Schmiegelow, F.K. and Mönkkönen, M. 2002. Habitat loss and fragmentation in dynamic landscapes: avian perspectives from the boreal forest. *Ecological Applications,* **12**, 375–389.

Schmiegelow, F.K., Machtans, C.S. and Hannon, S.J. 1997. Are boreal birds resilient to forest fragmentation? An experimental study of short-term community responses. *Ecology,* **78**, 1914–1932.

Schneider, D.C. 1994. *Quantitative Ecology. Temporal and Spatial Scaling.* Academic Press, San Diego, California.

Schoener, T.W. 1974. Resource partitioning in ecological communities. *Science,* **185**, 27–39.

Schoener, T.W. and Schoener, A. 1983. Distribution of vertebrates on some very small islands. I. Occurrence sequences of individual species. *Journal of Animal Ecology,* **52**, 209–235.

Schoener, T.W. and Spiller, D.A. 1992. Is extinction rate related to temporal variation in population size? An empirical answer for orb spiders. *American Naturalist,* **139**, 1176–1207.

Schonewald-Cox, C.M. and Buechner, M. 1990. Park protection and public roads. Pp. 373–395 in *Conservation Biology: The Theory and Practice of Nature Conservation, Preservation and Management.* P.L. Fiedler and S.K. Jain, eds. Chapman and Hall, New York.

Schorger, A.W. 1973. *The Passenger Pigeon: Its Natural History and Extinction.* University of Oklahoma Press, Norman.

Schtickzelle, N. and Baguette, M. 2003. Behavioural responses to habitat patch boundaries restrict dispersal and generate emigration–patch area relationships in fragmented landscapes. *Journal of Animal Ecology,* **72**, 533–545.

Schultz, C.B. 1998. Dispersal behaviour and its implications for reserve design in a rare Oregon butterfly. *Conservation Biology,* **12**, 284–292.

Schultz, C.B. and Crone, E.E. 2001. Edge-mediated dispersal behaviour in a prairie butterfly. *Ecology,* **82**, 1879–1892.

Schwartz, M.W., ed. 1997. *Conservation in Highly Fragmented Landscapes.* Chapman and Hall, New York.

Schwartz, M.W. and van Mantgem, P.J. 1997. The value of small reserves in chronically fragmented landscapes. Pp. 379–394 in *Conservation in Highly Fragmented Landscapes.* M.W. Schwartz, ed. Chapman and Hall, New York.

Scientific Panel on Ecosystem Based Forest Management. 2000. *Simplified Forest Management to*

Achieve Watershed and Forest Health: A Critique. Report to the National Wildlife Federation, Seattle, Washington.

Scott, J.M. 1999. Vulnerability of forested ecosystems in the Pacific Northwest to loss of area. Pp. 33–42 in *Forest Wildlife and Fragmentation: Management Implications.* J. Rochelle, L.A. Lehmann and J. Wisniewski, eds. Brill, Leiden, Germany.

Scott, J.M., Davis, F.W., McGhie, R.G., Wright, R.G., Groves, C. and Estes, J. 2001. Nature reserves: do they capture the full range of America's biological diversity? *Ecological Applications,* **11**, 999–1007.

Scotts, D.J. 1991. Old-growth forests: their ecological characteristics and value to forest-dependent fauna of southeast Australia. Pp. 147–159 in *Conservation of Australia's Forest Fauna.* D. Lunney, ed. Royal Zoological Society of New South Wales, Mosman.

Seabloom, E.W., Dobson, A.P. and Stoms, D.M. 2002. Extinction rates under nonrandom patterns of habitat loss. *Proceedings of the National Academy of Sciences,* **99**, 11229–11234.

Seabrook, W.A. and Dettmann, E.B. 1996. Roads as activity corridors for cane toads in Australia. *Journal of Wildlife Management,* **60**, 363–368.

Searle, S.R., Casella, G. and McCulloch, C.E. 1992. *Variance Components Analysis.* Wiley, New York.

Sekercioglu, C.H., Ehrlich, P.R., Daily, G.C., Aygen, D., Goehring, D. and Sandi, R.F. 2002. Disappearance of insectivorous birds from tropical forest fragments. *Proceedings of the National Academy of Sciences,* **99**, 263–267.

Sekercioglu, C.H., Daily, G.C. and Ehrlich, P.R. 2004. Ecosystem consequences of bird declines. *Proceedings of the National Academy of Sciences of the United States of America,* **101**, 18042–18047.

Semlitsch, R.D. and Bodie, J.R. 1998. Are small, isolated wetlands expendable? *Conservation Biology,* **12**, 1129–1133.

Semlitsch, R.D. and Bodie, J.R. 2003. Biological criteria for buffer zones around wetlands and riparian habitat for amphibians and reptiles. *Conservation Biology,* **17**, 1219–1228.

Sergio, F., Newton, I., and Marchesi, L. 2005. Top predators and biodiversity. *Nature,* **436**, 192.

Shafer, C.L. 1990. *Nature Reserves: Island Theory and Conservation Practice.* Smithsonian Institution Press, Washington, D.C.

Shaffer, M.L. 1981. Minimum population sizes for species conservation. *BioScience,* **31**, 131–134.

Shaffer, M.L. 1990. Population viability analysis. *Conservation Biology,* **4**, 39–40.

Shaffer, M.L. and Samson, F.B. 1985. Population size and extinction: a note on determining critical population size. *American Naturalist,* **125**, 144–152.

Sharitz, R.R., Boring, L.R., van Lear, D.H. and Pinder, J.E. 1992. Integrating ecological concepts with natural resource management of southern forests. *Ecological Applications,* **2**, 226–237.

Sharpe, D.M., Gunterspergen, G.R., Dunn, C.P., Leitner, L.A. and Stearns, F. 1987. Vegetation dynamics in a southern Wisconsin agricultural landscape. Pp. 137–155 in *Landscape Heterogeneity and Disturbance.* M.T. Turner, ed. Springer-Verlag, New York.

Shiel, D. and Burslem, F.R. 2003. Disturbing hypotheses in tropical forests. *Trends in Ecology and Evolution,* **18**, 18–26.

Shields, J. and Kavanagh, R. 1985. *Wildlife Research and Management in the Forestry Commission of N.S.W. A Review.* Technical Paper No. 32. Forestry Commission of NSW, Sydney.

Shine, R. and Mason, R.T. 2004. Patterns of mortality in a cold-climate population of garter snakes *(Thamnophis sirtalis parietalis). Biological Conservation,* **120**, 201–210.

Shipley, B. 2000. *Cause and Correlation in Biology.* Cambridge University Press, Cambridge, United Kingdom.

Shorrocks, B. and Sevenster, J.G. 1995. Explaining local species diversity. *Proceedings of the Royal Society of London, Series B,* **260**, 305–209.

Short, J. and Turner, B. 1994. A test of the vegetation mosaic hypothesis: an hypothesis to explain the decline and extinction of Australian mammals. *Conservation Biology,* **8**, 439–449.

Shrader-Frechette, K.S. and McCoy, E.D. 1993. *Method in Ecology: Strategies for Conservation.* Cambridge University Press, Cambridge.

Shriar, A.J. 2001. The dynamics of agricultural intensification and resource conservation in the buffer zone of the Maya biosphere reserve, Petén, Guatemala. *Human Ecology,* **29**, 27–48.

Siebert, S.F. 2002. From shade- to sun-grown perennial crops in Sulawesi, Indonesia: implications for biodiversity conservation and soil fertility. *Biodiversity and Conservation,* **11**, 1889–1902.

Sieving, K.E., Willson, M.F. and de Santo, T.L. 2000. Defining corridor functions for endemic birds in fragmented south-temperate rainforest. *Conservation Biology,* **14**, 1120–1132.

Siitonen, P., Lehtinen, A. and Siitonen, M. 2005. Effects of forest edges on the distribution, abundance and regional persistence of wood-rotting fungi. *Conservation Biology,* **19**, 250–260.

Silsbee, D.G. and Larson, G.L. 1983. A comparison of streams in logged and unlogged areas of the Great Smoky Mountains National Park. *Hydrobiology,* **102**, 99–111.

Simandl, J. 1992. The distribution of pine sawfly cocoons (Diprionidae) in Scots pine stands in relation to stand edge and tree base. *Forest Ecology and Management,* **54**, 193–203.

Simberloff, D. 1992. Do species–area curves predict extinction in fragmented forest? Pp. 75–89 in *Tropical Deforestation and Species Extinction.* T.C. Whitmore and J.A. Sayer, eds. Chapman and Hall, London.

Simberloff, D. 1995. Why do introduced species appear to devastate islands more than mainland species? *Pacific Science,* **49**, 87–97.

Simberloff, D. 2000. Extinction-proneness of island species: causes and management implications. *Raffles Bulletin of Zoology,* **48**, 1–10.

Simberloff, D. and Abele, L.G. 1982. Refuge design and island geographic theory: effects of fragmentation. *American Naturalist,* **120**, 41–45.

Simberloff, D. and Levin, B. 1985. Predictable sequences of species loss with decreasing island area: land birds in two archipelagoes. *New Zealand Journal of Ecology,* **8**, 11–20.

Simberloff, D. and Martin, J.L. 1991. Nestedness of insular avifaunas: simple summary statistics masking species patterns. *Ornis Fennica,* **68**, 178–192.

Simberloff, D., Parker, I.M. and Windle, P.N. 2005. Introduced species policy, management and future research needs. *Frontiers in Ecology and Environment,* **3**, 12–20.

Simberloff, D.A. 1986. The proximate causes of extinction. Pp. 259–276 in *Patterns and Processes in the History of Life.* D.M. Raup and D. Jablonski, eds. Springer-Verlag, Berlin.

Simberloff, D.A. 1988. The contribution of population and community biology to conservation science. *Annual Review of Ecology and Systematics,* **19**, 473–511.

Simberloff, D.A. 1998. Flagships, umbrellas, and keystones: is single-species management passe in the landscape era? *Biological Conservation,* **83**, 247–257.

Simberloff, D.A., Farr, J.A., Cox, J. and Mehlman, D.W. 1992. Movement corridors: conservation bargains or poor investments? *Conservation Biology,* **6**, 493–504.

Sisk, T., Haddad, N.M. and Ehrlich, P.R. 1997. Bird assemblages in patchy woodlands: modeling the effects of edge and matrix habitats. *Ecological Applications,* **7**, 1170–1180.

Sisk, T.D. and Margules, C.R. 1993. Habitat edges and restoration: methods for quantifying edge effects and predicting the results of restoration efforts. Pp. 57–69 in *Nature Conservation 3: Reconstruction of Fragmented Ecosystems.* D.A. Saunders, R.J. Hobbs and P. Ehrlich, eds. Surrey Beatty and Sons, Chipping Norton.

Sisk, T.D. and Zook, J. 1996. La influencia de la composición del paisaje en la distribución de *Catharus ustulatus* en migración por Costa Rica (Influence of landscape composition on the distribution of Swainson's thrush *Catharus ustulatus,* migrating through Costa Rica). *Vida Silvestre Neotropical,* **5**, 120–125.

Sizer, N. and Tanner, E.V. 1999. Responses of woody plant seedlings to edge formation in a lowland tropical rainforest, Amazonia. *Biological Conservation,* **91**, 135–142.

Sjorgen-Gulve, P. 1994. Distribution and extinction patterns within a northern metapopulation of the pool frog *Rana lessonae. Ecology,* **75**, 1357–1367.

Skole, D. and Tucker, C. 1993. Tropical deforestation and habitat fragmentation in the Amazon: satellite data from 1978 to 1988. *Science,* **260**, 1905–1910.

Skonhoft, A., Yoccoz, N.G., Stenseth, N., Gilliard, J. and Loison, A. 2002. Management of chamois *(Rupicapra rupicapra)* moving between a protected core area and a hunting area. *Ecological Applications,* **12**, 1199–1211.

Slabbekoorn, H. and Peet, M. 2003. Birds sing at a higher pitch in urban noise. *Nature,* **424**, 267.

Small, M.F. and Hunter, M.L. 1988. Forest fragmentation and avian nest predation in forested landscapes. *Oecologia,* **76**, 62–64.

Smith, A.P., Moore, D.M. and Andrews, S.P. 1992. *Proposed Forestry Operations in the Glen Innes Forest Management Area: Fauna Impact Statement.* Supplement to the Environmental Impact Statement. Report for the Forestry Commission of New South Wales by Austeco Pty. Ltd.

Smith, A.T. 1980. Temporal changes in insular populations of the pika *(Ochotona principes). Ecology,* **61**, 8–13.

Smith, A.T. and Peacock, M.M. 1990. Conspecific attraction and the determination of metapopulation colonization rates. *Conservation Biology,* **4**, 320–323.

Smith, D.S. and Hellmund, P.C., eds. 1993. *Ecology of Greenways: Design and Function of Linear Conservation Areas.* University of Minnesota Press, Minneapolis.

Smith, G.C. and Agnew, G. 2002. The value of "bat boxes" for attracting hollow-dependent fauna to farm forestry plantations in southeast Queensland. *Ecological Management and Restoration,* **3**, 37–46.

Smith, G.T., Arnold, G.W., Sarre, S., Abensperg-Traun, M. and Steven, D.E. 1996. The effects of habitat fragmentation and livestock-grazing on animal communities in remnants of gimlet *Eucalyptus salubris* woodland in the western Australian wheatbelt. II. Lizards. *Journal of Applied Ecology,* **33**, 1302–1310.

Smith, J.H. 2003. Land-cover assessment of conservation and buffer zones in the Bosawas Natural Resource Reserve of Nicaragua. *Environmental Management,* **31**, 252–262.

Smith, J.N. and Hellman, J.J. 2002. Population persistence in fragmented landscapes. *Trends in Ecology and Evolution,* **17**, 397–399.

Smith, J.N., Cook, T.L., Rothstein, S.I., Robinson, S.K. and Sealey, S.G., eds. 2000. *Ecology and Management of Cowbirds and Their Hosts.* University of Texas Press, Austin.

Smith, J.N., Taitt, M.J., Zanette, L. and Myers-Smith, I.H. 2003. How do brown-headed cowbirds *(Molothrus ater)* cause nest failures in song sparrows *(Melospiza melodia)?* A removal experiment. *Auk,* **120**, 772–783.

Smith, P.A. 1994. Autocorrelation in logistic regression modeling of species' distribution. *Global Ecology and Biogeography Letters,* **4**, 47–61.

Smith, R.J. 2000. *An Investigation into the Relationships between Anthropogenic Forest Disturbance Patterns, Population Viability and Landscape Indices.* MSc Thesis, Institute of Land and Food Resources, University of Melbourne, Australia.

Smyth, A.K., Lamb, D., Hall, L., McCallum, H., Moloney, D. and Smith, G. 2000. Towards scientifically valid management tools for sustainable forest management: species guilds versus model species. Pp. 118–124 in *Management for Sustainable Ecosystems.* P. Hale, A. Petrie, D. Moloney and P. Sattler, eds. Centre for Conservation Biology, University of Brisbane, Brisbane.

Soderquist, T.R. and Mac Nally, R. 2000. The conservation value of mesic gullies in dry forest landscapes: mammal populations in the box–ironbark ecosystem of southern Australia. *Biological Conservation,* **93**, 281–291.

Solomon, B.D. 1998. Impending recovery of Kirtland's warbler: case study in the effectiveness of the Endangered Species Act. *Environmental Management,* **22**, 9–17.

Soulé, M.E. 1985. What is conservation biology? *BioScience,* **35**, 727–734.

Soulé, M.E. and Gilpin, M.E. 1991. The theory of wildlife corridor capability. Pp. 3–8 in *Nature Conservation 2: The Role of Corridors.* D.A. Saunders and R.J. Hobbs, eds. Surrey Beatty and Sons, Chipping Norton.

Soulé, M.E. and J. Terborgh. 1999. Conserving nature at regional and continental scales: a scientific program for North America. *BioScience,* **49**, 809–817.

Soulé, M.E., Estes, J.A., Berger, J. and Del Rio, C.M. 2003. Ecological effectiveness: conservation goals for interactive species. *Conservation Biology,* **17**, 1238–1250.

Soulé, M.E., Mackey, B.G., Recher, H.F., Williams, J.E., Woinarski, J.C.Z., Driscoll, D., Dennison, W.C. and Jones, M.E. 2004. The role of connectivity in Australian conservation. *Pacific Conservation Biology,* **10**, 266–279.

Soulé, M.E., Estes, J.A., Miller, B. and Honnold, D.L. 2005. Strongly interacting species. conservation policy, management, and ethics. *BioScience,* **55**, 168–176.

Southwood, T.R.E. 1996. *Ecological Methods with Particular Reference to the Study of Insect Populations.* Methuen, London.

Soutter, M. and Musy, A. 1993. Evaluating the width of hydrological buffer zones between drained agricultural land and nature reserve areas. *Irrigation and Drainage Systems,* **7**, 151–160.

Spackman, S.C. and Hughes, J.W. 1995. Assessment of minimum stream corridor width for biological conservation: species richness and distribution along mid-order streams in Vermont, USA. *Biological Conservation,* **71**, 325–332.

Spellerberg, I. 1998. Ecological effects of roads and traffic: a literature review. *Global Ecology and Biogeography,* **7**, 317–333.

Spiegelhalter D., Thomas A., Best N. and Lunn D. 2003. WinBUGS User Manual Version 1.4. MRC Biostatistics Unit, Cambridge.

Spies, T.A. and Turner, M.G. 1999. Dynamic forest mosaics. Pp. 95–160 in *Managing Biodiversity in Forest Ecosystems.* M. Hunter III, ed. Cambridge University Press, Cambridge.

Spooner, P., Lunt, I. and Robinson, W. 2002. Is fencing enough? The short-term effects of stock exclusion in remnant grassy woodlands in southern NSW. *Ecological Management and Restoration,* **3**, 117–126.

Srivastava, D.S. and Lawton, J.H. 1998. Why more productive sites have more species: an experimental test of theory using tree-hole communities. *American Naturalist,* **152**, 510–529.

Srivastava, D.S., Kolasa, J., Bengtsson, J., Gonzalez, A., Lawler, S.P., Miller, T.E., Munguia, P., Romanuk, T., Schneider, D.C. and Trzcinski, M.K. 2004. Are natural microcosms useful model systems for ecology? *Trends in Ecology and Evolution,* **19**, 379–384.

Stacey, P.B. and Taper, M. 1992. Environmental variation and the persistence of small populations. *Ecological Applications,* **2**, 18–29.

Starfield, A.M. and Bleloch, A.L. 1992. *Building Models for Conservation and Wildlife Management.* Burgess International Group, Edina.

State of the Environment Advisory Council 1996. *State of the Environment Australia 1996.* CSIRO Publishing, Collingwood.

State of the Environment Report 2001. *Biodiversity.* Commonwealth of Australia, Canberra.

Steadman, D.W. 1995. Prehistoric extinctions of Pacific island birds: biodiversity meets zooarcheology. *Science,* **267**, 1123–1131.

Steadman, D.W. 1997a. Human-caused extinction of birds. Pp. 131–161 in *Biodiversity II: Understanding and Protecting Our Biological Resources.* M.L. Reaka-Kudla, W.E. Wilson and W.O. Wilson, eds. Joseph Henry Press, Washington, D.C.

Steadman, D.W. 1997b. The historic biogeography and community ecology of Polynesian pigeons and doves. *Journal of Biogeography,* **24**, 737–753.

Stenseth, N. and Lidicker, W., eds. 1992a. *Animal Dispersal.* Chapman and Hall, London.

Stenseth, N. and Lidicker, W., eds. 1992b. Appendix 1. Where do we stand methodologically about experimental design and methods of analysis in the study of dispersal? Pp. 295–315 in *Animal Dispersal.* N. Stenseth and W. Lidicker, eds. Chapman and Hall, London.

Stephens, S.E., Koons, D.N., Rotella, J.J. and Willey, D.W. 2003. Effects of habitat fragmentation on avian nesting success: a review of the evidence at multiple spatial scales. *Biological Conservation,* **115**, 101–110.

Stickel, L.F. 1946. The source of animals moving into a depopulated area. *Journal of Mammalogy,* **27**, 301–307.

Stirzaker, R., Vertessey, R. and Sarre, A., eds. 2002. *Trees, Water and Salt: An Australian Guide to Using Trees for Healthy Catchments and Productive Farms.* Joint Venture Agroforestry Program, Canberra.

Stone, L., Dayan, T. and Simberloff, D. 1996. Communitywide assembly patterns unmasked: the importance of species' differing geographical scales. *American Naturalist,* **148**, 997–1015.

Stouffer, P.C. and Bierregaard, R.O. 1995. Use of Amazonian forest fragments by understorey insectivorous birds. *Ecology,* **76**, 2429–2445.

Stow, A.J. and Sannucks, P. 2004. Inbreeding avoidance in Cunningham's skink *(Egernia cunninghamii)* in natural and fragmented habitat. *Molecular Ecology,* **13**, 443–447.

Stratford, J.A. and Robinson, W.D. 2005. Gulliver travel to the fragmented tropics: geographic variation in the mechanisms of avian extinction. *Frontiers in Ecology and Environment,* **3**, 91–98.

Strong, A.M. and Bancroft, G.T. 1994. Postfledging dispersal of white-crowned pigeons: implications for conservation of deciduous seasonal forests in the Florida Keys. *Conservation Biology,* **8**, 770–779.

Struhsaker, T.T., Struhsaker, P.J. and Siex, K.S. 2005. Conserving Africa's rain forests: problems in protected areas and possible solutions. *Biological Conservation,* **123**, 45–54.

Stuart, S.N., Chanson, J.S., Cox, N.A., Young, B.E., Rodrigues, A.S.L., Fischman, D.L. and Waller,

R.W. 2004. Status and trends of amphibian declines and extinctions worldwide. *Science,* **306**, 1783–1786.

Stuart-Smith, A.K. and Hayes, J.P. 2003. Influence of residual tree density on predation of artificial and natural songbird nests. *Forest Ecology and Management,* **183**, 159–176.

Styskel, E.W. 1983. Problems in snag management implementation: a case study. Pp. 24–27 in *Snag Habitat Management*. Proceedings of the Symposium, June 7–8 1983. Northern Arizona University, Flagstaff, Arizona. USDA, Rocky Mountain Forest and Range Experiment Station.

Suckling, G.C. 1982. Value of reserved habitat for mammal conservation in plantations. *Australian Forestry,* **45**, 19–27.

Sumner, J., Moritz, C. and Shine, R. 1999. Shrinking forest shrinks skink: morphological change in response to rainforest fragmentation in the prickly forest skink *(Gnypetoscincus queenslandiae)*. *Biological Conservation,* **91**, 159–167.

Sutherland, W. 1996. *Ecological Census Techniques: A Handbook*. Cambridge University Press, Cambridge.

Sutherland, W.J. 2000. *The Conservation Handbook. Research, Management and Policy.* Blackwell Science, Oxford.

Sweanor, L.L., Logan, K.A. and Hornocker, M.G. 2000. Cougar dispersal patterns, metapopulation dynamics and conservation. *Conservation Biology,* **14**, 798–808.

Sweeney, B.W., Bott, T.L., Jackson, J.K., Kaplan, L.A., Newbold, J.D., Standley, L.J., Hession, W.C. and Horowitz, R.J. 2004. Riparian deforestation, stream narrowing, and loss of stream ecosystem services. *Proceedings of the National Academy of Sciences,* **101**, 14132–14137.

Swenson, J.E., Alt, K.L. and Eng, R.L. 1986. Ecology of bald eagles in the Greater Yellowstone ecosystem. *Wildlife Monographs,* **95**, 1–46.

Swets J.A., Dawes, R M. and Monahan, J. 2000. Better decisions through science. *Scientific American,* **October 2000**, 82–87.

Swihart, R.K. and Slade, N.A. 1985. Testing for independence of observations in animal movements. *Ecology,* **66**, 1176–1184.

Szaro, R.C. and Jakle, M.D. 1985. Avian use of a desert riparian island and its adjacent scrub habitat. *Condor,* **87**, 511–519.

Tang, S.M. and Gustafson, E.J. 1997. Perception of scale in forest management planning: challenges and implications. *Landscape and Urban Planning,* **39**, 1–9.

Taylor, A.C., Tyndale-Biscoe, H. and Lindenmayer, D.B. 2006. Marsupial gliders "marooned" on islands of native bush in a sea of exotic pine? *Journal of Animal Ecology.*

Taylor, B.L. 1995. The reliability of using population viability analysis for risk classification of species. *Conservation Biology,* **9**, 551–558.

Taylor, P.D. and Merriam, G. 1995. Habitat fragmentation and parasitism of a forest damselfly. *Landscape Ecology,* **11**, 181–189.

Taylor, P.D., Fahrig, L., Henein, K. and Merriam, G. 1993. Connectivity is a vital element of landscape structure. *Oikos,* **68**, 571–573.

Taylor, R. 1979. How the Macquarie Island parakeet became extinct. *New Zealand Journal of Ecology,* **2**, 42–45.

Taylor, R.J., Bryant, S.L., Pemberton, D. and Norton, T.W. 1985. Mammals of the Upper Henty River region, Western Tasmania. *Papers and Proceedings of the Royal Society of Tasmania,* **119**, 7–15.

Telleria, J.L. and Santos, T. 1992. Spatiotemporal patterns of egg predation in forest islands: an experimental approach. *Biological Conservation,* **62**, 29–33.

Temple, S.A. 1986. The problem of avian extinctions. *Current Ornithology,* **3**, 453–485.

Temple, S.A. and Cary, J.R. 1988. Modelling dynamics of habitat interior bird populations in fragmented landscapes. *Conservation Biology,* **2**, 340–347.

Temple, S.A. and Wiens, J.A. 1989. Bird populations and environmental changes: can birds be bio-indicators? *American Birds,* **43**, 260–270.

Terborgh, J. 1974. Preservation of natural diversity: the problem of extinction prone species. *BioScience,* **24**, 715–722.

Terborgh, J. 1989. *Where Have All the Birds Gone?* Princeton University Press, Princeton, New Jersey.

Tewksbury, J.J., Levey, D.J., Haddad, N.M., Sargent, S., Orrock, J.L., Weldon, A., Danielson, B.J., Brinkerhoff, J., Damschen, E.I. and Townsend, P. 2002. Corridors affect plants, animals and their interaction in fragmented landscapes. *Proceedings of the National Academy of Science,* **99**, 12923–12926.

Tews, J., Brose, U., Grimm, V., Tielborger, K., Wichmann, M.C., Schwager, M. and Jeltsch, F. 2004. Animal species diversity driven by habitat heterogeneity/diversity: the importance of keystone structures. *Journal of Biogeography*, **31**, 79–92.

Thiollay, J.-M. 1998. Long-term dynamics of a tropical savanna bird community. *Biodiversity and Conservation*, **7**, 1291–1312.

Thomas, C.D. 1990. What do real population dynamics tell us about minimum viable population sizes. *Conservation Biology*, **4**, 324–327.

Thomas, C.D., Cameron, A., Green, R.E., Bakkenes, M., Beaumont, L.J., Collingham, Y.C., Erasmus, B.F., Siqueiria, M., Grainer, A., Hannah, L., Hughes, L., Huntley, B., Jaarsveld, A., Midgley, G.F., Miles, L., Heurta-Ortega, M.A., Townsend Peterson, A., Philips, O.L. and Williams, S.E. 2004. Extinction risk from climate change. *Nature*, **427**, 145–148.

Thomas, J.A. 1984a. The behaviour and habitat requirements of *Maculinea nausithous* (the dusky large blue butterfly) and *M. teleius* (the scarce large blue) in France. *Biological Conservation*, **28**, 325–347.

Thomas, J.A. 1984b. The conservation of butterflies in temperate countries: past efforts and lessons for the future. Pp. 333–353 in *The Biology of Butterflies*. R.I. Vane-Wright and P.R. Ackery, eds. Symposium of the Royal Entomological Society. Academic Press, London.

Thomas, J.A. and Morris, M.G. 1995. Rates and patterns of extinction among British invertebrates. Pp. 111–130 in *Extinction Rates*. J.H. Lawton and R.M. May, eds. Oxford University Press, Oxford.

Thomas, J.W., Forsmann, E.D., Lint, J.B., Meslow, E.C., Noon, B.R. and Verner, J. 1990. *A Conservation Strategy for the Northern Spotted Owl*. U.S. Government Printing Office, Portland, Oregon.

Thompson, I.D. and Angelstam, P. 1999. Special species. Pp. 434–459 in *Maintaining Biodiversity in Forest Ecosystems*. M.L. Hunter, ed. Cambridge University Press, Cambridge.

Thompson, I.D., Baker, J.A. and Ter-Mikaelian, M. 2003. A review of the long-term effects of postharvest silviculture on vertebrate wildlife, and predictive models with an emphasis on boreal forests in Ontario, Canada. *Forest Ecology and Management*, **177**, 441–469.

Thompson, W.L., White, G.C., and Gowan. 1998. *Monitoring Vertebrate Populations*. Academic Press, San Diego.

Threatened Species Network 2003. *Western Swamp Tortoise*. Fact Sheet. World Wide Fund for Nature, Perth.

Thrush, S.F., Hewitt, J.E., Cunnings, V.J., Green, M.O., Funnell, G.A. and Wilkinson, M.R. 2000. The generality of field experiments: interactions between local and broad-scale processes. *Ecology*, **81**, 399–415.

Tickle, P., Hafner, S., Lesslie, R., Lindenmayer, D.B., McAlpine, C., Mackey, B., Norman, P. and Phinn, S. 1998. *Scoping Study: Final Report.*. Montreal Indicator 1.1e. Fragmentation of forest types: identification of research priorities. A study prepared for the Forest and Wood Products Research and Development Corporation. Canberra, Australia.

Tilghman, N.G. 1989. Impacts of white-tailed deer on forest regeneration in northwestern Pennsylvania. *Journal of Wildlife Management*, **53**, 524–532.

Tilman, D. 1987. Secondary succession and the pattern of plant dominance along experimental nitrogen gradients. *Ecological Monographs*, **57**, 189–214.

Tilman, D., May, R.M., Lehman, C.L. and Nowak, M.A. 1994. Habitat destruction and the extinction debt. *Nature*, **371**, 65–66.

Tilman, D., Lehman, C.L. and Karieva, P. 1997. Population dynamics in spatial habits. Pp. 3–20 in *Spatial Ecology: The Role of Space in Population Dynamics and Interspecific Interactions*. D. Tilman and P. Karieva, eds. Princeton University Press, Princeton, New Jersey.

Tischendorf, L. and Fahrig, L. 2000a. On the usage and measurement of landscape connectivity. *Oikos*, **90**, 7–19.

Tischendorf, L. and Fahrig, L. 2000b. How should we measure landscape connectivity? *Landscape Ecology*, **15**, 235–254.

Tocher, M.D., Gascon, C. and Zimmerman, B.L. 1997. Fragmentation effects on a central American frog community: a ten-year study. Pp. 124–137 in *Tropical Forest Remnants. Ecology, Management and Conservation of Fragmented Communities*. W.F. Laurance and R.O. Bierregaard, eds. University of Chicago Press, Chicago.

Tommerup, I.C. and Bougher, N.L. 1999. The role of ectomycorrhizal fungi in nutrient cycling in temperate Australian woodlands. Pp. 190–224 in *Temperate Eucalypt Woodlands in Australia: Biology, Conservation, Management and Restoration*. R.J. Hobbs and C.J. Yates, eds. Surrey Beatty and Sons, Chipping Norton.

Travis, S.E., Slobodchikoff, C.N. and Keim, P. 1995. Ecological and demographic effects on intraspecific variation in the social system of prairie dogs. *Ecology*, **76**, 1794–1803.

Trayler, K.M. and Davis, J.A. 1998. Forestry issues and the vertical distribution of stream invertebrates in southwestern Australia. *Freshwater Biology*, **4**, 331–342.

Trombulak, S.C. and Frissell, C.A. 2000. Review of ecological effects of roads on terrestrial and aquatic communities. *Conservation Biology*, **14**, 18–30.

Tscharntke, T. 1992. Fragmentation of Phragmites habitats, minimum viable population size, habitat suitability, and local extinction of moths, midges, flies, aphids and birds. *Conservation Biology*, **6**, 530–536.

Tscharntke, T., Klein, A.M., Kruess, A., Steffan-Dewenter, I. and Thies, C. 2005. Landscape perspectives on agricultural intensification and biodiversity: ecosystem service management. *Ecology Letters*, **8**, 857–874.

Tubelis, D.P., Lindenmayer, D.B., Saunders, D.A., Cowling, A. and Nix, H.A. 2004. Landscape supplementation provided by an exotic matrix: implications for bird conservation and forest management in a softwood plantation system in southeastern Australia. *Oikos*, **107**, 634–644.

Tuchmann, E.T., Connaughton, K.P., Freedman, L.E., and Moriwaki, C.B. 1996. *The Northwest Forest Plan*. A report to the president and Congress. USDA Office of Forestry and Economic Assistance, Portland, Oregon, USA.

Turner, F.B., Jennrich, R.I. and Weintraub, J.D. 1969. Home ranges and body size of lizards. *Ecology*, **50**, 1076–1081.

Turner, I.M. 1996. Species loss in fragments of tropical rain forest: a review of the evidence. *Journal of Applied Ecology*, **33**, 200–209.

Turner, M., Gardner, R.H., and O'Neill, R.V. 2001. *Landscape Ecology in Theory and Practice: Pattern and Processes*. Springer, New York.

Turner, M.G. 1989. Landscape ecology: the effect of pattern on process. *Annual Review of Ecology and Systematics*, **20**, 171–197.

Turner, M.G., Pearson, S.M., Romme, W.H. and Wallace, L.L. 1997. Landscape heterogeneity and ungulate dynamics: what spatial scales are important? Pp. 331–348 in *Wildlife and Landscape Ecology: Effects of Pattern and Scale*. J. Bissonette, ed. Springer-Verlag, New York.

Turner, M.G., Romme, W.H. and Tinker, D.B. 2003. Surprises and lessons from the 1988 Yellowstone fires. *Frontiers in Ecology and Environment*, **1**, 351–358.

Tyndale-Biscoe, C.H. 1997. A fresh approach to quarantine. *Search*, **28**, 54–58.

Tyre, D., Possingham, H.P. and Lindenmayer, D.B. 2001. Inferring process from pattern: can territory occupancy provide information about life history parameters? *Ecological Applications*, **11**, 1051–1061.

Underwood, A.J. 1995. Ecological research, and research into environmental management. *Ecological Applications*, **5**, 232–247.

UNEP 1999. *Global Environmental Outlook 2000*. United Nations Environment Programme, Nairobi, Kenya.

United Nations Development Programme 2000. United Nations Development Programme, United Nations Environment Programme, World Bank, and World Resources Institute. *World Resources 2000–2001: People and Ecosystems: The Fraying Web of Life*. Elsevier Science, Amsterdam.

United States Fish and Wildlife Service 1980. *Habitat Evaluation Procedures*. Ecological Services Manual No. 102. U.S. Government Printing Office, Washington, D.C.

United States Fish and Wildlife Service 1981. *Standard for the Development of Habitat Suitability Index Models*. Ecological Services Manual No. 103. U.S. Government Printing Office, Washington, D.C.

Usher, M.B., ed. 1986. *Wildlife Conservation Evaluation*. Chapman and Hall, London.

Vallan, D. 2000. Influence of forest fragmentation on amphibian diversity in the nature reserve of Ambohitantely, highland Madagascar. *Biological Conservation*, **96**, 31–43.

van der Ree, R. 1999. Barbed wire fencing as a hazard for wildlife. *Victorian Naturalist*, **116**, 210–217.

van Dorp, D. and Opdam, P.F. 1987. Effects of patch size, isolation and regional abundance of forest bird communities. *Landscape Ecology*, **1**, 59–73.

van Horn, M.A., Gentry, R.M. and Faaborg, J. 1995. Patterns of ovenbird *(Seiurus aurocapillus)* pairing success in Missouri forest tracts. *Auk*, **112**, 98–106.

van Horne, B. 1983. Density as a misleading indicator of habitat quality. *Journal of Wildlife Management*, **47**, 893–901.

van Horne, B. and Wiens, J.A. 1991. Forest bird habitat suitability models and the development of general habitat models. Pp. 1–31 in Fisheries and Wildlife Research Report No. 8. U.S. Fish and Wildlife Service, Washington, D.C.

van Jaarsveld, A.S., Ferguson, J.H. and Bredenkamp, G.J. 1998. The Groenvaly grassland fragmentation experiment: design and initiation. *Agriculture, Ecosystems and Environment*, **68**, 139–150.

van Nieuwstadt, M.G., Sheil, D. and Kartawinata, K. 2001. The ecological consequences of logging in the burned forests of east Kalimantan, Indonesia. *Conservation Biology*, **15**, 1183–1186.

Várkonyi, G., Kuussaari, M. and Lappalainen, H. 2003. Use of corridors by boreal *Xestia* moths. *Oecologia*, **137**, 466–474.

Vellend, M. 2003. Habitat loss inhibits recovery of plant diversity as forests regrow. *Ecology*, **84**, 1158–1164.

Vera, F.W.M. 2000. *Grazing Ecology and Forest History*. CABI Publishing, New York.

Verts, B.J. and Carraway, L.N. 1986. Replacement in a population of *Perognathus parvus* subjected to removal trapping. *Journal of Mammalogy*, **67**, 201–205.

Vetaas, O.R. 1992. Microsite effects of trees and shrubs in dry savannas. *Journal of Vegetation Science*, **3**, 337–344.

Viana, V.M., Ervin, J., Donovan, R.Z., Elliott, C. and Gholz, H. 1996. *Certification of Forest Products*. Island Press, Covelo, California.

Vice, D.S., Engeman, R.M., and Vice, D.L. 2005. A comparison of three trap designs for capturing brown treesnakes on Guam. *Wildlife Research*, **32**, 355–359.

Vickery, J.A., Bradbury, R.B., Henderson, I.G., Eaton, M.A. and Grice, P.V. 2004. The role of agrienvironment schemes and farm management practices in reversing the decline of farmland birds in England. *Biological Conservation*, **119**, 19–39.

Villard, M.A. 1998. On forest-interior species, edge avoidance, area sensitivity, and dogmas in avian conservation. *Auk*, **115**, 801–805.

Villard, M.A. 2002. Habitat fragmentation: major conservation issue or intellectual attractor? *Ecological Applications*, **15**, 319–320.

Villard, M.A. and Taylor, P.D. 1994. Tolerance to habitat fragmentation influences the colonization of new habitat by forest birds. *Oecologia*, **98**, 393–401.

Villard, M.A., Martin, P.R. and Drummond, C.G. 1993. Habitat fragmentation and pairing success in the ovenbird *(Seiurus aurocapillus)*. *Auk*, **110**, 759–768.

Villard, M.A., Trzcinski, M.K. and Merriam, G. 1999. Fragmentation effects on forest birds: relative influence of woodland cover and configuration on landscape occupancy. *Conservation Biology*, **13**, 774–783.

Vincent, P.J. and Haworth, J.M. 1983. Poisson regression models of species abundance. *Journal of Biogeography*, **10**, 153–160.

Virkkala, R., Rajasarrkka, A., Vaisanen, R.A., Vickholm, M. and Virolainen, E. 1994. The significance of protected areas for the land birds of southern Finland. *Conservation Biology*, **8**, 532–544.

Viveiros de Castro, E.B. and Fernandez, F.A.S. 2004. Determinants of differential extinction probabilities of small mammals in Atlantic forest fragments. *Biological Conservation*, **119**, 73–80.

Von Uexküll, J. 1926. *Theoretical Biology*. Kegan Paul, Trench, Trubner and Co. Ltd, London.

Vos, C.C. and Chardon, J.P. 1998. Effects of habitat fragmentation and road density on the distribution patterns of the moor frog *Rana arvalis*. *Journal of Applied Ecology*, **35**, 44–56.

Wace, N. 1977. Assessment of the dispersal of plant species: the car-borne flora of Canberra. *Proceedings of the Ecological Society of Australia*, **10**, 166–186.

Wahlberg, N., Klemetti Selonen, V. and Hansik, I. 2002. Metapopulation structure and movements in five species of checkerspot butterflies. *Oecologia*, **130**, 33–43.

Wahungu, G.M., Catterall, C.P. and Olsen, M.F. 1999. Selective herbivory by red-necked pademelon *(Thylogale thetis)* at rainforest margins: factors affecting predation rates. *Austral Ecology*, **24**, 577–586.

Waide, R.B., Willig, M.R., Steiner, C.F., Mittelbach, G., Gough, L., Dodson, S.I., Juday, G.P. and Parmenter, R. 1999. The relationship between productivity and species richness. *Annual Review of Ecology and Systematics*, **30**, 257–300.

Walker, A. 1999. *Examination of the Barriers to Movements of Tasmanian Fish*. Honours thesis, School of Zoology, University of Tasmania. Hobart, Tasmania.

Walker, B. 1995. Conserving biological diversity through ecosystem resilience. *Conservation Biology*, **9**, 747–752.

Walker, B. 1998. The art and science of wildlife management. *Wildlife Research*, **25**, 1–9.

Walker, B. and Meyers, J.A. 2004. Thresholds in ecological and social–ecological systems: a developing database. *Ecology and Society*, **9**, 3. http://www.ecologyandsociety.org/vol9/iss2/art3.

Walker, B., Kinzig, A. and Langridge, J. 1999. Plant attribute diversity, resilience, and ecosystem function: the nature and significance of dominant and minor species. *Ecosystems*, **2**, 95–113.

Walker, B.H. 1992. Biodiversity and ecological redundancy. *Conservation Biology*, **6**, 18–23.

Walker, J., Bullen, F. and Williams, B.G. 1993. Ecohydrological changes in the Murray-Darling Basin. I. The number of trees cleared over two centuries. *Journal of Applied Ecology*, **30**, 265–273.

Wall, J. 1999. Fuelwood in Australia: impacts and opportunities. Pp. 372–381 in *Temperate Eucalypt Woodlands in Australia: Biology, Conservation, Management and Restoration*. R.J. Hobbs and C.J. Yates, eds. Surrey Beatty and Sons, Chipping Norton.

Wallace, K.J., Beecham, B.C. and Bone, B.H. 2003. *Managing Natural Biodiversity in the Western Australian Wheatbelt*. Department of Conservation and Land Management. November 2003. Perth, Western Australia.

Wallis, A., Stokes, D., Wescott, G. and McGee, T. 1997. Certification and labelling as a new tool for sustainable forest management. *Australian Journal of Environmental Management*, **4**, 224–238.

Walters, C.J. 1986. *Adaptive Management of Renewable Resources*. Macmillan, New York.

Walters, J.R., Ford, H.A. and Cooper, C.B. 1999. The ecological basis of sensitivity of brown treecreepers to habitat fragmentation: a preliminary assessment. *Biological Conservation*, **90**, 13–20.

Wang, K.M. and Hacker, R.B. 1997. Sustainability of rangeland pastoralism: a case study from the west Australian arid zone using stochastic optimal control theory. *Journal of Environmental Management*, **50**, 147–170.

Warburton, N.H. 1997. Structure and conservation of forest avifauna in isolated rainforest fragments in tropical Australia. Pp. 190–206 in *Tropical Forest Remnants. Ecology, Management and Conservation*. W.F. Laurance and R.O. Bierregaard, eds. University of Chicago Press, Chicago.

Watson, D.M. 2001. Mistletoe: a keystone resource in forests and woodlands worldwide. *Annual Review of Ecology and Systematics*, **32**, 219–249.

Watson, D.M. 2002. Effects of mistletoe on diversity: a case study from southern New South Wales. *Emu*, **102**, 275–281.

Watson, J., Warman, C., Todd, D. and Laboudallon, V. 1992. The Seychelles magpie-robin *Copsychus sechellarum:* ecology and conservation of an endangered species. *Biological Conservation*, **61**, 93–106.

Watson, J., Freudenberger, D. and Paull, D. 2001. An assessment of the focal-species approach for conserving birds in variegated landscapes in southeastern Australia. *Conservation Biology*, **15**, 1364–1373.

Watson, J.E., Whittaker, R.J. and Dawson, T.P. 2004. Habitat structure and proximity to forest edge affect the abundance and distribution of forest-dependent birds in tropical coastal forests of southeastern Madagascar. *Biological Conservation*, **120**, 315–331.

Watson, J.E., Whittaker, R.J. and Freudenberger, D. 2005. Bird community responses to habitat fragmentation: how consistent are they across landscapes? *Journal of Biogeography*, **32**, 1353–1370.

Weathers, K.C., Cadenasso, M.L. and Pickett, S.T. 2001. Forest edges as nutrient and pollutant concentrators: potential synergisms between fragmentation, forest canopies and the atmosphere. *Conservation Biology*, **15**, 1506–1514.

Webb, J.K. and Shine, R. 1998. Using thermal ecology to predict retreat-site selection by an endangered snake species. *Biological Conservation*, **86**, 233–242.

Webb, N.R. 1997. The development of criteria for ecological restoration. Pp. 133–158 in *Restoration Ecology and Sustainable Development*. K.M. Urbanska, N.R. Webb and P.J. Edwards, eds. Cambridge University Press, Cambridge.

Webb, N.R. and Haskins, L.E. 1980. An ecological surveys of heathlands in the Poole Basin, Dorset, England, in 1978. *Biological Conservation*, **17**, 153–165.

Webster, R. and Ahern, L. 1992. *Management for Conservation of the Superb Parrot* (Polytelis swainsonii) *in New South Wales and Victoria*. Department of Conservation and Natural Resources, Melbourne, Australia.

Wegner, J. 1994. *Ecological Landscape Variables for Monitoring and Management of Forest Biodiversity in Canada*. Report to Canadian Ministry of Natural Resources. GM Group, Ecological Land Management, Manotick, Canada.

Wegner, J.F. and Merriam, G. 1979. Movements by birds and small mammals between a wood and adjoining farmland habitats. *Journal of Applied Ecology*, **16**, 349–357.

Wei, F.W., Feng, Z.J., Wang, Z.W. and Hu, J.C. 2000. Habitat use and separation between the giant panda and the red panda. *Journal of Mammalogy*, **81**, 448–455.

Weibull, A.C., Ostman, O. and Granqvist, A. 2003. Species richness in agroecosystems: the effect of landscape, habitat and farm management. *Biodiversity and Conservation*, **12**, 1335–1355.

Weishampel, J.F., Shugart H.H. and Westman, W.E. 1997. Phenetic variation in insular populations of a rainforest centipede. Pp. 111–123 in *Tropical Forest Remnants. Ecology, Management and Conservation*. W.F. Laurance and R.O. Bierregaard, eds. University of Chicago Press, Chicago.

Welsh, A., Cunningham, R.B. and Lindenmayer, D.B. 1996. Modelling the abundance of rare species: statistical models for counts with extra zeros. *Ecological Modelling*, **88**, 297–308.

Welsh, A.H., Cunningham, R.B. and Chambers, R.L. 2000. Methodology for estimating the abundance of rare animals: seabird nesting on North East Herald Cay. *Biometrics*, **56**, 22–30.

Western, D. and Gichohi, A. 1993. Segregation effects and the impoverishment of savanna parks: the case for ecosystem viability analysis. *African Journal of Ecology*, **31**, 269–281.

Wethered, R. and Lawes, M.J. 2003. Matrix effects on bird assemblages in fragmented Afromontane forests in South Africa. *Biological Conservation*, **114**, 327–340.

Wethered, R. and Lawes, M.J. 2005. Nestedness of bird assemblages in fragmented Afromontane forest: the effect of plantation forestry in the matrix. *Biological Conservation*, **123**, 125–137.

Whelan, R.J. 1995. *The Ecology of Fire*. Cambridge Studies in Ecology. Cambridge University Press, Cambridge.

Whelan, R.J., Rodgerson, L., Dickman, C.R. and Sutherland, E.F. 2001. Critical life cycles of plants and animals: developing a process-based understanding of population changes in fire-prone landscapes. Pp. 94–124 in *Flammable Australia: The Fire Regimes and Biodiversity of a Continent*. R.A. Bradstock, J.A. Williams and M.A. Gill, eds. Cambridge University Press, Melbourne.

White, G.C. and Garrot, R.A. 1990. *Analysis of Wildlife Radio-Tracking Data*. Academic Press, San Diego, California.

Whittaker, R.H. 1954a. The ecology of serpentine soils. I. Introduction. *Ecology*, **35**, 258–259.

Whittaker, R.H. 1954b. The ecology of serpentine soils. IV. The vegetational response to serpentine soils. *Ecology*, **35**, 275–288.

Whittaker, R.J. 1998. *Island Biogeography: Ecology, Evolution and Conservation*. Oxford University Press, Oxford.

Whittaker, R.J., Willis, K.J. and Field. R. 2001. Scale and species richness: towards a general, hierarchical theory of species diversity. *Journal of Biogeography*, **28**, 453–470.

Whitten, S., Bennett, J., Moss, W., Handley, M. and Phillips, W. 2002. *Incentive Measures for Conserving Freshwater Ecosystems*. Environment Australia, Canberra.

Wiens, J. 1980. Concluding comments: are communities real? *Acta XVII Congress International Ornithological*, 1088–1089.

Wiens, J. 1994. Habitat fragmentation: island vs landscape perspectives on bird conservation. *Ibis*, **137**, S97–S104.

Wiens, J. 1995. Landscape mosaics and ecological theory. Pp. 1–26 in *Landscape Mosaics and Ecological Orocesses*. L. Hansson, L. Fahrig and G. Merriam, eds. Chapman and Hall, London.

Wiens, J. 1997. Wildlife in patchy environments: metapopulations, mosaics and management. Pp. 53–84 in *Metapopulations and Wildlife Conservation*. D.R. McCullogh, ed. Island Press, Covelo, California.

Wiens, J. 1999. The science and practice of landscape ecology. Pp. 37–383 in *Landscape Ecological Analy-*

sis. *Issues and Applications*. J.M. Klopatek and R.H. Gardner, eds. Springer, New York.

Wiens, J.A. 1989. Spatial scaling in ecology. *Functional Ecology*, **3**, 385–397.

Wiens, J.A., Stenseth, N.C. van Horne, B. and Ims, R.A. 1993. Ecological mechanisms and landscape ecology. *Oikos*, **66**, 369–380.

Wiens, J.A., Schooley, R.L. and Weekes, R.D. 1997. Patchy landscapes and animal movements: do beetles percolate? *Oikos*, **78**, 257–264.

Wiggett, D.R. and Boag, D.A. 1989. Intercolony natal dispersal in the Colombian ground squirrel. *Canadian Journal of Zoology*, **67**, 42–50.

Wilcove, D.S. 1985. Nest predation in forest tracts and the decline of migratory songbirds. *Ecology*, **66**, 1211–1214.

Wilcove, D.S., McLellen, C.H. and Dobson, A.P. 1986. Habitat fragmentation in the temperate zone. Pp. 237–256 in *Conservation Biology: The Science of Scarcity and Diversity*. M.E. Soulé, ed. Sinauer, Sunderland.

Wiles, G.J., Bart, J., Beck, R.E. and Aguon, C.F. 2003. Impacts of the brown tree snake: patterns of decline and species persistence in Guam's avifauna. *Conservation Biology*, **17**, 1350–1360.

Williams, B.K., Nichols, J.D. and Conroy, M.J. 2002. *Analysis and Management of Animal Populations*. Academic Press, San Diego, California.

Williams, J.E. and Gill, A.M. 1995. *The Impact of Fire Regimes on Native Forests in Eastern New South Wales*. Environmental Heritage Monograph Series No. 2. New South Wales National Parks and Wildlife Service, Sydney.

Williams, J.E. and West, C.J. 2000. Environmental weeds in Australia and New Zealand: issues and approaches to management. *Austral Ecology*, **25**, 425–444.

Williams, S.E. and Hero, J.M. 2001. Multiple determinants of Australian tropical frog biodiversity. *Biological Conservation*, **98**, 1–10.

Williamson, I. 1999. Competition between the larvae of the introduced cane toad *Bufo marinus* (Anura: Bufonidae) and native anurans from the Darling Downs area of southern Queensland. *Australian Journal of Ecology*, **24**, 636–643.

Wilson, A. and Lindenmayer, D.B. 1996. Wildlife corridors: pros and cons for wildlife conservation. *Australian Journal of Soil and Water Conservation*, **9**, 22–28.

Wilson, D.E. Cole, F.R., Nichols, J.D., Rudran, R. and Foster, M.S., eds. 1996. *Measuring and Monitoring Biological Diversity: Standard Methods for Mammals*. Smithsonian Institution Press, Washington.

Wilson, E.O. and Willis, E.O. 1978. Applied biogeography. Pp. 522–534 in *Ecology and Evolution of Communities*. M.L. Cody and J.M. Diamond, eds. Belknap Press of Harvard University, Cambridge, Massachusetts.

Wilson, J., and Wilson, R. 1978. Observations on the Seychelles magpie-robin *Copsychus sechellarum*. *British Ornithological Club Bulletin*, **98**, 15–21.

Wilson, J. 1978. Caves: changing ecosystems? *Studies in Speleology*, **3**, 35–42.

Wilson, J.B. 1995. Null models for assembly rules: the Jack Horner effect is more insidious than the Narcissus effect. *Oikos*, **72**, 139–144.

Wilson, J.B. and Roxburgh, S.H. 1994. A demonstration of guild-based assembly rules for a plant community, and determination of intrinsic guilds. *Oikos*, **69**, 267–276.

Wilson, J.B. and Whittaker, R.J. 1995. Assembly rules demonstrated in a saltmarsh community. *Journal of Ecology*, **83**, 801–807.

Wilson, J.B., Allen, R.B. and Lee, W.G. 1995. An assembly rule in the ground and herbaceous strata of a New Zealand rainforest. *Functional Ecology*, **9**, 61–64.

With, K.A. 1994. Ontogenic shifts in how grasshoppers interact with landscape structure: an analysis of movement patterns. *Functional Ecology*, **8**, 477–485.

With, K.A. and Crist, T.O. 1995. Critical thresholds in species' responses to landscape structure. *Ecology*, **76**, 2446–2459.

Witham, T.G., Morrow, P.A. and Potts, B.M. 1991. Conservation of hybrid plants. *Science*, **254**, 779–780.

Woinarski, J., Recher, H.F. and Majer, J.D. 1997. Vertebrates and eucalypt formations. Pp. 303–341 in *Eucalypt Ecology*. J. Williams and J. Woinarski, eds. Cambridge University Press, Melbourne, Australia.

Woinarski, J.C. and Recher, H.F. 1997. Impact and response: a review of the effects of fire on the Australian avifauna. *Pacific Conservation Biology*, **3**, 183–205.

Woinarski, J.C.Z. 1999. Fire and Australian birds: a review. Pp. 55–112 in *Australia's Biodiversity: Responses to Fire*. Biodiversity Technical Paper No. 1. A.M. Gill, J.C.Z. Woinarski and A. York, eds. Environment Australia, Canberra.

Wolf, M. and Batzli, G. 2004. Forest edge: high or low quality habitat for white-footed mice *(Peromyscus leucopus)? Ecology*, **85**, 756–769.

Wolfenbarger, D.O. 1946. Dispersion of small organisms. *American Midland Naturalist*, **35**, 1–152.

Wolff, J.O., Schauber, E.M. and Edge, W.D. 1997. Effects of habitat loss and fragmentation in the behaviour and demography of gray-tailed voles. *Conservation Biology*, **11**, 945–956.

Woodroffe, R. and Ginsberg, J.R. 1998. Edge effects and the extinction of populations inside protected areas. *Science*, **280**, 2126–2128.

Woodward, F.I. 1987. *Climate and Plant Distribution*. Cambridge University Press, Cambridge.

World Commission on Forests and Sustainable Development. 1999. *Our Forests Our Future*. Report of the World Commission on Forests and Sustainable Development. Cambridge University Press, Cambridge, England.

World Resources Institute 1992. *World Resources 1992–1993*. Oxford University Press, New York.

Worthen, W.B. 1996. Community composition and nested-subset analyses: basic descriptors for community ecology. *Oikos*, **76**, 417–426.

Wright, D.H. and Reeves, J.H. 1992. On the meaning and measurement of nestedness of species assemblages. *Oecologia*, **92**, 416–428.

Wright, D.H., Patterson, B.D., Mikkelson, G.M. Cutler, A. and Atmar, W. 1998. A comparative analysis of nested subset patterns of species composition. *Oecologia*, **113**, 1–20.

Yaffee, S.L. 1994. *The Wisdom of the Spotted Owl: Policy Lessons for a New Century*. Island Press, Washington, D.C.

Yahner, R.H. 1983. Seasonal dynamics, habitat relationships and management of avifauna in farmstead shelterbelts. *Journal of Wildlife Management*, **47**, 85–104.

Yamada, K., Ansari, M., Harrington, R., Morgan, D. and Burgman, M. 2004. Sindh ibex in Khirthar National Park, Pakistan. Pp. 214–228 in *Species Conservation and Management: Case Studies Using RAMAS GIS*. H.R. Akçakaya, M. Burgman, O. Kindvall, C.C. Wood, P. Sjogren-Gulve, J. Hatfield and M. McCarthy, eds. Oxford University Press, New York.

Yanai, R.D. 1991. Soil solution phosphorus dynamics in a whole-tree-harvested northern hardwood forest. *Soil Science Society of America Journal*, **55**, 1746–1752.

Yates, C.J., Hobbs, R.J. and True, D.T. 2000. The distribution and status of woodlands in Western Australia. Pp. 86–106 in *Temperate Woodlands in Australia: Biology, Conservation, Management and Restoration*. R.J. Hobbs and C. Yates, eds. Surrey Beatty and Sons, Chipping Norton.

Yates, E.D., Levia, D.F. and Williams, C.L. 2004. Recruitment of three nonnative invasive plants into a fragmented forest in southern Illinois. *Forest Ecology and Management*, **190**, 119–130.

Young, A., Boyle, T. and Brown, A. 1996. The population genetic consequences of habitat fragmentation for plants. *Trends in Ecology and Evolution*, **11**, 413–418.

Young, A. and Clarke, G., eds. 2000. *Genetics, Demography and Viability of Fragmented Populations*. Cambridge University Press, Cambridge.

Zackrisson, O. 1977. Influence of forest fires on the north Swedish boreal forest. *Oikos*, **29**, 22–32.

Zanette, L. and Jenkins, B. 2000. Nesting success and nest predators in forest fragments: a study using real and artificial nests. *Auk*, **117**, 445–454.

Zanette, L., Doyle, P. and Tremont, S.M. 2000. Food shortage in small fragments: evidence from an area-sensitive passerine. *Ecology*, **81**, 1654–1666.

Zanette, L., MacDougall-Shakleton, E., Clinchy, M. and Smith, J.N. 2005. Brown-headed cowbirds skew host offspring sex ratios. *Ecology*, **86**, 815–820.

Zartman, C.E. 2003. Habitat fragmentation impacts on epiphyllous bryophyte communities in Central Amazonia. *Ecology*, **84**, 948–954.

Zedler, J.B. 2003. Wetlands at your service: reducing impacts of agriculture at the watershed level. *Frontiers of Ecology and Environment*, **1**, 65–72.

Zeng, H., Peltola, H., Talkkari, A., Venäläinen, A., Strandman, H., Kellomäki, S. and Wang, K. 2004. Influence of clear-cutting on the risk of wind damage at forest edges. *Forest Ecology and Management*, **203**, 77–88.

Zimmerman, B.L. and Bierregaard, R.O. 1986. Relevance of equilibrium theory of island biogeography and species–area relations to conservation with a case study from Amazonia. *Journal of Biogeography*, **13**, 133-143.

Zuidema, P.A., Sayer, J. and Dijkman, W. 1996. Forest fragmentation and biodiversity: the case for intermediate-sized reserves. *Environmental Conservation*, **2**, 290–297.

Index

Pages with *f* denote figures and photographs; *t* denote tables.